THE LONGEVITY REVOLUTION

Books by the Author

Birren, J.E., R.N. Butler, S.W. Greenhouse, L. Sokoloff, and M.R. Yarrow. *Human Aging*, 1963.

Butler, R.N. and M.I. Lewis. *Aging and Mental Health*, 1973; with Sunderland, T. (5th ed.), 2003.

Butler, R.N. *Why Survive? Being Old in America*, Pulitzer Prize, 1976.

Butler, R.N. and M.I. Lewis. *Love and Sex after Sixty*, 4th ed., 2002.

Butler, R.N. and A.G. Bearn (eds.). *The Aging Process*, 1985.

Butler, R.N. and H.P. Gleason (eds.). *Productive Aging, Enhancing Vitality in Later Life*, 1985.

Warner, H., R.N. Butler, E. Schneider, and R. Sprott (eds.). *Modern Biological Theories of Aging*, 1987.

Kent, B. and R.N. Butler (eds.). *Human Aging Research*, 1987.

Bianchi, L., P. Holt, O.F.W. James, and R.N. Butler (eds.). *Aging in Liver and Gastrointestinal Tract*, 1987.

Butler, R.N., M. Schechter, and M. Oberlink (eds.). *The Promise of Productive Aging*, 1990.

Butler, R.N. and K. Kiikuni (eds.). *Who Is Responsible for My Old Age?* 1993.

Butler, R.N. and J.A. Brody (eds.). *Delaying the Onset of Late Life Dysfunction*, 1995.

Fillit, H.M. and R.N. Butler (eds.). *Cognitive Decline: Strategies for Prevention*, 1997.

Butler, R.N., L.K. Grossman, and M. Oberlink (eds.). *Life in an Older America*, 1999.

Butler, R.N. and C. Jasmin (eds.). *Longevity and Quality of Life*, 2000.

THE LONGEVITY REVOLUTION

The Benefits

and Challenges

of

Living a Long Life

ROBERT N. BUTLER, M.D.

PublicAffairs / NEW YORK

Published in the United States by PublicAffairs™,
a member of the Perseus Books Group.

PublicAffairs books are available at special discounts for bulk
purchases in the U.S. by corporations, institutions, and other
organizations. For more information, please contact the Special Markets Department at the Perseus Books Group, 2300
Chestnut Street, Suite 200, Philadelphia, PA 19103, call (800)
255-1514, or e-mail special.markets@perseusbooks.com.

Library of Congress Cataloging-in-Publication Data
Butler, Robert N., 1927-
The longevity revolution : the benefits and challenges of living
a long life / Robert N. Butler.—1st ed.
p. ; cm.
Includes bibliographical references and index.
ISBN 978-1-58648-553-5 (hardcover : alk. paper) 1. Longevity.
2. Longevity—Social aspects—United States. 3. Geriatrics—
United States. 4. Medical policy—United States. I. Title.
[DNLM: 1. Longevity. 2. Geriatrics. 3. Public Policy. 4. Quality
of Life. 5. Social Welfare. 6. Socioeconomic Factors. WT 116 B986L
2008]
QP85.B88 2008
612.6'8—dc22
2007048204

Text design and composition by Jenny Dossin

First Edition

10 9 8 7 6 5 4 3 2 1

To the Memory of My Late Wife
Myrna Irene Lewis

and

To Our Daughter
Alexandra Nicole Butler

"Man can will his future."

Rene Dubos
(20th Century Biologist, Scientist, and Author)

"Do not yearn after immortality, but exhaust the limits of the possible."

Pindar
Olympians
(Greek Poet, 5th Century)

Contents

PREFACE

In fewer than one hundred years, human beings made greater gains in life expectancy than in the preceding fifty centuries. From the Bronze Age to the end of the nineteenth century, life expectancy grew by only an estimated twenty-nine years—from about twenty to just under fifty years. But since the beginning of the twentieth century in the industrialized world, there has been an unprecedented gain of more than thirty years of average life expectancy from birth to over seventy-seven years of age. The aging of populations is occurring rapidly throughout the world. By 2025, the number of people in the United States age sixty-five and older will nearly double. Old age, once seen in 2 to 3 percent of a population, is now common and favors women, who outlive men by over five years in America today. Moreover, nearly 20 percent of the gain in life expectancy now applies to those sixty-five years of age.

I call this unprecedented demographic transformation the Longevity Revolution.[1] Once the privilege of the few, longevity has rapidly become the destiny of the many in both the developed and developing worlds. How did it come about that the fulfillment of ancient hopes to extend life—with genuine possibilities of more to come—has not been welcomed with total enthusiasm?

People of good will and deep concern over the future of our country, and other nations, have asked tough questions about the impact of the aging of our population and its advancing longevity. Can we afford old age as a society? Will Social Security and Medicare collapse under the pressure of growing numbers of retirees? It is altogether good that corporate leaders, economists, politicians, and commentators are raising these crucial issues. However, the unfortunate outcome of such criticism has been the spread of fear and gloom among the general public.

Despite this pervasive ambivalence, I remain optimistic. We have the tools to take advantage of this exceptional demographic shift. But it will require nothing less than a total transformation of both the personal experience of aging and of cultural attitudes.

I wrote *The Longevity Revolution* for the thoughtful public. Its central purpose is to describe the origins, challenges, and adjustments to advancing longevity and the aging of populations and to question contemporary assumptions about late life. Because I believe in the activism of an enlightened citizenry, the general thrust of this book is toward an agenda for action and the presentation of a body of knowledge to support it.

The book is divided into six sections. The first part describes the origins of the Longevity Revolution. Part II outlines the major challenges of a longer life, including ageism, the changing nature of the family, and the various disorders of longevity.

Part III offers a detailed overview of biomedical science related to aging. But doing science takes money, lots of money, so we will begin with a discussion of the pressing need for funding. The devastating reality of Alzheimer's disease is one of the greatest challenges facing scientists today. Despite advances in neuroscience, we do not know the causes of this disease nor have we discovered how to curb its progression in any meaningful way. We must launch a colossal public-private research initiative to combat Alzheimer's disease if we are to realize the full potential of a longer life. Finally, this section covers the basic biology of aging and longevity, from the evolutionary theories of aging to the discoveries that appear so promising.

Part IV is the heart of the book. Here I will offer a range of solutions to the challenges raised in previous chapters. I will discuss how societies can improve health promotion and health care as well as how they can finance these added years of life. The United States has a long way to go. Although Americans are living longer than ever, having reached 77.9

years on average, American life expectancy has slipped over the last two decades from eleventh place to forty-second. This puts the richest country in the world behind Jordan, Singapore, and the Cayman Islands, among other countries. The United States has not made a research investment in ageing. Only eleven American medical schools out of 145 have geriatrics departments; in England geriatrics is the number two specialty. Americans need a *health* program rather than a disease program.

All of this, to be sure, is an enormous undertaking and will require the development of a vigorous politics of aging and longevity. In particular, the baby boomers must exercise their political power to force changes in policy.

In Part V, I discuss some necessary cautions. There is a staggering inequality of longevity around the world, which is a serious obstacle to globalization. In Sierra Leone, a child can expect to live to be about thirty-four years old. In Japan, which has the world's highest life expectancy, a person will likely live to be eighty-two. Moreover, unless we deal with significant threats—from industrial pollution to diseases ranging from tuberculosis to AIDS to the epidemic of obesity—we could lose the longevity we have gained.

The final part of the book discusses quality-of-life issues and the future of longevity. If we continue to add years to our life span, will we make good use of them? Will the added years bring about greater maturity in our personal relationships and our relationship to the world?

Through this book, I hope to contribute to cultural, scientific, and social thought and, most of all, to encourage a national discussion about the challenges posed by the Longevity Revolution. My intentions are immediate and practical. The so-called baby boomers—the seventy-six million Americans born between 1946 and 1964—are, by some counts, the largest generation in American history. Their numbers surprised everyone, and consequently there were not enough diapers for them as babies, not enough schools for them as children, and not always enough good jobs for them as adults. In 2011, the oldest of this generation will turn sixty-five, and many worry that there will not be enough resources for them in old age.

I believe that the baby boomer generation is both a generation at risk for unhappy old age and the key to transforming the character of old age in America. The baby boomers are discovering old age through their parents, and they want a financially secure, vigorous, and healthy final chapter to life. They want to age better than older people do today.[2] And, since the fields of medicine and biological science, among others, helped create the Longevity Revolution, they must contribute to and make the adjustments necessary to accommodate and take advantage of this new phenomenon.[3]

The baby boomers are an enormously influential interest group. They can transform what it means to live a long life. The American penchant for crisis management rather than foresight and design will be challenged, I hope, by a thoughtful analysis of the risks and potential rewards of meeting longevity head on, and sooner rather than later. The baby boomers should not have to turn gray of head before we notice that we haven't made room for them in America's land of old age. At the same time, society must also prepare for generations X and Y, who in turn must do their part. (See Table 1.)

If we are to successfully meet the challenges of the Longevity Revolution we must have the audacity to question conventional wisdom. This book refutes many long-held beliefs about the effects of an aging population on society. It is *not* true, for example, that:

- Decreased birthrates are disadvantageous.

- Welfare state-type social protections are unsustainable.

- The aging population accounts for rising health costs.

- Excessive medical costs are associated with the end of life.

- The AARP is the most powerful lobby in Washington.

- Age prejudices have been ended by laws and legal actions.

- Older workers are unproductive.

- Old people receive more public and private support than children and youth.

While there are radical social transformations to be made if we are to make good use of the extra thirty years granted us, there has also been a healthy growth in science. It is what got us here, and it promises still greater longevity. Scientific advancements should and will add vigor and health throughout life, and not just at its end. The aging population increasingly consists of active, vigorous, robust older people. We must not take them for granted, but the trend can continue and it should be celebrated. Above all, I hope that this book will help convince people that our increased longevity constitutes a supreme achievement.

PART I: INTRODUCTION

CHAPTER 1

WHAT IS THE LONGEVITY REVOLUTION?

> *In a state of nature . . . no Arts, no Letters,*
> *no Society; and which is worst of all,*
> *continual fear, and danger of violent death;*
> *and the life of man, solitary, poor, nasty,*
> *brutish, and short.*
>
> **Thomas Hobbes (1588–1679)**
> *Leviathan,* **Part I, Chap. XIII (1658)**

Life through much of human history was indeed brutish and short—humans lived barely long enough to reproduce themselves. For the mere survival of the human race, a proportion of individuals had to live long enough to give birth and rear their young (Figure 1.1). Yet, it is sobering to note that, according to archeological estimates, half of all Neanderthals (the archetypal caveman living one hundred thousand to thirty-five thousand years ago) and Upper Paleolithic homo sapiens (beginning forty thousand years ago and including Cro-Magnon man) died by the time they were twenty, with only a few living beyond age fifty.

The Cro-Magnon era brought longer life expectancy for some, with the rare individual living beyond age sixty. This fledgling longevity, sporadic as it was, became possible as humans began to work together to create a better standard of living. Although they were less muscular, Cro-Magnons eventually replaced the intellectually outpaced Neanderthals. However, like their predecessors, the vast majority of Cro-Magnons continued to die at an average age of eighteen to twenty.

Before I continue, it is important to define the terms *life expectancy* and *life span* and distinguish between them. Life expectancy is based on the average number of years that each sex can expect to live, under specific conditions. Life span is the genetically determined length of life of a specific animal species under the best of environmental circumstances. It probably increased during early hominid development but

Figure 1.1—Average Length of Life from Ancient to Modern Times

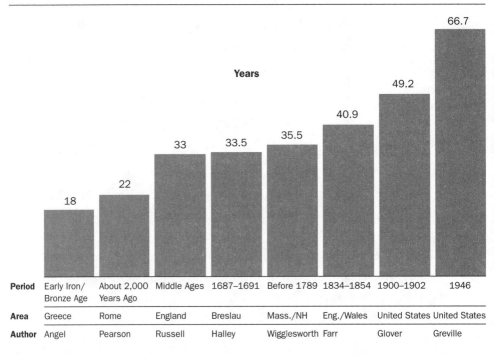

Period	Early Iron/ Bronze Age	About 2,000 Years Ago	Middle Ages	1687–1691	Before 1789	1834–1854	1900–1902	1946
Area	Greece	Rome	England	Breslau	Mass./NH	Eng./Wales	United States	United States
Author	Angel	Pearson	Russell	Halley	Wigglesworth	Farr	Glover	Greville

This bar graph shows how life expectancy has increased since the early Iron and Bronze Age. Adapted from L.I. Dublin, A.J. Lotka, M. Spiegelman: *Length of Life*. New York: Ronald Press, 1949.

has not, in all likelihood, increased since that time. Life expectancy is more malleable and is dependent upon a variety of factors that can change quickly, such as the conquest of diseases. For example, between 1900 and 2000, life expectancy in the United States increased over thirty years.[1]

Although early humans had bodily defenses, their immunities were probably specific to local pathogens, so when our forebears traveled they were exposed to new infectious diseases to which they lacked immunity. Illness was largely a mystery to be accounted for by spirits, gods, evil, and retribution, or, as Euripides wrote in *Medea*, "A throw of chance—and there goes Death bearing off your child into the unknown." Nature or the gods were blamed for accidents, plagues, pestilence, famine, and even wars.

The prospects for health and survival brightened a bit as hunting, fishing, and food gathering progressed to include the cultivation of plants and the domestication of animals during the Neolithic era. The Fertile Crescent was the site of the first technology essential to public health. Reliable fire-making techniques made cooked food and heated water possible. The development of pottery in the Neolithic period advanced more hygienic and convenient storage of food and water and the disposal of waste and garbage. Permanent population sites became common, and the overall number of human beings greatly increased. Nonetheless, the length of life that could be expected by any individual changed little, even though many more people were alive.

The Bronze Age opened the door to the first real improvements in human longevity, such as the manufacture of metal tools and weapons made of bronze, the rise of urbanization, the specialization of labor, the exploration and colonization of new territories in search of raw materials, and undoubtedly the production of surplus food. These developments provided the social and environmental supports for increases in life expectancy. The Iron Age (around 1200 BC) continued the trend toward permanent settlements, laying the foundation for social organization and advancing agriculture through the use of iron implements.

Imperial Rome at its height brought a relatively high standard of living and health to its more than one million inhabitants. Life expectancy was twenty-five years; however, it is important to note that this number includes a very high infant mortality rate. Those who survived beyond childhood had an average life expectancy of forty.[2] But by AD 180 Roman culture began its decline, and nothing remotely like it appeared again until the eighteenth and nineteenth centuries in Europe. Scholars believe there was a dip in life expectancy after the decline of the empire.

As humankind moved from prehistory to the early modern era of Western civilization,[3] life continued to be fragile. Infant and childhood mortality and the random and frequent deaths of adults—especially of women during childbirth—from infections, disease, and accidents were the norm. These commonplace and expectable deaths were punctuated by the devastations of plagues, failed harvests, famines, and scourges like scurvy and beriberi that seemingly sprang from nowhere. Typically, epidemics broke out as people became so weakened by hunger and malnutrition that they were predisposed to disease.

Although plague was first recorded in Athens in 430 BC, and five thousand died daily in Rome in an epidemic in the third century AD, the most infamous plague occurred in the late Middle Ages. The Black Death, as the bubonic form of the plague was known, began in Constantinople in 1334 and spread throughout Europe from 1348 until 1351. In fewer than twenty years, it is estimated to have killed perhaps three-quarters of the population of Europe and Asia.

It was not until Shibasaburo Kitasato and Alexander Yersin discovered the plague bacterium in 1894 that it became clear that the disease was carried by fleas from rats to humans and proliferated in dirty, garbage-strewn living situations. Plague continued into the nineteenth century in Europe; the last was in India in the early twentieth century, resulting in ten million deaths.

Epidemics continued into the twentieth century. Tuberculosis, known variously as consumption or the white plague, was particularly rampant and deadly. As late as 1930, eighty-eight thousand people in the United States died of tuberculosis. Although Robert Koch, a German physician and bacteriologist, discovered the tubercle bacillus in 1882, the disease was not controlled until the 1940s when streptomycin and other drugs became available.

Many scholars consider smallpox to have been more significant in its effect on populations and political developments than even the Black Death, because it struck all classes of society. Edward Jenner, an English physician, demonstrated how it could be prevented by a vaccination with cowpox virus. His discovery laid the foundation for the sciences of modern immunology and virology as well as the eventual elimination of smallpox. The last outbreak occurred in the 1970s in Somalia, where it was quickly suppressed. In May 1980 the World Health Organization officially declared its global eradication, marking perhaps the world's greatest public health achievement. Smallpox, a scourge throughout history, is now extinct *in nature*.

Polio could be the second disease of epidemic proportions to become extinct worldwide. As with smallpox, humans are its only known natural host. During its peak, from 1943 to 1956, polio infected some four hundred thousand Americans and killed about twenty-two thousand. The disease began to decline rapidly in the United States in 1955, after a mass immunization program with the Salk vaccine, followed by the Sabin vaccine in 1961.

Influenza, an old enemy, made its most spectacular showing between 1917 and 1919, killing at least twenty-one million worldwide and infecting half the world's population. Twice as many people perished in a few months' time as had been killed in World War I, with people dying faster than from any other disease. Half a million died in the United States. Eventually, the pandemic ended, probably because most survivors developed antibodies, producing what might be called a herd immunity and leaving few people for the virus to attack.

In assessing the impact of disease over the centuries, or even over the first half of the twentieth century, it is important to remember that death was not the only consequence. Permanent disability was widespread in children who survived the so-called children's diseases, such as whooping cough and German measles, and disability hastened death. In addition to the physical trauma, one can only imagine the emotional anxiety and fear that families felt, especially for their children, when diseases of mysterious origin, and for which no treatment was known, struck at random or with chilling predictability during epidemics. The life of a child was precarious; that of an adult only somewhat less so.

Nonetheless, the Industrial Revolution and the wealth it generated brought significant increases in longevity.[4] Beginning in the middle of the eighteenth century in England, Europe was transformed from a rural, agricultural, and handicraft economy to one dominated by mass production of manufactured goods, improved agriculture, and wider distribution. It became possible to feed a much larger population, many of whom were now working in urban areas. The significance of this transformation cannot be overstated. Robert Fogel, economist and Nobel Prize winner, estimates that prior to the Industrial Revolution in France and England, about one-fifth of working class people had a calorie intake that was inadequate to sustain them and that during the eighteenth and even the nineteenth century there was widespread and chronic malnutrition.[5] Fogel notes that the increase of longevity and stature[6] over the past two hundred years was due to the availability of more food, and he introduced the concept of "technophysio evolution" to describe these changes.[7]

In response to epidemics of yellow fever, cholera, smallpox, typhoid, and typhus, communities began to recognize the benefits of organized efforts to address health issues. Predating the germ theory of disease, social reformers, motivated by moral concern, contributed critically to public health measures.[8] In 1866, New York created the first state health department, with local boards in each town mandated to monitor serious health problems and attend to unsanitary living conditions. Other states followed, and organized public health efforts began.

Another element that contributed to the Longevity Revolution really began in the bedrooms of Europe (first, notably, in France) in the nineteenth century, when couples started to limit the number of children they conceived by purposefully abstaining from sexual intercourse in order to save the women from dying in childbirth. The resulting decline in birthrates produced two very important changes: the proportion of older persons and other age groups in the population increased and the longer time between births as well as fewer pregnancies per woman contributed to increased health and survival of both infants and mothers.

Meanwhile, by the beginning of the nineteenth century, another major element of the Longevity Revolution was taking shape, namely a revolution in medical science. In 1846, the general structure of the human body was almost fully known when an American surgeon, William Morton, opened the way to the field of surgery by using ether as a general anesthetic. At the same time, it was becoming clear that specific organisms caused infections. John Snow, a physician, unraveled the basis of contagion when he demonstrated that contaminated water flowing through a Broad Street pump was the cause of the 1866 London cholera epidemic. Subsequent improvements in water supply and sewage systems reduced both water-borne and food-borne diseases.

The real breakthrough came in the late nineteenth century when Louis Pasteur and Robert Koch developed and demonstrated the germ theory of disease, beginning a dramatic decline in death rates. With this discovery, health professionals and the general public finally understood how some diseases and infections were communicated. The work of Pasteur, a French chemist, led directly to pasteurization and the protection of millions of children and adults from disease transmitted through milk. At about the same time, Koch developed ingenious techniques for the study of bacteria that are still in use today. He established criteria, referred to as Koch's postulates, for proving the bacterial cause of a disease. In the process he discovered the microorganism causing tuberculosis as well as those causing wound infections and Asiatic cholera.

By the end of the nineteenth century, proof of the germ theory of disease began to transform medical care and hospital practices. Decades earlier, Hungarian physician Ignaz Semmelweiss was driven to insanity and suicide after his peers ridiculed his pioneering belief that midwives and other medical personnel delivering babies should thoroughly wash their hands and wear clean clothes to prevent "childbirth fever" (puerperal fever),[9] a major cause of death among women giving birth. Skeptics were unconvinced even by the evidence of greatly reduced deaths from infection in the Viennese hospital where Semmelweiss worked.[10] In 1865, the very year of Semmelweiss's death, English surgeon Joseph Lister demonstrated that heat sterilization of surgical instruments and the use of antiseptic agents on wounds could dramatically reduce infection. Cleanliness during childbirth and in medical care in general was adopted by the 1890s, unfortunately too late for Semmelweiss to know that he had been vindicated.

The developing science of endocrinology, the study of the body's hormonal system, also brought dramatic changes, exemplified by the important discovery of insulin by Frederick Banting and Charles Best, both Canadians, in 1921. Practically overnight, diabetics were saved from almost certain death and given the prospect of reasonably long and healthy lives.

Other drug discoveries led to seemingly miraculous cures, and Alexander Fleming's discovery of penicillin in 1928 was perhaps the most dramatic of all. Prior to penicillin, even minor cuts and bruises could have dire consequences. Many Americans past the age of sixty still clearly recall the days of childhood ear infections and other miseries that quickly became curable when penicillin came into general use in the 1940s. Young military personnel in World War II were among the first to benefit from its lifesaving effects.

The Industrial Revolution changed the world and touched nearly every aspect of life—from the social and cultural to the political, economic, and ecological—much of it without precedent. Wealth became more widely distributed and contributed to the growth of the middle classes, the famous bourgeoisie.[11] Political power reflected the shift in wealth and the needs of an industrializing society. Cities grew and workers organized, spawning labor movements and universal public education. The economic foundation for the modern Western welfare state was laid, and it, in turn, contributed to longevity.

It is sheer foolishness to imagine that we can extend life or sustain complex modern societies without substantial governmental participation. Many European countries realized the necessity of developing protections as well as social programs and services for the workers who were crowding urban areas. Laissez-faire capitalism slowly gave way to welfare capitalism. Worker compensation and safety were promoted in Germany, Austria, and Great Britain in the late 1880s, and by 1920 most of the United States had passed some form of relevant legislation. Unemployment benefits were made available in Europe in the latter part of the nineteenth and the first part of the twentieth century but were not legislated in the United States until the Social Security Act of 1935.

THE MODERN WELFARE STATE

In many countries the underlying purpose of social programs is not to be humane or altruistic. It is understood that death, accidents, and disease and their timing are unpredictable. Moreover, it is implied, if not always directly stated, that social supports and a floor of security for all are necessary to produce a healthy, educated, and productive workforce capable of maintaining economic productivity and buying power, and of making any necessary adjustments to changing economic conditions and technology. And as Otto von Bismarck so shrewdly recognized in the late nineteenth century, they protect against civil unrest and are potent vote-getters in democratic societies. Eventually, this long-term and essentially enlightened self-interest may penetrate the general American consciousness.

The success of post-1945 European welfare was due to social-democratic principles and policies and, interestingly, to the *success of capitalism*. Although the socialism and its electoral successes helped regulate capitalism humanely,[12] prosperity after World War II made implementation of social protections possible. Thus, although it was the strength of capitalism, not its failure, that brought us the welfare state, its failure could bring down the infrastructure of social support, as unemployment and dwindling prosperity put the taxation base of social protections at risk, ultimately jeopardizing our increased longevity.

At present, the industrialized world appears to be moving toward a five-legged stool of support for social protections. First, much more is expected of individuals. They are being called upon to alter measurably their health habits and to prudently save for the future. Second, the family is expected to take on more caregiving responsibilities, at the same time as some nations have instituted pro-family policies that provide family assistance and respite programs. Third, in the United States the civil society continues its important role and is continuing to grow. Philanthropy and volunteerism are gaining more support by the community at large.[13] Fourth, despite its continuing resistance to them, the business enterprise, too, along with labor, continues to provide some social protections. Finally, the government itself is endeavoring to reduce its vulnerability in times of social and individual financial crises, while retaining at least some measure of responsibility for the health and well-being of the people.

The United States was the twenty-eighth country to adopt a social security system,[14] specifically the beginnings of the guarantee of income maintenance in old age. But it was not until 1939 that the United States moved toward a *family-oriented, life-course social protection system* that eventually included disabled workers and survivors of deceased workers.

THE TWENTIETH AND TWENTY-FIRST CENTURIES

By the mid-twentieth century, the Longevity Revolution spawned what could be called the *new longevity* (Tables 1.1 and 1.2). Life expectancy after age sixty-five to age ninety tripled between 1940 and 1980. In 1940, only 7 percent of Americans had a chance of living to the age of ninety; by 1980 that percentage had risen to 24 percent. People over eighty are usually referred to as the fastest-growing age group of significant size, growing at 3.8 percent annually. In fact, centenarians are the most rapidly increasing age group. There are some seventy-two thousand centenarians today, and one Census Bureau projection predicts nearly one million by the middle of the twenty-first century! And we're not only living longer, we're living better. There has been a 60 percent drop in deaths from cardiovascular disease and stroke since 1950, as well as significant decreases in disability rates.[15] From 2000 to 2001 alone there was an overall decline in deaths of 1.7 percent, with a decline of 4.9 percent in those resulting from stroke and 3.8 percent from heart disease.

In the 1990s, the United States was in a "population lull," with a relatively small number of persons in their fifties and early sixties waiting to move into old age. This was because older women had relatively few children.[16] Influenced by the difficult years of the Great Depression in the 1930s, these older persons limited the number of children they produced—20 percent of women now over seventy-five had had no babies, and 20 percent had only one. But then came the baby boom generation, born after World War II. Now middle-aged, this generation will soon begin reaching old age, conventionally defined as beginning at sixty-five. (There is, however, nothing magical or scientific about this or any other number in defining old age.)

In the twentieth century we were offered realistic opportunities for health promotion and disease prevention through public health measures, healthy lifestyles, education, rising wealth, and workplace regulation, in addition to application of new knowledge, such as in understanding hypertension and atherosclerosis. The advent of possible means to *delay* aging and *extend* longevity and the growing encouragement of health promotion/disease prevention converge to offer a strategy that could be adapted by individuals and by society in the twenty-first century.

Progress made thus far creates rising expectations of still more profound advances in medical technology and in life expectancy. Coronary bypass grafts, balloon angiography with stents, drugs that lower blood pressure and cholesterol, knee and hip replacements, and cataract extractions with lens implants simply whet the appetite.

Indeed, some believe that humans can master their evolution. Among them is Aubrey de Grey of Cambridge University, who suggests a life expectancy of five thousand years by 2100.[17] The philosopher John Harris of Oxford views extraordinary longevity from another perspective when he considers the possibility of immortality and its consequences for humankind.[18]

A NEW SENSE OF THE LIFE CYCLE

For the first time in recorded history we are beginning to see the entire life cycle unfolding for a majority of the population in developed nations. Infancy, childhood, adolescence, early adulthood, middle age, and old age have become expectable stages in the lives of nearly all. The special characteristics, responsibilities, and needs of each stage, as well as the transitions that carry individuals from one stage to the next, are revealing themselves. Names have evolved. The term "adolescence" was introduced in 1904 by G. Stanley Hall, "teenager" in the United States in the 1920s, and "youth" as a special appellation in the 1960s by Kenneth Kenniston. "Empty nest" was coined in the 1960s to refer to the period somewhere in parents' midlife when children grow up and leave home. Also, from the 1960s, the term "prime of life generation" was applied to those fifty to sixty-four. "Young-old" and "old-old," terms originally intended by gerontologist Bernice Neugarten to differentiate among healthy and less well-functioning older persons, gradually evolved in common usage to mean persons sixty-five to seventy-five (the young-old) and seventy-five and above (the old-old).

We are beginning to get a grasp of the nature and complexities of old age itself. It is already clear that it is not a fixed and unchangeable condition. Nor are older people a homogeneous group. In fact, there is increasing variability among people as they grow older. Children are much more like one another than are "the elderly." This tremendous variability among the old is a function of the combination of genetics, health habits, health care, life styles, personality, personal history, occupation, chance, and, in some measure, luck or the lack thereof.

The concept of old age itself is undergoing constant redefinition; it is not the same as it was, nor will it be. As an example, the last third of life has typically been equated with decline and illness. But the aging population in the past few decades is increasingly represented by vigorous, robust older people. There has been a growing, "active" life expectancy with a reduction of disability. To be sure, this is not yet universal, but it is a portent of the future as we now see it. Already, over half of the "oldest old," the eighty-five-plus group, report no significant physical disability whatsoever. They can go about their everyday activities without any personal assistance.[19] Illness and disability rates among older people declined by 5 percent between 1982 and 1989, according to Manton et al. (1993), who used data from the National Long Term Care Survey.[20]

There has been a modest decline in the life expectancy differential for men and women. Two new factors have become influential: men's survival rates are increasing somewhat more rapidly than women's, probably reflecting a decrease in deaths from heart disease, and lung cancer rates are increasing faster for women, reflecting the greater numbers of women who began to smoke some thirty and forty years earlier. However, around the globe, women still live longer than men except in ten countries including Pakistan and Bangladesh, in part due to female infanticide. (See Table 1.3.)

Gender differences in life expectancy disappear at age 105, at which point both men and women have an equal chance of living on. The apparent reason is that those sturdy men who managed to reach 105 represent the fittest (and perhaps the luckiest) of their sex and thereby have outlived the disadvantages of being male. Eighty-five percent of centenarians are women, but the men in that group are in better shape physically and cognitively.

THE IMPACT OF RACE AND CLASS ON LONGEVITY

Social class (measured by income and education) is a major factor in how long we can expect to live. Urban, middle-class, salaried individuals have the longest life expectancy. Many more whites fall into this category than do persons of minority ethnic and racial backgrounds. Differences in life expectancy between blacks and whites narrowed from 7.6 years in 1970 to 5.6 years in 1983 and 1984. Since that time the difference has begun to widen again. In 1984, black life expectancy began declining steadily for the first time in eighty years. The gap in life expectancy remains significant across racial lines. In 2003, life expectancy for blacks (at birth) was 72.2 years (the average for males and females combined), compared to 77.7 years for whites (at birth).

Black men, on average, will not live to retirement. The mortality rate of a black man in Harlem is currently higher than that of a man in Bangladesh, one of the world's poorest nations. In Bangladesh, 55 percent of men live to age sixty-five, while in Harlem, only 40 percent do. The main causes of such "excess mortality" (a statistical term that hardly conveys the human dimension of the situation) are cardiovascular disease, cirrhosis of the liver, homicide, tumors, and drug dependency. Violence (a homicide rate six times greater among blacks than whites), substance abuse, AIDS, and poor health care underlie this shocking state of affairs.

Overall, blacks are disproportionately among the 47 million Americans who do not have health insurance coverage and who therefore often do not receive health care until they develop some acute problem that brings them to an emergency room. This means that much injury and disease that could be treated easily in early stages becomes full blown before treatment is given. One study reported in the *International Journal of Epidemiology* found that blacks, who number only 13 percent of the population, account for nearly *80 percent* of what are considered "premature deaths" in the United States. In that study, deaths occurred between the ages of fifteen and forty-four from disorders that are normally not fatal if treated early, such as appendicitis, asthma, bladder infections, and pneumonia.

THE LONGEVITY REVOLUTION IS WORLDWIDE

Growing longevity is a global geopolitical force. Japan took only twenty-four years to become an "aged society" (defined by the UN as one in which more than 14 percent of a population is over sixty). This is a faster pace than that in either Europe or the U.S. For instance, it took Germany forty-five and France 130 years to accomplish what Japan achieved.

As developing countries evolve, they will experience population shifts *unique* in the history of national development. In the past, countries have typically seen their first great gains in life expectancy occurring from improvements in infant, child, and maternal mortality, as well as from the control of infection and disease in the general population. For example, water purification and vaccination of children have been inexpensive and effective. And with the recent gains in prevention, treatment, and control of the illnesses of older people, the developing world will witness the increased survival of at least a proportion of the old who have access to such care at the same time that survival among the young is improving. We anticipate seeing simultaneous survival booms among the young and the old. By 2025, it is anticipated that 80 percent of all persons over sixty-five will be living in the developing world.

With the Longevity Revolution, the world enters a new and unprecedented stage of human development—the impact of which has been made greater because of its rapidity. We are no longer limited to a life view that must accommodate itself to the historic brevity of life, to random and premature illness and death, as Thomas Hobbes described it. The Longevity Revolution is a great intellectual and social as well as medical achievement and an opportunity that demands changes in outmoded mind-sets, attitudes, and socioeconomic arrangements. Many of our economic, political, ethical, health, and other institutions, such as education and work life, have been *rendered obsolete* by the added years of life for so many citizens. The social construct of old age, even the inner life and the activities of older persons, is now subject to a positive revision.

As Thomas Jefferson said, "Our laws and institutions must move forward hand in hand with the progress of the human mind." Just as societies have not yet adapted fully to the Industrial Revolution, with its pollution, unemployment, displacement, and other unsolved social and environmental issues, so too we have not yet fully adapted to the Longevity Revolution. But the revolution is fully launched. Adjustments are well under way. What will the twenty-first century bring?

PART II: CHALLENGES

CHAPTER 2

THE NEW LONGEVITY IS THE BIGGEST CHALLENGE

The improved standards of living, social protections, and health conditions brought about by the industrial-scientific revolution have helped make abundant what was once scarce: older people. Notwithstanding the challenges to government, business, and individuals presented by the increase in the disorders of longevity, such as dementia, physical frailty, and poverty, the vast majority of older persons and their families experience these extra years as a joyous and precious gift, and their contributions to society that follow are beyond estimation.

The world at large is now gaining an additional one million older persons each month. Both the proportions and absolute numbers are growing remarkably, and while the percentage of them is advancing more slowly in the developing world, because of its huge population it is outdistancing the developed world.

If the growing numbers of older persons were all financially independent and enjoying good health, there would be no challenge. A significant minority, however, is not so fortunate. The challenge is to assist those in need and to help people and society plan better. Were the United States to enjoy the strong economic growth it had in the 1950s and 1960s, there could be enough private and governmental revenue to support the new longevity.

But the United States and the world's governments will be saddled with a staggering amount of unfunded pension liabilities and massive health costs that will absorb the value of the economy's annual output. Lester C. Thurow, the economist, called America's old a "revolutionary class." By seeking advantages for itself, this class, Thurow believes, presents a danger to society.[1]

CHRONIC ILLNESS: CURSE AND COST

Chronic illness both saps productivity and drives up health costs. "Chronic conditions affect all ages," Catherine Hoffman and her colleagues have written.[2] Older people, however, require fewer resources than one might expect. "While the elderly were far more likely to have a chronic condition than other age groups, they actually accounted for only about a quarter of all people living in the community with chronic conditions. Working-age adults 18 to 64 years old accounted for 60 percent of all non-institutionalized persons with chronic conditions."[3]

Over 30 percent of Americans reported chronic conditions in 2000, but they accounted for more than 75 percent of medical cost: $1 trillion of $1.4 trillion.[4] They were also responsible for 96 percent of home care visits, 88 percent of prescription drug use, 76 percent of days spent in hospitals, and 72 percent of doctor visits.[5]

Over 90 percent of persons in nursing homes suffer from chronic conditions[6] and account for one of the most rapidly rising costs. About 43 percent of people over sixty-five will have at least one nursing home admission during their lifetime.[7] In 2002, the U.S. nursing home bill was $103.2 billion,[8] with public funding accounting for 64 percent and private funding for 36 percent of payments. Medicaid was the largest payer, paying for almost half of all nursing home services. While medical costs overall continue to increase faster than inflation (9.6 percent, more than twice the 3.6 percent increase in the GDP), nursing home expenditures grew more slowly than expenditures for other health services.

Annual growth in expenditures for nursing home care provided by freestanding facilities slowed to 4.1 percent in 2002, following growth of 5.7 percent in 2001. This correlates with slow growth in nursing home capacity, and a deceleration in the costs of supplies and services.[9]

By 1996, the Medicare Hospital Insurance Trust Fund (Part A) was in trouble. It was projected, erroneously, that it would be exhausted by 2001 unless significant reductions were made in payments to providers, new funds were made available, cuts were made in benefits, delivery of care was streamlined, or breakthroughs were made in research. But costs continue to advance. Moreover, it is anticipated that staggering numbers of victims of Alzheimer's disease and other dementias will overwhelm services in the twenty-first century unless prevention and treatment are found.

In 2002, persons over sixty-five accounted for $544 billion out of $1.6 trillion in health care expenditures.[10] Medicare outlays have increased from $191 billion in fiscal 1996 to $256 billion in 2004. Between 2003 and 2013, Medicaid and Medicare outlays for long-term care are projected to increase by 83 percent for nursing homes[11] and 105 percent for home care.[12]

OLD AGE, INCOME, AND WEALTH

Our society believes in two opposing myths regarding the financial situation of older persons: Old people are very rich *and* very poor. These contradictory views are due to the significant wealth inequality within the older population and to how researchers define wealth. It is true that Americans over the age of fifty (mostly those between fifty and sixty-four) control 70 percent of the nation's disposable income; however, like the rest of the country, where the top 1 percent control one-third of the nation's wealth and the richest 5 percent hold more than half of total wealth, the percentage of wealthy, older Americans is small. In 2002, of the employed individuals over sixty-five, only 7.9 percent earned more than fifty thousand dollars, and 24.7 percent earned more than twenty-five thousand; 31.5 percent of this group reported earning less than ten thousand dollars. The median income for an individual American sixty-five and older was $14,251 in 2002.

In 2000, the U.S. census reported that the median net worth of households with a householder aged sixty-five or older was $109,885 versus $55,000 for the total population, primarily as a result of higher home ownership among older persons (78 versus 66 percent). Home ownership is the largest asset for households with a householder over the age of sixty-five, accounting for $85,516 or 78 percent of their median net worth. However, these numbers do not reflect the lower median income of $23,486 for older households, compared to $50,010 for households whose primary householder is under the age of sixty-five.

The challenge of aging primarily affects women, who outlive men decisively. Most of the 30 percent of older persons who live alone are women, and they suffer from the greater risk of poverty, chronic illness, institutionalization, abuse, and vulnerability to crime. One-third of all women age seventy-five and older are financially dependent and need assistance in daily living. About 75 percent of nursing home residents are women.

CAN SOCIETY ADAPT?

Some people are pessimistic about the growing numbers of older persons. There are reasons for concern. Unlike other nations, the United States does not enjoy a national health program, and about forty-seven million children and adults (15.8 percent of the population) do not have insurance coverage. Hospital emergency rooms are an expensive and unpleasant place for the poor and uninsured to get health care. Furthermore, costs have escalated beyond inflation largely due to technology, administrative costs, and insurer profit, and only minimally, so far, from population aging. Health care, especially long-term care, is difficult to provide and is costly. It will remain a controversial and refractory challenge for years to come.

Baby boomers are not saving enough for their later years, and alarmists were claiming as early as the 1980s that Social Security was in a crisis. Reforms are necessary and the sooner they are set in motion the better. Many state and municipal pension systems are also in trouble. The private pension system is in an even greater crisis. Defined benefit private pensions are disappearing and are being replaced by defined contribution plans—401(k)s—which ultimately depend upon the stock market, with its inherent risks. These challenges are solvable but politically difficult.

Of course, we could face a public health and financial crisis if we do not solidify private pension plans and make Social Security sounder, if we cannot prevent or slow the progression of diseases, if we are unable to develop more effective methods of health care, and if we do not offer more appropriate living arrangements. The governmental and private sectors must collaborate quickly and effectively since the *velocity* of change adds to the demographic challenge.

As older persons increase in number, other notable societal and demographic changes complicate the adaptation to population aging. So-called big government is diminishing, and private enterprise for services is growing.

The need to assist those unable to care for themselves raises many issues. For one, women entering the workforce leave fewer caregivers behind, just as older persons are expressing the need to maintain their independence and not live with their families until and unless it becomes absolutely necessary.

Already, some industries are gearing up in response. Home care and the assisted living parts of the long-term care industry, for example, are growing rapidly. Of the nearly five million persons over eighty-five in the United States in the year 2000, many needed some help with one or more of the activities of daily living (ADLs), such as walking and bathing, or with instrumental activities of daily living (IADLs), such as paying bills and buying groceries. Since there are so many women in the workforce, paid and volunteer caregivers have become necessary. There will be new jobs in the "silver industries." How rapidly will the for-profit and not-for-profit service industries respond to the looming caregiving crisis?

And there are other challenges:

- Can we afford older persons? Are they a burden, given pension and health care costs? How much *can* and should society and the individual bear? How can individuals and societies finance longevity and not outlive their resources?

- Will costly physical and financial dependency take away resources from the young, creating intergenerational conflicts? How can we achieve an equitable distribution of resources across the generations?

- Will the overall aging of the population, with its burdens and costs, cause stagnation of the economy? And society at large? A weakening of the national will? In war and in peace? Or does increasing longevity and health create wealth?

- Will societies be overwhelmed by decrepitude and dementia? Or can the dementias and frailty be conquered?

- Will there be an excessive concentration of power in the hands of older persons? Will we live under a gerontocracy? Or will older people help reform government and be productive citizens?

- Will growing numbers of older people worsen the overpopulation of the already overpopulated globe? Or will concomitant declines in birthrates balance out the generations?

- How do we supply a full measure of independence and active choice throughout life and in late life?

- How can we avoid saddling future generations with burdens and liabilities while simultaneously creating a lasting infrastructure for the use of future generations, a legacy of value?

- How can we reorganize the health care system to be generous, yet affordable, while offering access and achieving outstanding performance standards?

- How can we promote the decline in disability rates through health promotion, accident prevention, worker safety, and greater access to health care?

- How can we assist paid and family caregiving? And build a long-term care system?

These and related questions require thoughtful answers and wise policies. Policymakers must consider the fact that it is *not* the revolution of longevity per se but other societal, technological, and economic factors that, from time to time, have caused economic stagnation. It is generally accepted that *wealth* generates health. I argue that the reverse is true. *Health, longevity, and aging engender wealth.*[13]

Some pundits exaggerate the costs that arise as a result of population aging—specifically pensions and health care. They present disturbing actuarial studies and appeal to the political establishment to avert an "aging crisis" by reducing public pensions and containing health care costs. By way of solutions, they offer added taxation, benefit cuts, and partial or complete privatization. They recommend greater individual responsibility, the competitive marketplace, and pronatalism (policies promoting the increase in birthrates) and immigration as ways to avert the so-called catastrophe of aging.

The work of some economists of diverse schools of thought, including those at the University of Chicago, Harvard, Yale, UCLA, the RAND Corporation, the University of Belfast, and the International Longevity Center, offer a different, more optimistic view.

David Cutler and Mark McClellan, for example, write, "The benefits from just lower infant mortality and better treatment of heart attacks have been sufficiently great that they alone are about equal to the entire cost of insurance for medical care over time."[14]

David Bloom and David Canning have demonstrated that nations that have a five-year advantage in life expectancy show significant increases in gross domestic product. It is revealing to reexamine the concept of the gross domestic product and to incorporate within it the advantages of improved health, conceptualized by economist William Nordhaus as "*health income*."[15] Nordhaus measures what he calls the "real output" of the health care industry and estimates the range of the dollar value of preventing an individual fatality at between $600,000 and $13.5 million. He settles on three million dollars as a reasonable figure.

Currently, conventional measures of national income and output exclude the value of improvements in the health of the population. Nordhaus developed a methodology to show how standard economic measures would change if they adequately reflected improvements in health status. He posits that the "value of increase in longevity in the last 100 years is about as large as the value of growth in non-health goods and services."

James Smith of the Rand Corporation has explained the role of health in creating wealth this way: He imagines someone who remains healthy all his life. This person does not miss school or work, he saves and invests, and he can remain productively engaged late in life.[16]

A second major reason that health creates wealth resides in the existence of the so-called silver industries. (See Table 2.1.) Older people constitute a powerful and growing market, variously called the silver industries, the mature market and the senior market, which are all as significant as the youth market of the baby boom 1960s. Indeed, they are more so. Longevity affects the entire life course, including the amount people spend on health and in the financial services industries. Optimism about the future encourages people to save and invest. The existence of life insurance and annuities illustrate how the future can be a powerful spur. People also seek medicines and surgical interventions to preserve their health and augment their longevity. And there is "luxury" spending—on grandchildren, and travel and recreation. Many states and cities in the South recruit older persons because of their value to the local economy.[17]

Robert Fogel writes,[18] "Public policy should not be aimed at suppressing the demand for health care. Expenditures on health care are driven by demand, which is spurred by income and by advances in biotech that make health interventions increasingly effective." He continues later, "The increasing share of global income spent on healthcare expenditures is not a calamity; it is a sign of the remarkable economic and social progress of our age."

Fogel argues that large expenditures on health care are a boon to the economy, and that the care industry will lead industry, just as railroads, automobiles, or computers did. If there are difficulties with the funding of health care, or with Social Security and pension schemes that were predicated on a lower life expectancy, the problems lie only with a "clumsy system of financing," Fogel argues.[19]

The fact that health and longevity generate wealth helps offset the fears of those who believe we cannot afford old age. But beyond concerns about maintaining productivity, we must address the underlying terror and distaste for aging, replete as it is with hysteria and anger, which ultimately becomes a self-afflicting prejudice. There is a related, seeming contradictory fear of longevity and boredom, and we must come to understand the personal and cultural denial of aging and the age prejudice that grips us.

We need new attitudes and policies to manage the new longevity. We could start by revaluing older persons, giving them productive roles in society, and acknowledging their value in the marketplace today no less so than the value of the youth market in the 1960s. Further, we need an alliance among the generations that confronts reality: The goals of older people and their children are similar. *Instead of blaming older people for health costs, we must remember the underlying reason for them—happily, we have postponed the high morbidity and mortality that commonly occurred among women in childbirth, their newborn and young children.* All else follows, including the better design of health care and pension systems and the support of medical research to end senility and frailty, and other scourges of old age, among other policy initiatives. When these are achieved, we will still face the challenge of the maturation of society itself, the construction of new attitudes and societal structures, to demonstrate the values of the new longevity. For why live longer if we cannot live well, enjoy the good life, and continue to contribute to society?

The new longevity is entering the national consciousness through the slow growth of provident thinking. While neither society nor the baby boom generation is prepared for their old age, is it too late to begin planning?[20]

Throughout the twentieth century there have already been myriad, effective societal adjustments to the increasing numbers and proportions of older people; in America, there is Social Security and Medicare. In addition, there are private pension programs such as 401(k)s, social service agencies, senior centers, the federal Administration on Aging, and the research programs of the National Institute on Aging (NIA). There is much to be done, but what has been accomplished gives us hope for eventual success.

CHAPTER 3

"THE GREEDY GEEZER": THE MANUFACTURED CLASH OF THE GENERATIONS

Intergenerational conflicts are part of world history. In 1968, worldwide youth uprisings marked the beginning of modern generational politics. Since then, politicians and journalists periodically have warned that a new and awesome clash of the generations is in the making because older people are draining society's resources at the expense of the young. In 1988, the cover of *The New Republic* depicted angry "greedy geezers" with garden trowels, golf clubs, and fishing rods in hands marching down a nameless hill in unison, presumably poised to attack and exploit society (see Figure 3.1).

If such a confrontation should indeed arise, it will result from poor planning, economic disparities, political manipulation, or media hype. So far there is no evidence of significant intergenerational conflicts in American society or elsewhere. Polling data from major survey companies such as Harris, Roper and Yankelovich bear this out.[1] *Children and youth are not the natural enemies of old people.* A few thoughtful observers, such as Robert Binstock, political scientist at Case Western Reserve University, predicted that "compassionate ageism" expressed in benefits and programs for older persons might backfire and lead to scapegoating of the old. Binstock also shrewdly refers to the dangers of "apocalyptic demography."[2]

Some politicians have worked especially hard to push the idea that the old are benefiting at the expense of the young. Begun in the 1980s, the Americans for Generational Equity (AGE), led then by former Senator David Durenburger (R-Minn.), former Congressman James Jones (D-Okla.), and former Colorado Governor Richard Lamm, expressed deep concerns for the baby boom generation *and posterity*. In a story in *The Washington Post,* Lamm said, "Simply put, America's elderly have become an intolerable burden on the economic system and the younger generation's future. In the name of compassion for the elderly, we have handcuffed the young, mortgaged their future, and drastically limited their hopes and aspirations."[3] Could it be that the leaders of AGE would use their organization as a political vehicle to exploit the concerns of the baby boomers in order to win votes? Another small organization, the Third Millennium, which speaks for Generation X (those born between 1965 and 1978), wants to cut Social Security and "stop paying the greens fees for well-heeled retirees."

The central objects of these intergenerational attacks are Social Security and Medicare, which enjoy great popularity among Americans,

Figure 3.1—Greedy Geezers   Permission by *The New Republic*

The New Republic / March 28, 1988
GREEDY GEEZERS

including young people. Moreover, we cannot trust the motives of those who perpetuate Social Security scare stories and declare that it is a state-run Ponzi operation. Paranoia has provoked many to believe that they will not receive their Social Security benefits when their time comes. In *The Return of Thrift: How the Collapse of the Middle Class Welfare State Will Reawaken Values in America,*[4] Phillip Longman envisioned the "imminent collapse of the middle-class welfare state." *This was twelve years ago, in 1996!*

Although the public does not generally realize it, Social Security serves younger people—through its life insurance, disability, and survival provisions. For example, nearly four million children and four million disabled persons receive Social Security. Moreover, the value of life insurance under Social Security for a 20-year-old is well over $200,000. Boomers and Generations X and Y (Generation X was born between 1965 and 1980 and Generation Y between 1981 and 1995) would do well to fight to preserve Social Security and Medicare rather than lament their presumed bankruptcy.

Although genuine issues of equity in our society have not been resolved,[5] it is *not* clear why and how they would be solved through attacks upon the old or by the reduction of what has been achieved on their behalf. The underlying issues are the allocation of a society's resources and the responsibilities of its constituents; that is, the private, nonprofit, and governmental sectors. The actual clash is not between the generations but, in the extreme, between positions taken by advocates of social protections and those of a pure laissez-faire system. The growing disparities of wealth and income are most disturbing.

INTERGENERATIONAL EQUITY AND ACCOUNTING

Discussion of intergenerational equity must be based on facts, not emotions. To discover if it indeed exists, all resources that go to the young and old must be placed on the accountant's table, including public and private resources. Public resources that derive from federal, state, and local taxes need to be differentiated. Having made such a tally, Merton Bernstein, a professor of law at Washington University, found that public funds for children and those for older persons are about equal. However, he did so in 1991 when $362 billion was spent on the aged, $392 billion on the young.[6] The costs of education, social security, and health care (through Medicaid) provided for children just about balanced the costs of health care (Medicaid and Medicare) and the Social Security that older persons received.[7] These calculations should be periodically updated. This is not to suggest that society is providing enough for either children or old people. But the necessary discussions and decisions must be based upon the same data.

Even excluding inheritances, most intergenerational transfers of money flow from older persons to their adult children. This is true of services as well.[8] Eventually, of course, children (and governments) inherit all remaining resources, including the intellectual infrastructure (such as museums, libraries, schools and universities—knowledge and, of course, the physical infrastructure such as buildings and bridges) created by the previous generations.[9] Besides its debts, we must also consider that the older generation saddles posterity with much that is good.

Generational accounting ought to be included in the U.S. budget, but only if it comprehensively integrates public and private resources—capital investments and the material, social, and intellectual landscape—in a serious effort to analyze intergenerational equity that includes society's obligations to the unborn. We must address the ancient question of what responsibility the rich have to the poor, and the healthy to the ill. How level is the playing field? Is each newborn given the opportunity to build a good life and contribute to society? What are the advantages of the universal risk pool?

To counter the potential for age divisiveness, the Gerontological Society of America sponsored an excellent book-length study, "Ties That Bind," that emphasized intergenerational interdependence.[10] In addition, the National Council on the Aging and the Child Welfare League of America formed Generations United, a coalition actively promoting the common interests of children, youth, families, *and* older persons.

The Children's Defense Fund joined in. Older persons are parents and grandparents, after all, and they share what burdens exist. There are over 2.42 million grandparents in midlife and older who are already providing primary care for their grandchildren. The majority of grandparents raising grandchildren are between the ages of fifty-five and sixty-four. But about 8.1 percent are seventy and older. About five million children are the financial dependents of grandparents.[11]

If funds such as Social Security are taken from the old, it is unlikely that they would be transferred to children. In fact, it would be illegal to do so directly without a countervailing reduction in Social Security income taxes. But if help for children were coupled with reductions in entitlements for the old by congressional action, there would be a retrogressive and undesirable return of many older persons to the impoverished status so many experienced only a few decades ago. Social class, income, ethnicity, and race adversely affect American children and American society in general. Growing economic disparity is highly undesirable for a variety of obvious reasons, among them business reasons, since prosperity depends upon consumer power.

As will be discussed in Chapter 14, the need to reform Social Security is realistic, as is the addition of a new program of mandated or voluntary supplementary private investments and savings to augment income in the later years. But there are dangers in replacing universal entitlements by means testing. Whatever is saved is not worth the political weakening of Social Security, risking its transformation into a program for the poor and making it, thereby, a poor program. Those who resent the millionaires who receive Social Security which they "don't need" can work to ensure a more effective and fair progressive taxation of millionaires. Income testing (i.e., taxation) is preferable to means-testing and less humiliating. Moreover, the well-to-do themselves who do not need or want the money could voluntarily donate their Social Security payments to their favorite charitable activities, to the government, or to Social Security itself.

However, it should be understood that inherited wealth will *not* be evenly distributed but will rather reflect the growing economic disparity in our society. The second wave of baby boomers (born in 1955 and beyond) and the succeeding cohorts (Generations X and Y) potentially face a lower standard of living than have their parents. The majority of working Americans earn annual family incomes of fifty thousand dollars per year.[12]

Another deeper, psychological element driving the alleged generational rebellion against the old is the dread of aging. Ageism and a youth quest are prominent in our society. Even older people are seeking corrective cosmetic and medical procedures. Do they hate themselves? Have they been driven to do so by a society hostile to the accompaniments of age? "Old is beautiful," says Maggie Kuhn of the Gray Panthers. Kathleen Woodward, director of the Stimpson Center for the Humanities at the University of Washington, has observed that our very language polarizes youth and age, yet each person is "multi-aged" because of their memories.

Traditionally, the old annoy the young and vice versa. The *Wall Street Journal* has reported that older people in Beijing were irritating the young, presumably creating a "generational gap" by reviving the "Yang Ge" folk dance brought to the nation's capital by the Communist People's Liberation Army in 1949. The decibel level was driving the young mad.[13] On the other hand, the decibel level of rock and roll may madden the old.

INTERGENERATIONAL DEPENDENCY

A dependency or support ratio is much discussed in the context of intergenerational relations. It is usually defined as the ratio of persons sixty-five and over per one hundred between eighteen and sixty-four. This number is extraordinarily misleading because the notion that all persons over sixty-four are dependent upon those between eighteen and sixty-four is false. More accurate is *the overall total dependency or support ratio.* This is the ratio of persons under eighteen and over sixty-five per hundred who are between eighteen and sixty-four. In the

United States this ratio is actually declining to less than it was in 1970 because of declining birthrates (see Table 3.1). Because not all persons between eighteen and sixty-four are employed—some may be pursuing their education or unemployable due to physical and mental illness—the dependency ratio should include all those who are working and those who are not. In 1964, there were seventy million working and 124 million not working. In 2004 there were 139.6 million working and 108.2 million not working.[14] Therefore, in the 1960s, as economist Richard C. Leone points out, "the worker to non-worker ratio was actually worse than today."

Older persons in society provide major support to their adult children and grandchildren in the form of direct financial assistance or the sharing of housing and providing child care and other services. Many older persons have savings and pensions or they still work. Economic and policy analyses ought to tell us how much an older person and a child (through maturity) costs on average.[15] We should examine personal needs at different stages of the life cycle. Society would be better served if more financial help were available to young persons during family formation and career development. Payment structures could better fit the stages of life. *Clearly, we do need generational economics.* But of a very comprehensive kind.

The crucial factor underlying the so-called dependency ratio is productivity. The idea that only so many jobs exist is a myth, for history has demonstrated otherwise. Women, for example, have been absorbed into the labor market.

International Comparisons

Table 3.2 informs us that nations that are not as wealthy as the United States have much higher percentages of older persons. Nonetheless, some are able or choose to provide even more generous programs for their older citizens.[16] I do not want to be like the journalism major who wrote beautifully about a soccer match but forgot to include the score. While Europe's welfare states are undergoing reevaluation in the light of changing economic conditions, I have been impressed by the near absence of ad-hominem attacks upon the unemployed and older people.

To help offset granny bashing, the Greedy Geezer image, and other forms of scapegoating, the United States should embark on a bold effort to bring the generations closer. One way to do this would be to inaugurate a children's initiative. Organizations such as AARP (formerly, the American Association of Retired Persons) and the Alliance for Retired Americans (formerly, the National Council of Senior Citizens) could lead the way. A national children's initiative is an undertaking older people should consider.

In a Grandpeople's Children's Initiative, older persons could responsibly focus on the multiple and increasing medical, social, and educational needs of children of all social, economic, racial, and ethnic groups of society. They could spend time with children, taking care of them and teaching them skills. More than that, enlightened self-interest demands that middle-aged and older generations address the needs of children who as working adults will support them later (e.g., through the pay-as-you-go Social Security system and caregiving), provided that children have the opportunities necessary for a productive future. This is the unspoken intergenerational contract.

There are those who fear a holocaust—intentionally hastening the deaths of older people, especially by neglect—for older people in this century. This includes individuals from diverse backgrounds, such as Anthony DeBono, Maltese physician and chair of the 1982 UN World Assembly of Aging; Claude Lanzmann, French author and filmmaker who made *Shoah;* and Yoshio Gyoten, Japanese physician and TV commentator. While I understand these concerns and regard them as serious, I am confident that the challenges are soluble if life is seen in its true character, as a dynamic, ongoing process. We are all beholden to the life cycle; inevitably, the young become the old who were once young. No one generation or stage of life is working against another.

The ultimate answer to all questions about generational inequity is the intergenerational contract—what parents do for their children, children must do for their parents—writ large. So long as we remember the natural cycle of the generations in all its profound and rich ramifications, we can deal with the necessarily creative ups and downs in intergenerational relations: the struggle by the young for identity and autonomy and the hopes of the old for a decent ending to life. Neither group need suffer.

CHAPTER 4

AGEISM: ANOTHER FORM OF BIGOTRY

Just as racism and sexism are based on ethnicity and gender, ageism is a form of systematic stereotyping and discrimination against people simply because they are old. As a group, older people are categorized as rigid in thought and manner, old-fashioned in morality and skills. They are boring, stingy, cranky, demanding, avaricious, bossy, ugly, dirty, and useless.

An ageist younger generation sees older people as different from itself; it subtly ceases to identify with its elders as human beings.[1] Old men become geezers, old goats, gaffers, fogies, coots, gerries, fossils, and codgers, and old women are gophers and geese. A crone, hag, or witch is a withered old woman.[2]

Ageism takes shape in stereotypes and myths, outright disdain and dislike, sarcasm and scorn, subtle avoidance, and discriminatory practices in housing, employment, pension arrangements, health care,[3] and other services. Older persons are subject to physical, emotional, social, sexual, and financial abuse. They are the focus of prejudice regarding their capacity for work and sexual intimacy,[4] which Freud described as the two most important human activities. Taking away the validation of work or purposeful activities and demeaning the capacity for love are surely the most profound forms of age prejudice.

Historically, older persons have been venerated in most societies and cultures *in word*, although not always in deed. In fact, to be old or disabled was always a liability for practical reasons. Nomadic groups from North Africa to Alaska abandoned their old when the welfare of the entire tribe or group was at stake.

The term *ageism*, which I introduced in 1968,[5] is now part of the English language. It is identical to any other prejudice in its consequences. The older person feels ignored or is not taken seriously and is patronized. Anthropologist Barbara Myerhoff speaks about "death by invisibility" when she describes an older woman who, "unseen," was "accidentally" killed by a bicyclist.[6]

This invisibility extends to emergencies, such as the tragic case of September 11, 2001, in New York City. Animal activists evacuated dogs and cats within twenty-four hours after the World Trade Center was attacked, while disabled or older persons were abandoned in their apartments for up to seven days before ad hoc medical teams arrived to rescue them.[7] Older persons were also invisible in the devastation caused by Hurricane Katrina in New Orleans.

Reminiscent of the great social scientist George Mead's concept of the "looking glass self,"[8] older persons may turn ageist prejudice inward, absorbing, accepting, and identifying with the discrimination. Some examples:

- Simone De Beauvoir, author of *The Coming of Age*, described her disgust at growing old, although she wrote lovingly of her own mother's aging in *A Very Easy Death*.

- Comedian George Burns noted the unfortunate tendency of old people to conform to their stereotype—what he called the old person's "act"—by learning to shuffle about and decondition in a kind of identification and collaboration with the ageist society that demeans them.

- Yale psychologist Becca R. Levy reports that constant bombardment of negative stereotypes increases blood pressure. *Ageism can make an older person sick.*[9]

Advertisements and greeting cards depict older persons as forgetful, dependent, childlike, and—perhaps the ultimate insult in our society—sexless. Conversely, older people who continue to have sexual desires are dirty old men and ridiculous old women.

Wrinkles, crow's feet, liver spots, and dull skin are disparaged in our youth-dominated culture and exploited by the cosmetics industry and plastic surgeons. Women who have relied upon their appearance for self-definition and men and women who have depended upon a youthful appearance in their work are up against overwhelming odds. The clock does not stop. When does one cease to be beautiful and start on the journey to being over-the-hill? How many women past fifty can look like model Lauren Hutton or Susan Sarandon? How many men and women can overcome disability with elegance and style?

A study conducted by the American Academy of Facial, Plastic and Reconstructive Surgery revealed that baby boomers have received nearly a quarter of a million face-lifts and other cosmetic surgeries. Most of these patients were over fifty. The Associated Press has quoted Karen Seccombe, a University of Florida sociologist, who said, "The thought of saggy breasts, hair loss or wrinkles doesn't sit well with people who have grown up emphasizing fitness and youth."

The film and television industries help to perpetuate ageism.[10]

- Less than 2 percent of prime-time television characters are sixty-five or older, although this group is 12.7 percent of the population.

- 11 percent of male characters between fifty and sixty-four are categorized as old versus 22 percent of female characters.

- 75 percent of male characters sixty-five and older are characterized as old versus 83 percent of female characters of the same age.

- Only one-third of older characters are women.

- Middle-aged and older white males have joined women and minorities on the sidelines, as white men under forty get most of the jobs writing for television and film. Employment and earning prospects for older writers have declined relative to those for younger writers.

- According to one study, approximately 70 percent of older men and more than 80 percent of older women seen on television are portrayed disrespectfully, treated with little if any courtesy, and often looked at as "bad."

- Although Americans who are forty and over are 42 percent of the American population, more than twice as many roles are cast with actors who are under forty.

But there is some good news, too. By the 1990s, soap operas such as *The Guiding Light* were presenting older characters having more love affairs and not just worrying about their children. In 1994 *New York Magazine* put Paul Newman on the cover, calling him "The Sexiest (70 year old) Man Alive," and *More* magazine offered women over forty an alternative to those that cater to women in their twenties. Older models began to make their appearance in general women's magazines, too. One widely circulated magazine advertisement in 1994 described "Betty Mettler, age 101, Noxzema user since 1925."

OUR CULTURE'S FEAR OF GROWING OLD

As Tolstoy noted, "*Old age is the most unexpected of all the things that happen to a man.*"

The underlying basis of ageism is the dread and fear of growing older, becoming ill and dependent, and approaching death. People are afraid, and that leads to profound ambivalence. The young dread aging, and the old envy youth.[11] Behind ageism is corrosive narcissism, the inability to accept our fate, for indeed we are all in love with our youthful selves, as is reflected in the yearning behind the expression "salad days."

Although undoubtedly universal,[12] ageism in the United States is probably fueled by the worship of youth in a still-young country dominated by the myth of the unending frontier. In 1965, the Who, a British rock group, sang, "I hope I die before I get old," while in America "you never say die."[13] Hollywood veils older actresses with gauzy lens filters. Moreover, age carries less authority.

The powerful imagery of the birth and adoration of the infant Jesus, and the journey of the Magi to see the Christ child, describes a birth of hope. How this contrasts with the final years of life! Children are seen as the future; older people, the past. Grimm's fairy tales depict gnarled and evil old women cursing innocent and beautiful youths with spells and afflictions.

Denial is a close cousin of ageism; in effect, it eliminates aging from consciousness.

One of the striking facts of human life is the intensity with which people avoid aging. Narcissistic preoccupation with our own aging and demise and perhaps, according to Freud, the inability of the unconscious to accept death make it difficult for society as a whole to deal with the challenges of aging. Note our gallows humor at birthdays, the money we spend on cosmetic surgery, and the popularity of anti-aging medicine. This was not always the case. In Europe in generations past, young men in high positions wore wigs they had powdered white in an attempt to appear older and, by implication, wiser. Today, men flock to cosmetic surgeons and colorists to preserve the illusion of youth.

AGEISM AND ECONOMICS

Is ageism in large measure a function of economics? Are older people no longer contributing to society? Are they a drain on the economy? As the numbers and proportions of older persons increase in the industrialized world and productivity becomes the essential measure of an individual's worth, the status of older persons has declined.

Although no mathematical relationship is likely, there appears to be a threshold in societies at which point the number or proportion of older persons is seen as a burden. Veneration is replaced by contempt. In primitive societies, the role of the older man as the oral historian and wise counselor who knows how conditions have changed over time was valued, especially if he was in control of land. With the movement away from agrarian and toward industrial societies, this power base eroded, except on the rare occasions when the older person was able to maintain power, wealth, and income.

The severe cutback in services following the $750 billion tax cut inaugurated by the Reagan administration in 1981 brought steady criticism that Social Security and Medicare provide entitlements for older people that are denied the young. I have called this the politics of austerity, leading in part to what has been termed the "New Ageism."[14] The new ageism is concocted from a dangerous brew that envies some old people for their improved financial status while resenting poor old people for being tax burdens and those who are not poor for making Social Security so costly. Two letters to the editor of *Newsweek* (April 3, 1995) exemplify this agenda. One described AARP as "ruthless in its pursuit of power, privilege and special interests for its members." The other writer wrote, "I submit that people over 65 should not be allowed to vote." (The shame that these attacks engender in their target is a measure of their success.)

In a frontal assault against the welfare state and its specific social protections, the new ageism derides people sixty-five and older for their avaricious nature. They ignore the fact that older people must be 8 to 10 percent poorer than an adult under sixty-five to be officially counted as poor by the government.[15] In truth, poverty among older people has not been ended, despite progress. As Table 4.1 shows, nearly 25 percent have incomes of less than $39 per day.

The poverty index is outdated. It was estimated in 1964 as the amount of income necessary to meet essential needs, based on the

Department of Agriculture's economy food plan for emergency use designed to keep a healthy person alive and functioning reasonably for thirty days. At that time, people spent approximately one-third of their income on food. Today food accounts for about 10 percent of one's income; housing, transportation, and medical care are the major costs.

Table 4.2 shows the limitations of the cost-of-living adjustment under Social Security in the face of the increased costs of Medicare and inflation.

Although older persons are about as likely to be poor as younger persons, income and assets are distributed more unevenly among older persons and concentrate highly among a tiny percentage of the rich old. As mentioned, the old, especially older women, are, on the whole, the poorest of the poor. Those eighty-five and older have the lowest income and the greatest incidence of chronic illness. They are more likely to require medical services and medications but less able to afford the care.

Only about 8 percent of people over sixty-five have annual incomes in excess of fifty thousand dollars, and they are often still employed. According to the U.S. Census Bureau, 18 percent of men and 10 percent of women sixty-five and older are currently in the civilian labor force.[16] The dismissal of talented, functional older people from the workforce on the grounds that there is a limited availability of work and that old people take away jobs from the young (the so-called fixed lump of labor noted in the last chapter) is based on an inaccurate assumption and is costly.

Widows are the primary victims of poverty in later years and thus bear the brunt of ageism's assault. Women's luck in living longer than men has, paradoxically, compounded their problems. Of older women, 41 percent are near-poor,[17] contrasted with 17 percent of older men. The fact that 75 percent of poor older persons are women reflects their lower wage levels during their working years, inadequate and inequitable Social Security coverage, and the increased risk of financial devastation from widowhood. Over half of widows become poor after their husbands die, probably due to consuming medical and funeral expenses and lost pension income. When New Ageists talk about denying health care or cutting entitlements to the old they are really talking about denying these benefits to poor old women.

ELDER ABUSE

Elder abuse is a widespread phenomenon that affects older adults who live in rich and poor nations alike. In the United States alone it is believed that as many as 1.2 million older adults are physically abused or neglected each year. Elder abuse takes many forms, including physical, emotional, financial, and sexual abuse—often by family members. It may involve neglect, such as the failure to provide food, shelter, clothing, medical care, and personal hygiene, as well as narcotic over-medication.

In 2004, UN Secretary General Kofi Annan released a report on the abuse of older persons that mentioned practices such as the ostracism of older women, which occurs in some societies when they are used as scapegoats for natural disasters, epidemics, or other catastrophes. The report stated: "Women have been ostracized, tortured, maimed or even killed if they failed to flee the community."

The World Health Organization (WHO) reported that 36 percent of nursing home staff in the U.S. reported having witnessed at least one incident of physical abuse of an older patient in the previous year, and 10 percent admitted having committed at least one act of physical abuse themselves. This represents sexism as well as ageism, for about 75 percent of nursing home residents are women. Other statistics are equally alarming:[18]

- 1 million to 3 million Americans sixty-five and older have been injured, exploited, or otherwise mistreated by someone on whom they depend for care or protection.

- Estimates of the frequency of elder abuse range from 2 percent to 10 percent.

- Only one out of six incidents of elder abuse, neglect, exploitation, and self-neglect is brought to the attention of authorities.

- Only twenty-one states report that they maintain an elder abuse registry/database on perpetrators of substantiated cases, and less than half of states maintain a central abuse registry.

- It is estimated that each year five million older Americans are victims of financial exploitation, but only 4 percent of cases are reported. Many of these cases involve the unauthorized use of older person's assets and the transferring of power of attorney for an older person's assets without written consent.

- Of the nearly $1 billion National Institute on Aging budget, only $1.7 million goes to NIA elder abuse and neglect research funding.

AGEISM IN HEALTH-CARE SETTINGS

Ageism can be invoked by aesthetic revulsion. Especially when weakened by disease, older persons can be disheveled, unwashed, and appear ugly and decaying. Some older persons "let themselves go" and unwittingly add fuel to the fire. Sphincters loosen, depositing stains and smells. Ear and nose hairs grow more quickly in older bodies, as does the cartilage, causing the nose and ears to enlarge. Some profound and common disorders of old age—mobility problems, dementia, and incontinence—are unattractive and provoke a negative response.

When older men or women are malodorous, scabrous, or disturbing in dress and language, they can scare, disgust, and discomfort younger people. Such older persons become untouchable. (Touch is powerful and therapeutic. Some older persons living alone have not been touched for years.)

Medical schools unwittingly promote the virus of ageism.[19] Fresh out of college, young students are confronted with aging and death and their own personal anxieties about both. They are left to their own devices to insulate themselves from anxiety and pain about disease, disability, disfigurement, and dying. A cadaver that requires dissection is usually the first older person medical students encounter, and they are not ordinarily provided with effective group or individual counseling, either at the time of dissection or later, upon the death of their first patient.

Defense mechanisms like gallows humor, cynicism, denial, the invention of negative language, and facetiousness are common. Long hours in medical training lead to angry exhaustion and feelings of being "put upon."[20]

It was in medical school that I first become conscious of the medical profession's prejudice toward age. For the first time I heard such insulting epithets as "crock," which was used to describe middle-aged women and older patients who were labeled hypochondriacal because they had no apparent organic basis for their complaints, as well as having many symptoms, and "GOMER" (Get Out of My Emergency Room).

The hidden curriculum in medical schools undermines student's idealism and can compromise their education. For example, in some studies up to thirty-five percent of doctors erroneously consider an increase in blood pressure to be a normal process of aging. In physical diagnosis courses, medical students meet older people who are stripped of their individuality and seen as archives or museums of pathology, rather than as human beings. Men and women in their eighties are particularly valuable in these sessions because they often have a plethora of symptoms and conditions about which the student must learn.

In addition, few medical school graduates will practice geriatrics, and practicing physicians often do not invest the same amount of time dealing with older patients. Medicare expenditures per capita steadily decline as people grow older. In fact, a UCLA study reported that, as people enter their forties, physicians spend less time with them per encounter. Logically, it should be the reverse since medical problems tend to increase as we grow older, and the ramifications are sobering. Sixty percent of adults over sixty-five do not receive recommended preventive services, and 40 percent do not receive vaccines for flu and pneumonia. They receive even less preventive care for high blood pressure and cholesterol.

Some doctors question why they should even bother treating certain problems of the aged; after all, the patients are old. Is it worth treating them? Their problems are irreversible, unexciting, and unprofitable. Their lives are over.

Between 1955 and 1966, Morris Rocklin, a volunteer in the NIMH Human Aging Study, was studied until he turned 101 years of age. Rocklin complained about his painful right knee to his physician, who said, "What do you expect at your age?" To this typical statement by a physician, Rocklin replied indignantly, "So why doesn't my left leg hurt?" The symmetry of the human body offers a good test of the realities of medical ageism. Rocklin's oft-quoted response has been used by many geriatricians to educate medical students on the topic of ageism.

NURSING HOMES: AGEIST SCANDAL

Nursing homes are licensed by the states and must meet federal standards to participate in Medicaid or Medicare. About 95 percent of the nation's sixteen thousand nursing homes (which house 1.5 million men and women) participate in those programs. According to a government study conducted in 2002,[21] nine of ten nursing homes in the United States lack adequate staff, and nurse's aides provide 90 percent of the care. In most nursing homes, the report said, a patient needs an average of 4.1 hours of care each day—2.8 hours from nurse's aides and 1.3 hours from registered nurses or licensed practical nurses.

In 2000, over 91 percent of nursing homes had nurse aide staffing levels that fell below the thresholds identified as minimally necessary to provide the needed care.[22] In response, the Department of Health and Human Services concluded that "it is not currently feasible" for the federal government to require that homes achieve a minimum ratio of nursing staff to patients. Nursing homes would have to hire 77,000 to

137,000 registered nurses, 22,000 to 27,000 licensed practical nurses, and 181,000 to 310,000 nurse's aides. This would take $7.6 billion a year, an 8 percent increase over current spending. The solution given by the Bush administration was to encourage nursing homes to adopt better management techniques so nurse's aides can achieve high productivity and, ultimately, to rely on market forces.

AGE-BASED HEALTH CARE RATIONING

Medical ageism is prevalent in preventive tests for cancer[23] and treatment of a variety of illnesses, some of them life threatening. For example:

- Only 10 percent of people sixty-five and over receive appropriate screening tests for bone density, colorectal and prostate cancer, and glaucoma. This despite the fact that the average age of colorectal cancer patients is seventy, that more than 70 percent of prostate cancer is diagnosed in men over sixty-five, and that people over sixty are six times more likely to suffer from glaucoma.

- Chemotherapy is underused in the treatment of breast cancer patients over sixty-five, even though for many of these patients it may improve survival.

- In a cost-cutting effort to reduce supposedly unnecessary medical tests in 1998, the American Cancer Society and government health agencies determined that if an older woman had no abnormalities in a Pap smear for three years in a row, she could be tested less often. Yet over 25 percent of cases of cervical cancer occur in women over sixty-five!

- The pelvic examination is often deferred because many doctors, especially men, do not like to do it. (In both men and women, the rectal examination may meet a similar fate.)

- Although deaths due to ischemic heart disease disproportionately affect persons over sixty-five (85 percent in the United States and 87 percent in France), few national comparisons focus on older people. For example, WHO's MONICA Project—the important international longitudinal study that monitors cross-national trends in cardiovascular disease—focuses on death before sixty-five.

- A patient under the age of seventy-five who is admitted with a heart attack is six times more likely to receive blood clot-dissolving drugs such as streptokinase than a patient over seventy-five, even though data indicate the value of streptokinase in improving the chances for survival of older patients.

- Advanced surgery for Parkinson's disease is less available to older persons.

Unless older people are knowledgeable or have strong advocates, even the more affluent members of our society experience age-determined limits in medical care. Parenthetically, when malpractice suits are won, older persons usually receive lower monetary awards.

Ageism in Research Protocols

Although there are legal requirements that women and racial minorities be appropriately represented in clinical studies and clinical trials funded by the National Institutes of Health, no such regulation exists regarding older persons, yet they are the largest patient population and the most frequent users of prescription drugs. In 1989, the Food and Drug Administration did issue guidelines, not requirements, recommending that older persons be appropriately represented in studies of drugs they are likely to use. Not much has changed, however;[24] 40 percent of clinical trials between 1991 and 2000 excluded people over seventy-five.

FIGHTING AGEISM

We can treat the psychosocial disease I call "ageism"[25] by working to transform cultural sensibility and legislative initiatives. But ageism remains both gross and subtle, and omnipresent, despite the fact that prejudice against age is a prejudice against everyone. We all chance to become its ultimate victims as longevity increases.

A key intervention against ageism comes from the recognition that older people themselves are an economic power. Japan has the most rapidly growing population of older persons in the world, as well as the highest life expectancy. When its Ministry of International Trade and Industry became excited by the "silver community concept"—the idea of establishing communities for their older citizens in other countries—many in Japan reacted negatively. When Spain heard about the plan, however, it was interested, for it realized that this novel concept could be a source of jobs.

If silver communities are economically valuable for Spain, then they are economically valuable for Japan itself, for the United States, and indeed potentially for all countries. There is a lot of gold in geriatrics, as *The Wall Street Journal* once reported, when one considers the vital connection between producers and consumers. Thus, the so-called high cost of health and social services produces jobs and consumption. Looked at another way, the health care enterprise is the largest producer of jobs, and we cannot forget its contribution to the gross domestic product.

The same can be said of Social Security. Indeed, Social Security offers stability to the economy when it is slowed. Pension funds (other than Social Security), one of the largest sources of capital formation in our country, own about one-quarter of American stocks and bonds.

Older persons may be spared some manifestations of ageism by virtue of being seen as a market in a capitalist society. And with money comes power. They need to be productive and develop a philosophy of responsible aging if we are to fight ageism. Survival is closely associated with individuals' views of themselves, as well as their sense of continued usefulness. In 1963, the National Institute of Mental Health Human

Aging Studies (1955–1966), a multidisciplinary 11-year longitudinal study of community-residing older men, found that major factors in a persons' experience of aging and adaptation to it were disease, social adversity, economic deprivation, personal losses, and cultural devaluation, as well as personality and previous life experience.[26]

Research Intervention

Heavy investment in biomedical, behavioral, and social research is the ultimate cost containment, the ultimate disease prevention, and the ultimate service. When we eliminate frailty and dementia, we will have revealed a very different old age. When we are able to prevent and treat Alzheimer's disease, the polio of geriatrics, we will empty half of our country's nursing home beds, which I consider the iron lungs of geriatrics. By investing the necessary dollars in research we can gain freedom from senility and demolish the myth of the inevitability of senility and debility in old age, while providing for an infusion of new funding for aging and longevity research (a topic I will cover in depth in Chapter 7). A better understanding of what accounts for the difference in life expectancy between men and women, and the development of means to assist men to catch up with women by living longer, will also do much to overcome many of the problems of age, as well as ageism.

LEGISLATION TO PROTECT OLDER WORKERS

Legislation, such as the Age Discrimination in Employment Act (ADEA), as well as other social protections, including entitlements and long-term care, are antidotes to ageism.

Title VII of the Civil Rights Act of 1964, which prohibits discrimination on the basis of race, color, national origin, sex, or religion, did not cover older persons. It should have and should be amended to do so now. The act instructed then Secretary of Labor Willard Wirtz to conduct a study and provide recommendations on "legislation to prevent arbitrary discrimination in employment because of age." The resulting Wirtz report in 1965 recommended legislation that resulted in the Age Discrimination in Employment Act.

The ADEA, which covers employers with twenty or more employees, was enacted in 1967 and protected individuals between forty and sixty-five. In 1974, an amendment to the ADEA extended the application of the law to public employees (it had previously applied to only private employees). This provision, as it affects state employees, has since been found to be unconstitutional by the Supreme Court, noting that it is an infringement on states' rights. Many states have their own age discrimination statutes for their employees.

In 1978, an amendment to ADEA passed that extended the protected age group from age sixty-five to seventy. The main purpose was to eliminate mandatory retirement before the age of seventy, a compromise with those who wanted to eliminate mandatory retirement entirely. Rep. Claude Pepper was the key figure in this effort. In 1986, further amendments were passed, sponsored by Rep. Pepper. This bill eliminated the age limit of seventy, thereby extending coverage to all individuals over forty and eliminating mandatory retirement for almost all workers.

In 1990, the Older Workers Benefits Protection Act was passed, which amended ADEA to specifically prohibit discrimination against older workers in all employee benefits. The level of benefits or cost of the benefits must be the same for older workers as younger workers. It also set out a process, unfortunately, by which an employee can voluntarily waive his/her rights under ADEA upon termination, the interpretation of which has become a source of lawsuits.

In 2005, the Supreme Court ruled that employees do not have to prove specific intention or deliberate bias to claim discrimination in pay or benefits. Employees may need to show only that a policy has a disparate negative impact on older workers, regardless of employer motivation. On the other hand, employers can claim "reasonable factors other than age" for failing to promote or for discharging older employees.

Unfortunately, the ADEA is not being enforced effectively by the U.S. Equal Employment Opportunity Commission (EEOC),[27] and the

Figure 4.1—Images of Old Age in America, 1790–Present

Source: W. Andrew Achenbaum. Institute of Gerentology, University of Michigan, and Wayne State University, 1978.

Department of Labor has not demonstrated profound concern for the older worker. Moreover, as noted, the Supreme Court has ruled that this federal law need not be enforced for state employees.

Reforms in Language and Imagery

It is time to change the language and imagery of old age in the media and to sensitize journalists and writers about the language of ageism. For example, some studies suggest that older persons prefer to be called older persons,[28] not the elderly. Nor do they generally like euphemisms such as senior citizens and golden agers. We need a media and advertising watch to track the images of age presented in America, similar to one created by Maggie Kuhn, who invented the Gray Panthers, a national grassroots organization advancing the causes of young and old people. As mentioned, older people are rarely seen on TV and in film, and when they are, their appearance is often not positive.[29]

We must challenge the advertising, news, and entertainment industries (so closely linked) to end ageist stereotypes and alter the climate toward older persons in a positive manner. There are wonderful films and novels that can help us understand better the personal experience of aging, such as the great Italian film *Umberto D.*

Madison Avenue, the locus of advertising, is said to be dominated by twenty- and thirty-year-olds. Certainly, their attention is focused on people eighteen to thirty-nine (or forty-nine) because, they reason erroneously, once people make their product choices early in life they stick with them and because older people have no money anyway. Sometimes ageist ads backfire, such as the TV spot in which a crotchety, wheelchair-bound grandmother goes ballistic at a family reunion when she finds they are out of Coke. Consumers complained that the ad was "mean spirited."[30]

Educational and Human Rights Interventions

Health promotion programs and senior mentors introduce medical students to healthy older persons, which is critical to changing the mind-set of medical students that older persons are all demented and decrepit. Along similar lines, the introduction of children in primary school to the perspectives of human development can make them aware of the life course and its stages.

The United Nations has not addressed the human rights of older persons. In fact, they were not considered in the Universal Declaration of Human Rights of 1948, and today groups such as Human Rights Watch do not focus on their plight. (Although it takes up the issues of women's rights, it does not even note those affecting older women.) References to the rights of older persons were made at the 1982 World Assembly on Aging, but no significant actions have been taken to deal with abuses (in nursing homes, for example, or in employment discrimination). At the 2002 UN World Assembly on Aging, I introduced a Declaration of the Human Rights of Older Persons.[31]

French theologian Georges Bernanos[32] wrote, "The worst, the most corrupting lies are problems poorly stated." Let us then state our problems as they really exist, putting various myths and distortions into their proper perspectives. Knowledge is one antidote to prejudice. To treat the prejudice of ageism, we need to realize what is true about older persons.

It is time to redefine the onset and character of "old age." The Myths and Reality of Aging (MRA) study, a 2000 reprise of a 1974 Harris poll, supported by the National Council on the Aging and the International Longevity Center, revealed that old age is now perceived as a time of continuing vitality. Forty-four percent of Americans over sixty-five described the present as "the best years of my life," compared to 32 percent in the 1974 survey. A *New Yorker* cartoon used the data with the caption "Good news, honey—seventy is the new fifty."

It is time to alter our deep-seated cultural sensibility and work to overcome our fear, our shunned responsibility, and harmful avoidance and denial of age. Strict legislation, as well as legal and police action against age discrimination and abuse, are essential but insufficient. We must help people deal with their fears of aging, dependency, and death, and develop a sense of the life course as a whole.

John Glenn said upon his safe return to Earth at age seventy-two, "Old people have dreams, too." Certainly, after his historic and courageous space flight no one can look at their grandparents in quite the same way again.

Perhaps Eleanor Roosevelt said it best: "Beautiful young people are accidents of nature, beautiful old people are works of art."

CHAPTER 5

THE CHANGING FAMILY AND LONGEVITY

Until the Industrial Revolution, the home was the center of both production and consumption. Women might have worked at the spinning wheel while men tilled the soil in nearby fields. The Industrial Revolution separated the family from the workplace and husbands and wives from each other. It is generally assumed to mark the end of the family as a self-contained socioeconomic unit. But did it?

Two centuries after the rise of the Industrial Revolution, new technologies are retraining people to work at home again. Today, in open and diverse societies, the rigid, traditional, patriarchal concept of family is being overturned, and new biological techniques are revolutionizing reproduction. Nonetheless, the family remains the basic unit of society. It is still the primary caretaker of its members, including its older members. The family also contributes significantly to the prospects for the longevity of its children through both heredity and upbringing.

Although rooted in biology,[1] the human family is more than a biological unit. It is a powerful, cohesive, emotional, economic, and social centripetal force that binds people together. A family and its home represent security and hope, a crucible and a citadel. As Robert Frost wrote in *The Death of the Hired Man*, "Home is a place where, when you have to go there, they have to take you in."

Beginning in the twentieth century, people could, for the first time in human history, reasonably expect to live long enough to become grandparents. Four- and five-generation families are alive at the same time, when once the best we could anticipate was three generations. About 70 percent of married people age fifty-one to eighty-one are members of four-generation families, and 25 percent are part of three generations.

EVOLUTION OF THE FAMILY

Many Americans believe that as a society we have lost the extended family, that the nuclear family has emerged to replace it, and that traditional values and the close-knit, vertically extended family have been lost. Stanford University sociologist William J. Goode calls this a belief in the "classical family of Western nostalgia."

In fact, the lamented family of the past had many children and was largely a two-generational family, horizontal in character with many brothers and sisters, uncles and aunts; boarders were sometimes counted as family. It was also a family routinely affected by the deaths of its younger members, as many mothers died in childbirth and young children did not live to adulthood.

Even the ideal of mother, father, and two children was not commonplace in American life. In the nineteenth century nearly 25 percent of children saw one of their parents die before they reached eighteen. Remarriage and stepparenting were common. With an average life expectancy under fifty, neither grandparents nor the biological, extended family existed in great numbers. Even in the early part of the twentieth century, a ten-year-old had only about a 40 percent chance of having two of the possible four grandparents alive. By the 1990s, these chances had doubled to about 80 percent. In the twenty-first century, about 50 percent of persons over sixty-five have great-grandchildren.

The conventional definition of a family encompasses those who consider themselves economically or emotionally related to each other by blood (consanguinity) or by marriage (conjugality). It was not really until the 1940s and 1950s that the ideal of the nuclear family and its survival came into being, and then only for a short time. In the 1960s the families of the white American middle and upper classes underwent dramatic changes, this time marked by divorce (see Table 5.1). Divorces rose during the 1960s and 70s but fell in 1986 to the lowest level since 1975. Nonetheless, the high divorce rate remains, and there is a declining remarriage rate. Nearly ten percent (21.6 million) of the population fifteen and older are divorced, of whom 12.5 percent are divorced before the age of thirty-five. The marriage age is rising, from 20.8 years for women and 23.2 for men in 1970 to 25.3 years and 27.1 years, respectively, in 2003. Marriage continued to decline in America in 2004. In 2003, 32 percent of men and 25 percent of women fifteen and older (64.3 million) had never married, up from 28 percent and 22 percent for men and women, respectively, in 1970.[2]

In 2003, only 23 percent of the U.S. population was married couples with children under 18—the conventional idea of the nuclear family. Now there exists a wide spectrum of nonfamily households: single individuals, unmarried couples, groups with no family blood relationship, single-parent families arising from unwanted teenage pregnancies or from pregnancies by unwed adult professional women, gay couples (some with children), and divorced single parents, both men and women.[3]

Are we witnessing the destruction of the family? *The American Heritage Dictionary* defines *family* as "two or more people who share goals and values, have long-term commitments to one another, and reside usually in the same dwelling place." In other words, one could argue that new arrangements that have arisen by mutual decision and the desire for sustained companionship also constitute families. These families, which may or may not share a common blood lineage, nonetheless share common values.

Despite these changes, the married-couple household remains the norm, and 70 percent of children in the United States and 80 percent in Great Britain live in families with two parents. Families may differ, even be unstable, but the core concept of family prevails.

Some argue that social dysfunction in the United States is due to the absence of "a strong family unit." But the American family has survived remarkably well under continuing, new, and varied forms of great pressure—the African-American family despite slavery, immigrant families despite stresses and strains associated with immigration and the adoption of a new land, and other families despite recurrent economic and social duress. The family in its broadened definition and diversification has proven to be quite resilient. Its structure has been changing from the very beginning of human time, adaptive through centuries of famines, plagues, and wars and in good times as well as bad.

The new consanguineous family, consisting of three or more generations, is rich with possibility. Increased longevity alone offers new family combinations and new roles for grandparents. But it has occurred at a time of great historical changes in the character of families, with unprecedented social changes introduced with the Industrial Revolution and continuing through the rise of feminism.

HOW IS THE FAMILY ADAPTING TO THE LONGEVITY REVOLUTION?

There is much controversy, even ill will, in some quarters toward the new kinds of family,[4] which do not conform to the so-called traditional American family. We see many blended, reconstituted, or step families due no longer largely to death but to high divorce rates and remarriage. Unrelated individuals, gay couples, foster children, and unmarried, cohabiting couples (more than ten million households in 2003) are drawn together by emotional and financial commitment and interdependence. The growing number of stepfamilies and single parent families in American society has provoked cultural, social, economic, and legal issues that must be worked out. According to Paul C. Glick, a former senior demographer at the Census Bureau, one child in four will become a stepchild before reaching the age of eighteen. The 2000 census showed that 5.8 million children live with their grandparents (including, of course, young grandparents). Of the grandparents who live with grandchildren, 4.5 million maintain their own households.[5]

In the longevous society, the roles and functions of grandparents are bound to expand, and it has been theorized that grandparents have an evolutionary role in the protection and education of children. The extended dependency of humans benefits from the survival of grandparents.

Peter Uhlenberg, a Duke University demographer, has pointed out that increasing survival prospects for infants encourages stronger emotional bonds between parents and children. Moreover, fewer deaths among adults between twenty and fifty has reduced the number of orphans.

The number of living grandparents per children has also increased. And as Uhlenberg says, "Decreasing mortality has increased the number of years that marriages survive without being disrupted by death. This change has probably contributed to the increase in divorce."[6] Indeed, the length of marriages has not changed significantly in centuries in the United States. Have divorces become a substitute for death and widowhood?

Decreasing numbers of infant and child death also allow more careful planning of family size and encourage a reduction in fertility. Thus, population has been stabilized (see Chapter 17).

HOW DO FAMILY STRUCTURE AND CIRCUMSTANCES CONTRIBUTE TO LONGEVITY?

We are confronted with another social myth about the American family, that it abandons its older members. There is little basis for this notion, as pioneering gerontologists Ethel Shanas and Gordon Streib discovered in the 1960s.[7] Generally, family members who grow old are not sent to nursing homes; in fact, they live close by their children. Another statistic that defies the stereotype is that some 2.4 million Americans, mostly women, provide unpaid care for some five million older people at home.[8]

Much is made of the burden of the old upon the young and of the drainage of public funds from young to old. However, as mentioned, studies reveal[9] that private funds move from old to young throughout life in caring for and educating children, supporting grandchildren, and, ultimately, providing inheritance. In many nations, middle-aged or older adults serve as banks and loan money to help their younger family members get started.

Many young people return home to live with their parents, especially during hard economic times. In 1990–1991, 59 percent of men and 47 percent of women between eighteen and twenty-four depended on their parents for housing. Forty percent of men and 28 percent of women twenty-five to thirty-four were also dependent on their parents. On the other hand, filial maturity, a concept introduced by Margaret Blenkner in 1965, means that children must achieve autonomy with respect to their parents and that their parents must accept the right of their children to live their own lives.[10]

The family is the locus for the initially defined prospects for the longevity of the individual—genes and the conditions of gestation and birth,[11] and how she or he is nursed, nourished, acculturated, and made adaptive. For example, prematurity occurs in epidemic proportions among the poor, leading to infant deaths or, with survival, long-term disabilities and a shorter life.

Family violence begins early. Often a perpetrator has been a victim, and is repeating vicious cycles. Socioeconomic upheaval and unemployment can also lead to conflict, violence, or alcohol and drug abuse. Alcohol, shockingly, is a problem in 25 percent of American families and is frequently associated with family abuse. One in six children lives in a family with alcoholics. One-fourth to one-half of all children born of alcoholics are genetically predisposed to alcoholism, which sharply curtails longevity. We energetically make substance abuse a crime and do so little about alcohol abuse!

The reality of elder abuse has shocked Americans. It often rises with economic pressures just like the abuse of a child or a wife. So do other forms of psychopathology. Elder abuse is a high price to pay for longevity.

Families, in general, accept responsibility and effectively deal with life's contingencies, including supporting the retired and the old. Individual families may suffer, however, since life's tragedies are not distributed uniformly. There are limits to what an individual family, however gifted, determined, and energetic, can do to protect its members. (That is why the concepts of insurance and of friendly societies emerged historically as private sector initiatives.) Eventually, private insurance companies, and especially governments, assumed roles of providing social protections.

Family life is related to longevity and family ties enhance life all along its course. Love and affection promote longevity. Marital status, for example, is associated with a longer life, certainly because of the mutual assistance marriage can provide, but also because of the emotional ties. Survival following heart attacks is improved when the victim is married. Marriage is especially good for men, in that it has been shown that married men live longer. On the other hand, there is some evidence that intensive caregiving without respite is stressful and may result in illness and earlier death; men especially are more vulnerable than women to dying during the first year after the death of a spouse. From the perspective of longevity, marriage may be less important for women.

WOMEN ARE KEY

Women are the key family figures in caregiving, kin keeping, health care decision-making, and shaping the prospects for longevity of family members. In short, family issues mainly involve women. They are the pioneers of aging and longevity, and the predominant representatives of the final stage of life. Family and women's issues must occupy a central place when the quality and length of life are considered.

Industry is now increasingly dependent upon women in the workforce, and families, of course, remain dependent on women's work. Fifty-nine percent of married couples are now both wage earners. In 21 percent of these couples, the wife earns more than the husband. Twelve percent of members of the armed forces, 42 percent of law students, and about 50 percent of medical students are now women. These are a few indicators of how their lot is changing.

The conservative movement in the United States, especially its religious constituency, believes the family is in peril, its unity disintegrating. Conservatives are troubled by changes in the character of the family and believe that women should stay at home and care for children and older people. They believe we should return to the era of obedient daughters, forbearing women, and sacrificing mothers, to the days when daughters simply obeyed fathers (and mothers) and wives obeyed husbands. Conservatives do not spell out how the return to the past would be financed and otherwise accomplished. In 2005, a nationwide survey by the *New York Times* (and CBS News) of teenagers found that boys, not girls, favor the conservative view of the family. Boys considered themselves better than girls (girls saw boys as equals) and believed strongly in the traditional marriage in which the wife stays home, cleans the house, cooks, and rears the children, while the husband is responsible for earning a living and doing household chores. Even the sons of working mothers believed in this 1950s view of marriage. Girls felt equally strongly about a career and egalitarian marriage. In fact, they were less interested in marriage and opposed the power relationship favoring boys. Fifty-five percent of them would consider single parentage. Conservatives, meanwhile, want more control over parenting, including school curricula (to teach creationism, not just evolution, and keep out sex education).

Longevity and the Feminist Movement

For years, feminist leaders neglected the plight of older women. Understandably, the feminist agenda was large, and intense, emphasizing reproductive rights and economic equality. As feminists themselves aged, they began to grapple with the issues of older women, such as the need to ensure that women have survivor benefits on the pensions of husbands and to gain equality in Social Security. Betty Friedan, the creative mother of modern feminism, understood the issues of age and wrote of the vital possibilities in *The Fountain of Age* (1993). Nonetheless, even when Medicare and Social Security have been severely threatened, the National Organization for Women and individual leaders in the women's movement have not participated as vigorously as their defense warranted.

Family Planning

Family planning is an instrument of public policy, but it is also an issue for individuals. Not simply birth control, it involves the maintenance of reproductive health in the workplace, prenatal and postnatal health care to ensure a vigorous and healthy child, avoiding unwelcome children, solving problems of infertility, and encouraging family stability, not promiscuity.

Moreover, family planning is not simply a tool to control family size, but for the timing of birth and its coordination with family resources and, by extension, with the overall age structure of society. It shares, with death and migration, the ultimate means for establishing that age structure. Couples can provide a sense of mastery through an improved ability to make choices. We do not need to be caught up in the Malthusian destruction of families and individuals. We can build the right mix of age structure using the greater capacity for family planning.

The biomedical revolution has certainly given new power to the family to control human biology, health, and leisure, together with the changing social perspectives needed to facilitate family and population stability. Although life expectancy for American women has increased thirty years in the twentieth century, the maximum reproductive age has, however, been unaffected. Yet, career women can now delay motherhood and be encouraged by the fact that with a healthy body and womb they may be able to carry their own or any fertilized egg through in vitro fertilization. Post-menopausal women can now have babies through a process of hormone therapy that prepares them to receive the embryos and uses younger woman's ovaries to produce eggs. Women can also have their own embryos frozen for later use, raising the issue of the rights of embryos and potentially leading to court battles over custody. It is conceivable that twins and siblings of the same age could be harvested ("born") at different times.

One can't help but wonder if the changing family structures will have the capacity to continue to support older members and give children a solid sendoff in the era of longevity. Will families forgo the fifth commandment "to honor thy mother and father"? Will affection be limited by cost to the family or the government? Families may be sorely tested.

FAMILY VALUES

The family transmits culture and values. It provides economic support, preserving and passing on property. It builds businesses. Property and wealth help to maintain it. At its best, it protects and educates children and fosters character development, autonomy, trust, confidence, civic responsibility, moral virtue, and, ideally, the capacity for nonviolent conflict resolution. The family also inhibits sexuality throughout much of childhood and adolescence. Its manner of controlling sexuality is a powerful determinant of later parent-child, marital, and cohabiting relationships. Religious, social, and cultural sexual taboos and restrictions protect against childhood (teenage) pregnancy and sexually transmitted diseases (STDs) as well as preserve the family and protect women against exploitation. Finally, the family that brings life also supports its members in their dying.

An estimated 49 percent of pregnancies in the United States are unplanned or unwanted; almost half of these (1.29 million in 2002) are terminated by abortion. Unintended pregnancies often occur at the beginning and the end of the childbearing years. Some of the worst complications occur for the very young (especially prematurity) and for pregnancies occurring later in life. On the other hand, babies who are wanted and planned are likely to be well cared for, enhancing their longevity.[12]

In Europe, remarkable changes are occurring with regard to marriage and parenthood. The trend, particularly in the Nordic countries, France, and Great Britain, is marked by cohabitation rather than marriage and the birth of children out of wedlock. These phenomena are independent of social class. In 1999, 62 percent of all births in Iceland were to unwed parents. In Norway that figure was 49 percent; in France, 41 percent; in Great Britain, 38 percent. Even in Catholic Ireland, where divorce did not become legal until 1995, 31 percent of births were out of wedlock. This is a comparable figure to the phenomenon in the United States. Classic family structures are giving way because of changed attitudes toward religion and independence from the state. European welfare policies are increasingly more compatible with cohabitation and single parenthood, including single motherhood. Children are treated the same whether their parents are married, cohabiting, separated, divorced, or single. They have the same inheritance rights. But as a result of these trends there are often highly complicated living arrangements and custodial agreements.

Family values in the best sense means the preservation of the integrity of the family against disintegration, divorce,[13] child abuse, disease, and poverty. (The last of these is often the cause of the first.) Most of all, it means fostering the development and care of family members, especially children. But the passions of those concerned with family values have focused on divorce, contraception, homosexuality, abortion, and school prayer.

Today, the age of first marriage is rising and the high divorce rate is continuing, thereby increasing the numbers of single-parent families. As a result, in the context of worldwide economic competition, we must introduce wide-ranging reforms to support the family.

Historically, children participated actively in the family, working in the fields, caring for their younger siblings, and providing household service: the eldest child would often cook for the family. *In the past, having a child was important because he would care for the parents when they became old. A child was a substitute for Social Security.* And there was a time when children worked in the factories built by the Industrial Revolution.

These tangible functions of children have declined in industrialized societies, making it difficult to compute the economic value of a child. For example, how were the decisions made that legislated the $500 claim per child on a tax return or the $90 a week that a child might receive under welfare?

It has been estimated that bringing up an American child can cost between $150,000 and $300,000, and the number goes even higher if unusual care or advanced education is part of the total cost. Although careful studies are not available, the family's financial burden of care for its children appears greater than for its elders. Vivianna Celizer's *Pricing the Priceless Child* (Princeton University Press, 1994) observed that by the 1950s a child's value in many families had become largely emotional.

The debate in the 1980s and 1990s over providing *unpaid* family leave for new mothers involved the concerns of businesses together with societal attitudes about the family. Of course, unpaid leave is not useful for those who cannot afford it. The United States is the only industrial democracy that does not mandate paid family leave. This includes Japan and European countries that offer full or partial wages. Most European countries, including Austria, France, Germany, Sweden, and Finland, now provide paid and job-protected leaves of at least five months. Eight of the first twelve European community nations offered paternal leave as well as maternity leave. European government programs are often geared to encourage population growth, which has declined throughout most of Europe.

Twenty-five states in the United States had enacted limited parental or family leave bills by 1992. Seven covered only state or public employees. In 1993 the federal Family and Medical Leave Act was passed. It provides up to twelve weeks of unpaid leave during any twelve-month period for workers who have been with a covered employer (one that employs five or more employees for each working day during each of twenty or more calendar workweeks in the current or preceding calendar year) for at least twelve months. Unpaid leave applies to the birth and care of the employee's newborn child or newly adopted children, the care of members of the worker's immediate family (spouse, child, or parent) with a serious health condition, or a worker's own serious health condition.

Comprehensive universal day-care legislation would help move the country in a new and realistic direction. In 1999, 29.2 percent of the 10.5 million children under six with working mothers were cared for in organized child-care facilities.[14] Yet, there is high staff turnover in day-care centers because wages are barely above poverty level. Many centers are unlicensed and unregulated. Plumbers, but not nannies, are licensed. Some centers are insecure, unsafe, and unsanitary. Poor care threatens the cognitive and emotional development of children. There are few twenty-four-hour centers even though shift work is increasing. Furthermore, family child care, not day-care centers, is the form of day care most widely used by working mothers, and even home-based day care by relatives may be inadequate. (Of course, care at home by mothers is not perfect either.) If we wish to really enhance the quality of life for children and preserve the family, a lot of money will have to be spent by parents, government, and business. Only a few U.S. firms, such as IBM and Stride-Rite, offer on-site day-care centers. We should create a private-public system of child-care services paid for by setting fees on a sliding scale based on income, not segregated by income levels. The French family day-care system works this way. Day care is essential for working parents and a productive America.

Other goals include:

- Strengthening the family's capacity to give children a healthy start and educating families to understand the influence of early care and nurture upon longevity and quality of life in old age.

- Establishing a national network of child-care centers (incorporating Head Start) with public-private financing using a sliding scale available to children beginning at three. Providing decent pay and pro-fessionalizing child-care workers.

- Relieving the caregiving burden through a variety of respite services from day care to elder care to disabled care to appropriate compen-sation equivalent for those who provide that care, both men and women. Building a national network of family service centers.

- Creating a long-term care system emphasizing community-based and home care for the impaired of all ages. This should include training, career growth, and decent compensation and benefits for paid caregivers.

- Incorporating family studies into a national population laboratory with longitudinal studies.

- Building effective mental health services, including family preser-vation.

- Creating a family contract regarding the sharing and transfer of money, goods, and personal services at different stages of the life cycle with implicit and explicit agreements, e.g., bequests for care in one's old age and the education of grandchildren.

- Building family-friendly policies such as comprehensive universal day care and paid family leave and doing away with the marriage penalty in our tax laws.

- Mounting scientific initiatives to better understand reproduction and to find natural means of contraception that would be an effec-tive means of reducing unwanted pregnancy *and* abortion.

- Strengthening pro-family policies to resolve workplace/family con-flicts and provide necessary leisure to mitigate stress.

- Meeting special needs of older people living alone through the development of shared housing and other living arrangements as well as services.

- Reforming legal arrangements and laws that protect the appropriate interests of grandparents in their grandchildren (e.g., visiting rights) and supporting situations in which grandparents are directly caring for their grandchildren.

The family of the twenty-first century is under enormous and continuing pressure and is so profoundly changed that unless its important functions are preserved and expanded more effectively, the health and vigor of our population, and its longevity, will be adversely affected— one more way we could lose the gains of the Longevity Revolution.

CHAPTER 6

THE DISORDERS OF LONGEVITY

As far back as the fifth century BC, Hippocrates, the father of medicine, observed the differences between acute and chronic disease. Today, with better control of acute infections and the lengthening of life, chronic, rather than acute, diseases dominate medicine, especially in the developed world.[1] Increasingly, people who have been disabled from their earliest years survive into old age. They include the developmentally disabled, victims of accidents, and persons afflicted with a variety of genetic and environmentally based diseases. Also included are the escalating numbers of impaired but surviving premature babies from neonatal units, who have continuing needs for health care throughout their lives. Many children suffer significant psychological traumas and stresses that affect their later lives. In addition, many people become disabled by chronic disease in their middle and later years. This convergence is a cause for national concern as we experience growing longevity. In 1995 in the United States, 38 percent of the population (an estimated one hundred million) suffered from chronic disease. This included 25 percent of children, 35 percent of young adults, 61 percent of the middle-aged, and 81 percent of older persons.

Chronic diseases usually take a long time to develop and frequently begin early in life—in utero, at birth, or in childhood.[2] In addition to genetic influences, many chronic diseases are related to the amount (or intensity) and the duration of exposure to one or more pathogens, including infectious agents,[3] toxins, carcinogens, and mutagens. The stage is often set by infectious diseases and malnutrition in childhood and other types of medical and socioeconomic stressors early in life, but unhealthy lifestyles, such as behaviors that cause heart disease, affect rich and poor alike.

Arguably, then, because many diseases generally considered geriatric did not suddenly emerge in old age but have a long history, the term "disorders of longevity" is more appropriate. Further, since many disorders of longevity started long before the onset of old age, it is desirable to intervene as early as possible. Osteoporosis, for example, may be seen as a pediatric disease that originates in pubescence and adolescence, when, ideally, bone density is achieved and "banked." In part, the disease is the cumulative outcome of a lack of anti-gravity exercise, calcium, and vitamin D. Heart disease and even type II diabetes,[4] particularly when associated with obesity, may take root in infancy and later childhood. Similarly, atherosclerosis may be seen as a disorder of longevity. Atherosclerotic streaks can be found in the aortas of toddlers during autopsy.

Aging itself is a risk factor for certain chronic diseases![5] While it is tempting to consider aging itself as a chronic disease, this is not accurate. It is more useful to consider aging as a series of risk factors that provide fertile soil for numerous events. In other words, the accumulation of normal aging characteristics creates a threshold beyond which a person becomes increasingly susceptible to an array of pathological outcomes.

Disorders like Alzheimer's disease, spinal stenosis,[6] and temporal arthritis are age-related, and a variety of diseases such as atherosclerosis may hasten aging. Moreover, factors related to aging alter the presentation, course, and outcome of a number of diseases and conditions. For example, older people demonstrate physiological functions such as immune responses, perception of pain, and temperature regulation that differ from those typically exhibited by younger persons. Two simple illustrations: Perhaps a third of persons over sixty who have a heart attack do not experience chest pain; older persons with appendicitis may not exhibit the characteristic symptoms of a fever and an elevated white blood cell count.

In 1900, the three leading causes of death in the United States were pneumonia and/or influenza, tuberculosis, and intestinal infections—all of them acute, infectious, and contagious. But by 1910, heart disease had already moved from fourth to first place as a cause of death, and by 1986, not a single acute infectious disease was among the top five. The better control of infection, a central focus of public health and medicine in the first half of the twentieth century, contributed to increasing numbers of people living longer. In 1981, with the appearance of the acquired immune deficiency syndrome (AIDS), acute infection returned with a vengeance. Yet as a result of advances in treatment, AIDS, too, has become a chronic infectious disease.[7] As more new therapies are introduced and death is deferred, AIDS will be seen in more and more older persons. Already today, some 10 percent of AIDS patients are over fifty-five.[8] (Table 6.1 lists causes of death in people sixty-five and older, and Table 6.2 lists them in the population as a whole.)

In the 1950s and 1960s a group of us purposely selected for study a "super healthy" population of older people in order to separate aging from disease. We found much that was attributed to aging is, in fact, a consequence of disease, social adversity, and even personality.[9] Before that time, most studies of older persons had been carried out mainly in chronic-disease populations. There was little information concerning normal or healthy aging. A group at Duke University also carried out important studies of healthy aging populations.

AGE-DEPENDENT

Age-dependent or age-related diseases occur at specific times in life if an individual is vulnerable. Examples are schizophrenia in adolescence, Huntington's disease between ages thirty-five and fifty, and amyotrophic lateral sclerosis (commonly known as Lou Gehrig's disease) in persons between forty and seventy. The onset of multiple sclerosis occurs between the midteen years and age fifty. Slow viruses or prions (proteinaceous infectious particles) also cause diseases that fall into the age-dependent category. These include kuru (sometimes called laughing sickness), Creutzfeldt-Jakob disease (CJD), and Gerstmann-Straussler syndrome. While these are rare conditions, there is reason to believe that understanding their underlying cause may advance study of many diseases.

Aging compromises homeostasis, the body's capacity to maintain the status quo. An excellent example of the progressively lower reserve and function of older persons can be found in their decreased capacity for thermo-regulation, requiring protection against exposure to extreme temperature levels. In the presence of multiple age-related social and medical conditions, homeostasis can be further reduced. Here again, aging itself is a risk factor.

Specific cell types in the body have varied vulnerabilities to both aging and diseases, and whether cells continue to actively divide (mitosis)[10] or do not divide (postmitosis)[11] is relevant to aging and disease.

The Gompertz Law[12]

British actuary Benjamin Gompertz propounded a theory known as the Gompertz Law of Mortality, in which he states that there is an increase in the rate of death as age increases after sexual maturation until roughly the ninth or tenth decade. Specifically, he found that over this significant portion of the age span, the death rate increases by a constant factor for successive equal age intervals (about every seven years). The Gompertz Law is ultimately based on biological events that occur within our bodies as we grow older and that predispose us to death and, of course, to various causes of death.

Other aspects of the origins of the disorders of longevity are evident in the evolutionary theory of aging of Sir Peter Medawar, the late Nobel Prize winner, and the concept of *pleiotropy.* "Reproductive success or perpetuation of the species is the key objective of evolution, and natural selection defers the accumulation of deleterious genetic components which may produce late-life diseases."[13] (This will be further discussed in Chapter 9.)

The Role of Society and Self in Aging

Until recently, the medical terminology for conditions of older persons commonly contained the adjective "senile"—senile osteoporosis, senile macular degeneration, and senile dementia—to identify them as belonging to the diseases of later life. The pejorative connotation and the sense of inevitability[14] did little to inspire either active diagnosis and treatment or research of these conditions, but it did convey the painful influence of aging as a predisposing factor.

Also significant is the older person's self-perception. Unfortunately, older persons (and doctors) often write off treatable conditions as aging. This is a serious issue that demands active questioning by the patient and the physician.

Diabetes and Aging

In some ways diabetes may be a most appropriate model of aging. Glycation (the cross-linking of glucose and proteins) occurs in both diabetes and aging. Cases of type II diabetes, which is the inability to effectively produce and utilize insulin, the hormone needed to metabolize glucose, increase with age. Today about twenty-one million people have diabetes, and perhaps the same number are probably prediabetic. Each year nearly seventy thousand people die from diabetes, and type II diabetes is increasing and occurs in children as young as ten, in association with growing obesity. By 2004 diabetes had reached crisis proportions. Twenty percent of dollars spent on health go to diabetes. How does obesity lead to diabetes? Suspects include fatty acids released by fat cells. While many genes probably play a role, lifestyle behavior is especially important.

The role of genetics and behavior is well illustrated by the Pimas of Arizona, who have the highest incidence of type II diabetes in the world: 50 percent of adults have the disease, and insulin resistance is common. Significantly, on the other hand, in studies of lifestyle habits, researchers found that the Pimas living in Mexico, following a traditional diet rich in squash, melons, and legumes, had a very low rate of diabetes.[15]

MEDICATED SURVIVAL

Historically, the National Center for Health Statistics has reported increased disability for each decade of life, in part because people who lived longer were dependent upon medical and surgical interventions, such as anti-hypertensive medications and angioplasty, and in part because neonatal critical care units, burn and trauma centers, and emergency medical technicians saved the lives of many who in the past would have died but who now in all likelihood were left with permanent disabilities.

However, recent studies[16] of older U.S. populations, as already noted, report that *disability rates are falling in the later decades of life* relative to the increased numbers of surviving older persons. Moreover, according to a Census Bureau study led by Cynthia Taeuber, the growth in the nursing-home population has slowed, in part because of declining disability but also because of assistive medical technology and greater availability of alternatives to nursing homes, such as assisted living in the community.

It appears that increased life expectancy is *not* accompanied by an *equal* increase of disability and institutionalization. This is extremely important since, clearly, people do not simply desire to live a longer life but also want to live in robust health. Analyses supported by the National Institute on Aging of the National Long-Term Care Survey (NLTCS) and the National Health Interview Survey[17] have shown that although the older population increased by 14.7 percent between 1982 (when the survey began) and 1989, the number of disabled older adults increased by only 9.2 percent. Even in adults over eighty-five there has been a decline in both rates and levels of disability since the 1980s through 1999—the most recent data available (see Figure 6.1 as of 2001). Not only are there declines in disability rates of 1 percent to 2 percent per year, but the incidence of specific chronic diseases such as high blood pressure, arthritis, and emphysema is falling.

These findings suggest the possibility of reductions in health costs by postponing disability. Some gerontologists refer to this result as the compression of morbidity, a construct that was articulated by James Fries[18] (Table 6.3 shows healthy life expectancy; a similar idea in some WHO member states). As life expectancy increases, he hypothesizes, disability-free years increase, and disabilities occur at an older age. The declines are substantial, for example, 2.6 percent annually from 1994 to 1999. But although the numbers are encouraging, it is not clear that the

Figure 6.1—Number of Chronically Disabled Americans Age 65 and Over (in Millions)

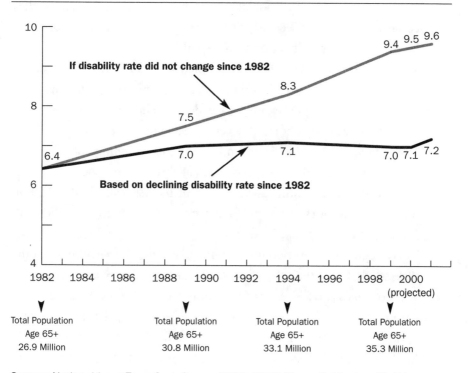

Source: National Long Term Care Survey 1982–1999 (Kenneth Manton, Ph.D.).

accumulation of disability over a lifetime has been reduced, nor have the causes of the declining rates been firmly established.[19]

One principal cause of declining disability rates across generations could be as simple as reducing injury rates as a result of workplace rules required by the Occupational Safety and Health Act of 1972, implemented by the Occupational Safety and Health Agency (OSHA), and the 20 percent decline in manufacturing jobs since 1980.

Of similar significance may be the anti-hypertension campaign and other public health efforts that help account for the 60 percent reduction in deaths from heart disease and stroke since 1960. Other factors may include increased public education about the applications of biomedical and pharmaceutical research (e.g., the use of statins to lower cholesterol levels), decreased infections, and healthier lifestyles that include a sensible diet, exercise, and tobacco cessation.

Declining disabilities could reduce Medicare, Medicaid, and other public health as well as out-of-pocket medical costs. Moreover, they could reduce Social Security costs if people remain in the workforce longer and lower indirect costs as a result of care received from family and friends that forces them to limit their workforce productivity.

Alvar Svanborg, Swedish geriatrician at the University of Gothenburg, conducted longitudinal studies in which he and his team extensively studied succeeding cohorts of seventy-year-olds (e.g., born 1900–1904, 1905–1909, 1910–1914). He found that each succeeding cohort was in greater health than the one preceding it, likely due to the social and medical resources that became universally available, especially in the 1920s in Sweden. Svanborg's work predated compatible findings by Kenneth Manton, Hiroshi Shibata, and others.

We begin to see the potential for an old age that is free of many of the illnesses that now tend to define it, and can envision what might occur when old age is no longer a sickness-oriented time of life. Freedom from chronic disease—cancer, heart disease, diabetes, stroke, arthritis, Parkinson's, Alzheimer's, and others—would release human energy that has been expended day by day, year after year, individually and collectively, in the struggle against illness and infirmity. The human body has been forced to serve as a fortress against outside invaders, namely those causes of disease that affect individuals who have been weakened by ill health, aging processes, malnutrition, and psychological states of depression and stress. Older people and their families have become sentinels, ever watchful for signs of weakness or vulnerability to disease. Freedom from constantly monitoring and fighting disease should bring about an energy dividend. It is not so far-fetched to imagine a newly vigorous and energetic third of life, one in which productivity and pleasure, participation and reflection can be practiced by a majority of older persons who remain physically and mentally able until very close to the end of life. To a considerable degree, this is already happening.[20]

Nonetheless, we are not built to last (see Figure 6.2).[21] Despite our best efforts, the human body is not constructed for unending operation. Table 6.4 tabulates the actual external (nongenetic) causes of death in the United States, as first estimated by James M. McGinnis and William H. Foege and later by Ali N. Mokdad. Obviously, the passage of time is a factor—increasing the duration and intensity of exposure to many of the causes. Left out of the McGinnis-Foege listing and the later follow-up is death due to hospital errors, nosocomial infections, and adverse reactions to prescription and over-the-counter drugs.[22]

Aging not only encompasses the basic biological changes that predispose a body to disease and disability, but it also signifies the passage of time that can lengthen one's exposure to pathogenic agents, the effects of use or disuse of joints, and the likelihood of unintended injuries or accidents. Even under the best of health conditions and the most favorable of lifestyles, our pumps, levers, tendons, joints, and valves wear out. Our body's methods of detoxification and our immune capacity are limited. And since the evolutionary plan is concerned only with reproductive success, the body's postreproductive fate is of little import to the perpetuation of the species, however troubling it is to us as individuals.[23] Hips and knees wear out as a result of traumatic or excessive exercise or sports injuries or even simply an active life. On the other hand, a lack of exercise and use contributes to disorders of longevity as well. Muscle thinning—the condition known as sarcopenia—occurs in proportion to inadequate muscle-strengthening exercises, and its impact is considerable. Sarcopenia contributes to falls and, because of declining lean body mass, untoward drug effects which occur because fat replaces

Figure 6.2—If Humans Were Built to Last

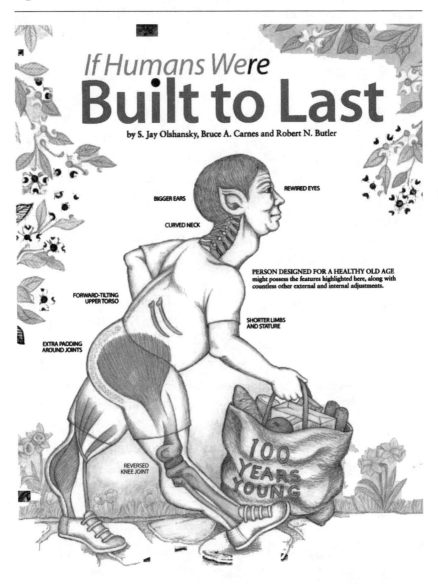

Source: *Scientific American* (2001) *284*: 50–55. Used with permission.

muscle protein, and fat-soluble medications such as diazepam (Valium R.) linger much longer in the body.

These limitations should quiet our hubris. However romantic, even heroic, our views of the human body and our nearly theological conception of its perfection and whatever mechanisms the body possesses to resist infections, to moderate oxidation, glycation, and other damaging processes, and to repair DNA damage—none of them proves to be enough, and our bodies can never catch up.

Ultimately, frailty, like dementia, becomes a huge problem. Dr. Linda Fried[24] has worked to define frailty—a syndrome of weight loss, weakness, exhaustion, slowed walking speed, and low activity—proposing that frailty is a medical syndrome that may have a multisystem root, including inflammatory, hormonal, and musculoskeletal bases. Does sarcopenia help explain frailty, or is it a component of the syndrome? Is loss of energy critical? Does weakness follow mitochondrial dysfunction? (Mitochondria are the energy centers or power plants in the cell whose functional capacity may decline with aging.) At the functional level, Fried developed screening tests to help care providers predict whether their older patients will develop frailty in the near future, and epidemiologist Jack Guralnik[25] found that quadricep (thigh) strength is a predictor of frailty.

UNINTENDED INJURIES OR ACCIDENTS RISE WITH AGE

Accidents remind us of the fragility of life. They are usually sudden, without warning, and are often terrifying both at the time and in retrospect. To die by accident rather than disease is difficult to accept. According to the Centers for Disease Control (CDC), more than ten thousand Americans over sixty-five died in 2000 as a result of falls. Among the many causes are a diminished sense of location in space (proprioception) and depth perception, mobility, and balance problems, as well as cardiovascular disease, visual impairments, and medication effects.

Unintentional injuries or accidents are connected not only with frailty in general but with subtle specific changes in muscle mass or sarcopenia, flexibility, and balance, as well as slowed reaction time and visual and hearing loss. They constitute the fifth leading cause of death in the United States—some ninety thousand deaths annually—and are likely to loom even larger in the future when longevity is further increased.[26]

Today, heart disease, cancer, and cerebral vascular accident (commonly known as CVA or stroke) are the three major chronic, degenerative, killer diseases.[27] They are all disorders of longevity and account for 58.7 percent of all deaths in the United States. But accidents take away more years of life than these diseases combined.[28] Until recently, these major killers were responsible for most reported deaths, but incidences have declined or stabilized over the last three decades, with the age-adjusted yearly death rate dropping 1.6 percent between 2000 and 2001. Between 1980 and 2001, the overall age-adjusted death rate declined 17.8 percent from heart disease, cancer, and CVA. (In 2001, the total age-adjusted rate for these three diseases was 501.7 deaths per hundred thousand.) In 2004, efforts to control circulatory disease continued to be the most successful in the overall push to improve life expectancy. Table 6.5 details chronic conditions by age and gender that compromise quality and length of life.

Older Persons and the Flu

According to the CDC, susceptibility to influenza increases with age, and because of the greater proportion of the population over sixty-five, the death rate from influenza rose markedly in the 1990s. Although the virus was no deadlier than in previous years, there was an average of thirty-six thousand flu deaths a year in the 1990s, as compared to twenty thousand a year in previous decades, with 90 percent among people sixty-five and older. In 2001, influenza and pneumonia together accounted for sixty-two thousand deaths, 2.6 percent of the total.

The virus was especially deadly in people over eighty-five, who are thirty-two times more likely to die. Although 65 percent of people age sixty-five and older are immunized, it is less effective in this population than in younger persons.

An Avian flu pandemic is now a possibility that confronts the world. Many experts believe it is a question of when, not if, and so far it has been more deadly than the 1917–1919 flu. The fatality rate for the avian flu is over 50 percent compared to about 2 percent for the 1917–1919 flu.

Overoptimism

We cannot assume that the shift from acute to chronic disease indicates the conquest of infectious disease, or that chronic disease is mainly a matter of genes and/or lifestyle. Moreover, the underlying biological changes of aging enhance susceptibility to infectious agents, genes, and lifestyle.[29]

Studies have suggested that recurrent infections play an important part in the development of a number of aging-related ailments. For example:

- Infections of the sinuses, lungs, and urinary tract may increase the risk of atherosclerosis by clogging arteries.

- Infectious agents may play a part in heart disease and stroke.

- Antibiotics have been reported to reduce further heart attacks.

- Most stomach cancers are due to the bacterium Helicobacter pylori.[30]

- Hepatitis B is associated with liver cancer.

- Human papillomavirus is associated with cervical cancer.

- One investigator has suggested that chlamydia plays an etiologic role in Alzheimer's disease.

• Medicare records reveal that the rate of septicemia in the older population more than doubled from 1986 to 1997, and in 1997 the rate
was found to increase with age.

WHAT IS A DISEASE?

Disease is a fluid concept influenced by societal and cultural attitudes
that change with time and in response to new scientific and medical
discoveries. As Lynn Payer observed,[31] culture itself also influences
medical diagnoses and practices. Our growing knowledge of genes will
also dramatically alter how we define, prevent, and treat disease.[32]

The current genocentric view of disease is limited, however, and
progress in understanding complex diseases depends on much more.
For example, many factors contribute to the development of heart disease, cancer, diabetes, and psychiatric illness. These diseases cannot be
ascribed solely to mutations in a single gene or to a single environmental factor. Rather they arise from the combined action of many genes,
environmental factors, aging, and behavior. About 70 percent of stroke,
colon cancer, coronary heart disease, and type II diabetes is due to
these factors and are potentially preventable by lifestyle modifications.
Further, the genes are difficult to identify because they typically exert
small effects on disease risk, and their effects are probably modified by
other unrelated genes as well as environmental factors.[33] For example,
lupus patients are genetically predisposed to have a hyperactive immune system, but their symptoms are also influenced by environmental factors, including viruses, drugs, and sunlight.

Disease as a concept, as well as specific diseases, will be redefined in
the genomics era. We must take care to avoid genetic reductionism and
to emphasize that genetic abnormalities differ in their impact, as do
the effects of environmental factors. Moreover, normal variations in
genes (polymorphisms) and detrimental changes (mutations) are not
easily distinguished.[34] To be labeled a disease, the condition must have
adverse consequences. Overly enthusiastic expectations regarding the

benefits of genetic research distort research priorities and spending for health.

Changes in disease classifications are not new. For instance, history shows differing emphasis with respect to the gastrointestinal tract. Between 1900 and 1940, the theory of bowel intoxication was at its height. Elie Metchnikoff, the Nobel-winning immunologist who introduced the word *gerontology*, believed bowel intoxication helped explain aging. Conditions such as obesity have not always been considered dangerous. We now know obesity is a serious risk factor for many life-threatening diseases. However, while the FDA and NIH now consider obesity a disease, it was only in 2004 that Medicaid and Medicare began to consider payment for anti-obesity medications or counseling programs—although they have long paid for the unfortunate results!

Not only are there fashions in diagnostic entities but in treatments as well, many of which turn out to be useless. Kristin Barker, a sociologist, refers to "diagnostic entrepreneurs."[35] Obviously, pharmaceutical companies seek to define various conditions as diseases or lifestyle conditions when they find or hope to find possible treatments to expand their markets. Of the 1,240 drugs that were licensed between 1975 and 1996, only 379 were for treatment of diseases as traditionally defined. (Just thirteen were for diseases primarily afflicting residents of tropical and poor countries.) Many of the remaining 861 were lifestyle drugs—treating conditions like baldness, severe axillary hyperhidrosis (excessive sweating), and erectile dysfunction, expanding the conventional view of diseases to issues of *quality* and *quantity* of life while creating profitable targets for drug companies.

PSYCHOLOGICAL DISORDERS OF LONGEVITY

We have stressed the medical disorders of longevity, but there are psychological ones as well—specific fear of aging (gerontophobia),[36] depression and suicide, anxiety, anger, and paranoia. Jacob Brody, an epidemiologist, wisely included what he called "social situations" in his category of aging-dependent conditions. He listed "signs of depression, social isolation and living alone, widowhood, and institutionalization." In addition to the dementias, which can result in destruction of memories and personhood, I would emphasize the frequency of symptoms of depression as well as diagnosable clinical depression. More than 20 percent of suicides are committed by persons sixty-five and older, most of them men—related to increases in depression and dementia.

As people grow older they frequently try to come to terms with the life they have led. Evaluative life review[37] in late life can be the basis of depression and despair as well as result in the successful resolution of conflicts. I believe suicide, in some cases, is the result of a profound sense that one's life has had no meaning or purpose.

New concepts and measures of stress further contribute to understanding various psychological conditions. For example, allostatic load, a theory developed by Bruce McEwen, refers to the ways that the body copes with psychosocial, environmental, and physical challenges, including work and family environments, lifestyle, and disturbances in normal sleep patterns.[38] Depression and lack of control are major factors in the allostatic load that are believed to result in conditions such as hypertension, obesity, and atherosclerosis as well as to lead individuals into social isolation, hostility, depression, and other conditions.[39] Socioeconomic status affects allostatic load. Stress may be as powerful a determinant of mortality as smoking, exposure to carcinogens, and many genetic risk factors.

With many acute diseases now in abeyance and with inroads being made against chronic disease, what about the issue of competing risks? Some have argued that elimination of one disease simply allows another to take its place. For example, it was once believed that the elimination of measles would result in deaths from other causes among those children who were so-called natural weaklings. However, this has not occurred. Jacob Brody, who has studied this hypothesis, says, "The concept of competing risk suggests that life is threatened by a continual pressure from a variety of sources." But he found no evidence of competing risk phenomena. For example, as death from cardiovascular disease declined over the period of his review, the incidence of cancer, the second leading cause of mortality, did not increase. Rather, life was prolonged. Of course, everyone dies ultimately; therefore, the final outcome of death is 100 percent certain. (Death, as it were, is the ultimate disorder of longevity.) So any disease that declines in incidence is replaced, in a sense, by death from other diseases, but these deaths *simply occur later*. Are causes of death independent after all? Are there individual genetic or common universal genetic factors that are reflected in the Gompertz curve? Put differently, are there intrinsic events that override all our efforts to curb disease and death—a kind of spiral of interdependency—namely, "intrinsic mortality" as described by Carnes, Olshansky, and Siegel?[40] Of course, there are extrinsic, independent causes of death such as obesity and smoking that bring death before its Gompertzian expectation. Another example might be heart disease influenced by behavioral and environmental conditions; or infections, such as AIDS. Nonetheless, intrinsic aging is a powerful factor. As already emphasized, we must always consider the complex interaction of environmental, social, behavioral, genetic, and intrinsic aging in the evaluation of disease.

WHAT MAKES SOME PEOPLE SURVIVE?

Too little is known about centenarians—people who survive beyond the period when Alzheimer's disease, cancer, and other disorders of longevity would have presumably already occurred.[41] Study of Danish twins indicates that genetic factors account for only about 25 percent of the variance in longevity. This study might be similar to the study of HIV survivors. What makes some people survive? Genes? Behavior? Environment? All three?

Thomas Perls, a student of centenarians, has stated that "the older you get, the healthier you've been."[42] Thirteen percent of the centenarians he has studied live independently, disease-free, at home. So-called longevity-enabling genes may be best conceived as genes that operate throughout life, providing for a vigorous life, not only for years added on at the end.

The Role Genes Play in the Disorders of Longevity

Certainly, we need to study more intensely the impact of the biological changes of aging in the maintenance of health and the genesis of disease.[43] We must understand better the profound interaction of the environment, behavior, aging, and genes. For example, a gene is expressed or activated in the environment, perhaps influenced by toxins, and with the passage of time affected by what some have called "accelerated aging." At times, however, it is difficult to determine the importance of genes, for example, when there is a late-onset disease such as Parkinson's or Alzheimer's. How do we isolate the role of the genetic contribution when the family history can be difficult or impossible to obtain because of the early death of parents, grandparents, or siblings? And, unless autopsy data are available, a family or pedigree analysis may be based upon inaccurate information.

The application of the Human Genome Project in producing diagnostic tests will make the role of genes much clearer, and proteomics, the large-scale study of the structure and function of proteins, is the necessary next step that may offer preventatives and treatments.

HAS ANYONE EVER DIED OF OLD AGE?[44]

In the 1930s German pathologist Ludwig Aschoff, reporting on the results of his extensive autopsy studies, claimed he was always able to find diseases to account for deaths. But Robert Kohn, a pathologist at Case Western, reported deaths that could not be explained by disease but rather were due to aging per se. (He used the term *senescence*.[45]) Regrettably, today the percentage of autopsies done in American hospitals has fallen dramatically from 50 percent to 10 percent of all who die, in part because insurers (including Medicare) will not pay for them and in part because doctors do not support the practice out of fear of malpractice claims. Their fears are partially well founded. Studies have shown that up to 40 percent of patient clinical diagnoses are inaccurate upon postmortem examination; in many cases the correct diagnosis would have modified treatment. (Clearly, death certificates are of limited value.)

On the other hand, by limiting the number of autopsies, we perpetuate these errors by preventing medical students and residents, as well as practicing doctors, from receiving the benefits of learning from the bodies of patients who have died of the diseases they are studying. I believe that Medicare and private insurers should pay for a randomized national sample of the dead (perhaps 15 percent of the 2.4-plus million annual deaths). (An autopsy costs between a thousand and three thousand dollars. Of course, autopsies should not be done where there are religious objections.)

In this era of longevity, it is important that we utilize all the modern tools of genetics and molecular biology to better understand the pathology of aging and senescence itself, as well as the disorders of longevity. Clinical pathological correlation at the organ, tissue, and cellular, as well as molecular, levels is also necessary. Longitudinal studies of the same samples over time are invaluable.[46] They contribute to developing a pathology of aging since rich comprehensive data (including autopsies) have been collected on these volunteer samples.

What is accelerated aging? A disease of accelerated aging would be especially interesting to the gerontologist who seeks precise biomarkers or biological signs of aging. The search for such biomarkers so far has proved elusive. Could accelerated aging be due to decreased availability of the energy (by mitochondria) that empowers metabolism? Might aging vary in speed based on individual genetic differences? Or is aging due to inadequate maintenance, such as by free-radical scavenging or failing repair mechanisms?

Thomas Perls[47] has noted that a greater number of centenarians had their babies in their forties. Just as accelerated aging is a possibility, so too is slowed aging.

Is rejuvenation of the whole human organism step by step by replenishing of cells, tissues, and organs—that is, by regeneration (embryonic stem cells)—the medicine in our future? Possibly, but it is much too early to be sure.

A strategy to delay the onset and slow the progression of fatal and nonfatal conditions is increasingly realistic. The goal is to delay the onset of aging itself—to postpone it, simply to slow it. To ensure quality of life and true health span or quality time, we must solve or at least ameliorate the myriad disorders of longevity. This requires a life-span perspective, with further investment in the health of pregnant women and their offspring as fetuses, infants, and children. An additional quantity of life is not enough. Access to comprehensive primary and specialty medical care should be assured. Furthermore, intensified research to improve quality of life is critical if society is to enjoy the fruits of the new longevity. Perhaps most of all, when we consider the actual causes of death, there is much we can already do.

We can now summarize the disorders of longevity (Table 6.6) which gives us a lifetime perspective regarding the common conditions found in old age—and suggests potential preventive and therapeutic interventions along the way. Indeed, where possible, prevention must begin at the very begining of life. Where precaution is not presently possible, such as in the case of Alzheimer's disease, major investments in research must be made, as will be discussed in the next chapter.

PART III: SCIENCE

CHAPTER 7

HUMAN INTEREST IN LONGEVITY: THE POWER OF RESEARCH

Five thousand years ago, according to Chinese legend, an aging emperor traveled in search of eternal life. He had studied medicine, taken drugs of immortality, and bathed in hot springs to restore his youth. Thousands of years after the Yellow Emperor the average life today still lasts 78 years, 28,470 days, 683,280 hours, or 2.4 million seconds. Immortality remains elusive.

Some centuries after the Yellow Emperor, in ancient Greece, ambrosia, food of the gods, was thought to conquer mortality. Dionysus was the God of wine and immortality, but the Greek ambrosia has yet to be discovered.

In more recent times, Faust, the German magician and alchemist of legend, sold his soul to the devil in exchange for power, knowledge, and youth. Some disparaged Faust, but others such as Goethe saw him as a seeker.

However intense the search, immortality is still nowhere in sight.

A 2001 survey conducted by the Alliance for Aging Research found that 63 percent of Americans polled by telephone would like to live to be one hundred (85 percent by online participation), although few expect to reach that goal. When asked how long they think they will live, 90 percent believed they would live at least to age eighty. Their enthusiasm for longevity was matched by their belief that individuals have some control over how long they will live. Sixty-seven percent felt there were things they could do to improve their chances of becoming centenarians at the same time as they were in favor of government support of research to help maintain vital old age.

The survey suggests that people vary in their views of what constitutes middle age and old age. According to people over fifty, middle age is achieved when one is over fifty. Two-thirds of baby boomers define older persons as seventy or older. (More recent, comparable surveys showed similar results.) According to a 2003 Harris poll,[1] nearly two-thirds of people between eighteen and twenty-nine would like to remain at about their present age, if they could live forever in good health. Among those in their early thirties, two of five would like to stay in that age range, though an almost equal number would like to be younger. However, among people in their forties, only one in five would like to live forever at that age, and two-thirds would prefer to be younger. A plurality of those fifty to sixty-four would like to be in their thirties, and those who are sixty-five and over would like to be in their forties.

What if people could be assured favorable conditions: a vigorous, healthy life expectancy, the presence of loved ones, reasonable socioeconomic circumstances, a sense of being valued by society, a continuing feeling of control over their lives? Given such assurances, would even more people wish to live a long time? I tested this assumption with a small longevity quiz among thirty persons of varied walks and stages of life. Obviously, it was not a nationally representative scientific sample, but, as expected, people's responses were that the better the conditions of their lives, the longer they wanted to live.

Most people underestimate the length of time they will live. A 1985 poll of six hundred people by the Association of Private Pension and Welfare Plans and the ERISA[2] Industry Committee (ERIC) found that most workers ages thirty-five to sixty-four underestimate their life ex-

pectancy by as much as five years or more. To test the validity of this response I conducted simple experiments that compared people's personal estimates of their life expectancy with the database estimates
from life insurance tables (see Table 7.1). A significant number did not
realize how long they could reasonably expect to live. Many seemed to
base their estimates on their own parents' and grandparents' experiences or on nothing at all.

The consequences of such underestimations can be sobering, especially for women, who may lack a realistic plan for the added years of
financial support they will need for the creative use of their time or for
the wise planning needed to accommodate a longer life. Individuals
who model their later years based on the example of previous generations may find themselves with unanticipated and even unwanted time
on their hands.

Current life insurance estimates may also underestimate life expectancy. In the twentieth century, every estimate of the absolute numbers and proportions of older persons made in the United States was
too low. What is really being underestimated is the impact of medical
and social progress on life expectancy. Such progress is likely to continue, possibly even at an accelerated rate. At the very least, this progress
is unpredictable and dynamic, with a history of outperforming expectations.

From ancient times to the search for the Fountain of Youth, to books
on life extension and maximum life span, longevity has always been a
topic of considerable fascination. People are interested in longevity for
different reasons. Some may be frightened of death and worried about
the hereafter, or worried about whether there is a life after death, and
they want to stay alive out of a desperate fear of the unknown. Others
may have a genuine interest in having more time to experience life itself. These are the people who would increase their desired life expectancy if they could be assured of a relatively pleasant time of it.

It is surprising how many people in later life, even up to the time of
dying, respond in the negative when asked if they would like to live
their lives over. Obviously, we must look to different points of view to

explain why many feel once is enough. For example, are the world's horrors just too painful to observe any longer? Might some individuals not be able to bear thinking that they had not had successful lives, or be fearful of having a second chance?

Does the desire for life and fear of death change over the course of life? One-third of those eighteen to twenty-nine say they are afraid to die, but that decreases to 16 percent among those fifty and older. How well do people prepare for death? Only one in four adults now has some written document such as a living will, or even a last will and testament, both of which require realistic preparation for dying and death. At a very deep and primitive level, there appears to be an instinctive drive toward holding on to life, seemingly contradicting the claim that most older people don't want a second chance. I have been impressed by the tenacity with which many of the very old and very sick fight to remain alive. Is it motivated by the fear of dying and death? Is it difficult to let go?

On the other hand, there are longevists who are altruistic, substantive, and vigorous persons, who enjoy their lives and wish to continue making their contributions. They have self-respect and are open to new experiences, precluding the boredom that precipitates anxious, even desperate actions or ennui.

In 1966, physician and historian Gerald Gruman defined prolongevity as "the belief that it is possible and desirable to extend significantly the length of life by human action."[3] He detailed two ancient notions—that there were long-lived people in some distant time and place (such as the fictional image of Shangri-la and the still-discussed Hunza near Tibet, the Caucasus in Georgia, and Vilcabamba in Ecuador) and that there are elixirs (such as the Fountain of Youth, as sought by Ponce De Leon) that will promote longevity.[4] Gruman's philosophic study follows the beliefs of prolongevity throughout history to modern times, including the emphasis upon human progress and the perfectibility of humankind. We have experienced a revolution in longevity but, to date, only negligible signs of human perfectibility and maturity.

CAN WE EXPAND LONGEVITY IN THE FUTURE?

How much longevity would we gain if we conquered the ten leading causes of death? Estimates suggest that conquest of cardiovascular disease, which accounts for nearly half of all deaths in the United States annually, would net us 10.2 additional years. The elimination of cancer would add 2.4, perhaps three years. Cures for other diseases would provide negligible increases in life expectancy.

The National Center for Health Statistics calculates lost years and premature death, defined as death occurring before sixty-five, and since 1995 it has also provided statistics of lost years and premature death before seventy-five. The concepts of lost years and premature death are productivity-oriented notions and are part of the language of health economists and rehabilitation planners who sought to obtain research or treatment funding by demonstrating years lost to national productivity as a result of disease and disability.

As Olshansky and Carnes have shown,[5] fresh, significant increases in longevity will have to come principally from reductions in mortality after age sixty-five because currently 85 percent of deaths occur then. (We would, of course, celebrate any mortality rate reductions for younger people, such as the infamous infant mortality rate in our black and Hispanic populations in America and of infants in the developing world.)

If the optimists are correct, a steady introduction of new treatments through research will add life expectancy through targeting diseases. If Olshansky and others are correct, to exceed the oft-stated limit of eighty-five years would require breakthroughs in retarding the processes of aging. All perspectives, of course, support further investment in fundamental research, whether disease-by-disease and/or aging research per se, with breakthroughs in aging research contributing to success in the former.

We know that short-term interventions (statins, Pap smears, flu shots, and blood pressure testing) are effective and measurably cost-saving. It is more difficult to assay the long-term effects of prevention in the reduction of heart and other diseases and, thereby, on mortality rates. But national dietary reforms, universal immunization schedules, accident prevention campaigns, and physical fitness programs, among other national efforts, if carried out, would undoubtedly extend longevity. Such efforts depend upon both individual and public responsibility, the latter being the expansion of the mission of national, state, and local public health-care programs and effectiveness of occupational, environmental, safety, and health efforts.

A simple example of an obstacle to be overcome is the absence of universal immunization in the United States. Many of our children do not receive the measles vaccine and a bare majority of our older population receive pneumococcal vaccine and annual flu shots. A second example is the absence of a truly major national accident prevention campaign, although accidents or unintended injuries are among the top ten causes of death.

Another source of added longevity available to us requires resolution of the high incidence of addictive disorders, including especially alcohol abuse, which, in turn, contributes to accidental deaths. We have achieved some commendable success against tobacco addiction, which, together with alcohol, may account for up to two-thirds of the difference in life expectancy between men and women. Homicide and suicide also count among the top ten causes of death. Homicide is strikingly higher in America than in other industrialized countries.

Can we extend longevity further? Yes, and beyond the practical and scientific research and preventive approaches already mentioned, there remains the possibility, still seen as largely far-fetched, that we can break through the age barrier, the maximum inherent life span, to extend the basic longevity of humankind. But short of such a major success, we must not forget that longevity is associated with education, improved economic status, health behavior, and access to health care.

People in Western societies especially, and perhaps humanity at large, are unwilling to accept natural barriers and are always struggling to overcome limits to mental and physical performance by breaking sports records, climbing mountains, exploring space, and adding years to life.

Yet there are limits. Try as we might, we cannot run a zero mile!

On the other hand, senility is not inevitable, natural, and irreversible, nor is three-score-ten a limit. Aging does not equate with disease; while natural, it, too, is not necessarily beyond intervention.

SHOULD WE EXPAND LONGEVITY?

Human interest in longevity is not always the same as the love of life. Some are simply greedy for an extended life but have no capacity to enjoy it. And those with a passion for life may be heedless about its perpetuity.

As can be seen, questions exist regarding the feasibility of increasing life expectancy, at least in the short term, especially through the disease-by-disease approach, unless there are either breakthroughs in retarding the processes of aging or in influencing genes that regulate longevity. So from a practical point of view, my question—Should we expand longevity?—is probably moot. But supposing there are breakthroughs, and soon? I see great social advantages. I believe both the individual and society will benefit from a more predictable and extended life to conduct the kind of human and social activities that, paraphrasing George Bernard Shaw, *simply require time and maturity.*

Would a prolongation of life reduce incentive? Would it destroy the motivation of scientists, artists, and others, freed from the threat of early death driving them on? Will the increased numbers and proportions of older persons provoke economic stagnation?

There are those who oppose efforts to extend life, among them Daniel Callahan and Leon R. Kass, who regard them as unnatural. Kass, chair of President George W. Bush's Council on Bioethics, has been especially influential, for instance, in opposing the use of embryonic stem cells. There are those who believe that we risk adversely affecting human authenticity, the nature of the family, marriage, sexual relations, parenting, aging, and the meaning of human experience if we tamper with the so-called natural order.

Throughout Western history, we see the recurrent fear of a prolonged life if it is associated with debility. In ancient Greek religion, Eos, the goddess of dawn, fell in love with a mortal, Tithonus. She beseeched Zeus to grant her love eternal life, which wily Zeus granted. However, Eos had failed to request continued health and vitality for her lover. He, therefore, continued growing eternally older and more impaired. Jonathan Swift, the Irish satirist, described the decrepit immortals, the Struldbrugs, in *Gulliver's Travels*. This imagery of inadvertent decrepitude in association with prolonged life affected Congress' deliberations in creating the legislation that produced the National Institute on Aging in 1974, and it adds discomfort in contemporary discussions of decision making at the end of life. I see the Tithonus complex as referring to those who are so fearful of decrepitude that they resist investment in the possibility of extending life.

Yes, fear of death is perhaps the greatest motivator of human behavior, at times prompting a desire to understand, seek knowledge, and contribute to others. Death and dying are great fears from which individually, however, we learn nothing. Ultimately, modern man accepts death with a sense of foreboding and horror, but anticipates that we may be able to prevent and conquer disease, even aging, and delay the inevitability of death.

RESEARCH IN THE UNITED STATES

A 1995 Harris poll found that 65 percent of Americans oppose cuts in medical research while 73 percent would pay higher taxes for research and 61 percent would urge Congress to provide tax incentives for private industry to conduct medical research. Sixty percent are willing to designate tax refund dollars for medical research. Over 90 percent endorsed maintaining the U.S. position as worldwide leader in medical research.

Nonetheless, after 2004, members of Congress had less of the budget available for making such choices because it was consumed by defense, entitlements, and other untouchables, while at the same time taxes were drastically reduced. In FY2007, there has been virtually no increase in the NIH budget. Will the American public tolerate a significant reduction in medical research? Or will the National Institutes of Health undergo a renaissance?

No institution in the world better illustrates the human desire to conquer disease and extend life than the National Institutes of Health.[6] The NIH is the research arm of the U.S. Public Health Service and is concerned with health as well as disease. Its mission is surely relevant to the Longevity Revolution: "Science in pursuit of fundamental knowledge about the nature and behavior of living systems and the application of that knowledge to extend healthy life and reduce burdens of illness and disability."

Of the NIH, Dr. Lewis Thomas has written enthusiastically: "All by itself, this magnificent institution stands as the most brilliant social invention in the Twentieth Century, anywhere. It has done something unique, imaginative, useful and altogether right."[7] Indeed, the NIH, over a hundred years old, is a great international treasure of biomedical science. Its intramural program and its support of scientists in American (and foreign) universities and medical schools have set the pace for medical research on a worldwide basis. As of 2007, the NIH had supported 114 American winners of the Nobel Prize who served both in its intramural program and/or received extramural support. Shockingly, although a majority of the public is familiar with NASA, only a small percentage know about the NIH.

The National Institute on Aging is part of the National Institutes of Health in Bethesda, Maryland. Only about two hundred million dollars of NIA's budget (20 percent of its total) was spent on the basic biology of aging in fiscal year 2006. The NIA's budget is less then 3 percent of the total NIH budget (about $29 billion) despite the revolution in longevity that has occurred and is likely to continue. It is also time for a nationwide private-sector effort to mobilize the resources of the private nonprofit and corporate sectors as well as individual philanthropists to both support aging research and to push our government to continue and expand support of the NIH and NIA.

The NIH works as follows: governmental officials do not decide who receives grants to conduct research. Study sections composed of scientists from universities do so. This is called peer review, and it has worked uncommonly well. Proposals are either rejected or approved and given scores. However, in 2007 only about 15 to 20 percent of approved grants were funded, depending on the institute.

I believe that at least 30 percent of approved grants (if not more) should be funded. Biomedical science obviously cannot thrive without the continual recruitment of young people, who are certainly the lifeblood of any scientific enterprise. When there are cutbacks in research and research training it is not surprising that grant applications from younger university scientists may decline or go unfunded. When funds are tight, review committees act too cautiously and conservatively. Funds should be available to support risky research.

Moreover, the nation's scientific infrastructure, especially its laboratories, need rebuilding. Furthermore, we must develop more cost-effective ways to conduct expensive but essential large clinical trials and long-term longitudinal human studies. Improved collaboration among industry, academia, and the NIH is desirable since we need to more rapidly apply scientific findings that assist human welfare, though. the commercialization of academia is an issue that must be confronted.

Economic Value of Research

In 2003, the U.S. invested $95 billion in biomedical research from public and private sources, including 57 percent from industry and 28 percent from NIH.[8] There are appropriate concerns that the natural sciences, engineering, and the physical sciences have not enjoyed the same funding boost as has biomedical science. For example, the National Science Foundation and NASA have not grown significantly in recent years. This fact also reduces valuable interdisciplinary study.[9]

Research supports hundreds of thousands of skilled jobs at universities, academic medical centers, and companies, large and small, across the nation. In 2004, the pharmaceutical industry employed 293,000 people[10] and generated sales of over $200 billion. In 2003, 1,473 bio-technology firms employed 198,300 people and generated $39 billion in sales.[11]

Science is a major economic stimulus to the economy. Indeed, the NIH received a 5.8 percent increase for fiscal year 1997 because the Republicans, joining Democrats, concluded that biomedical research resulted in economic growth. The NASDAQ stock market is advertised as the future, and its stocks fund in technology and biotechnology. Despite the bursting of the so-called new economy bubble of 2001, research and development are still the future.

Equally impressive, but still measured using conventional practices, are the cost savings of diagnostic and treatment procedures for particular diseases. For example, a seventeen-year program which invested only $56 million in research on testicular cancer has led to a 91 percent cure rate and an annual savings of $166 million.[12]

Increases in life expectancy in the 1970s and 1980s were worth $57 trillion to Americans—a figure six times larger than the entire output of tangible goods and services last year.

THE POLITICS OF ANGUISH

The architects of advocacy groups, patient activism, and foundations in this country have often been families who have lost family members to a particular disease. The Wexler family represents one remarkable example.

Nancy Wexler, a psychologist, lost her mother, three maternal uncles, and her maternal grandfather to Huntington's disease, a fatal neurological disorder that afflicts about thirty thousand Americans annually and puts an estimated 150,000 others at risk. Wexler's father, a Los Angeles psychoanalyst, created the Hereditary Disease Foundation in 1968, and it became the focus of the family. Nancy Wexler wrote that "the struggle against hereditary disease has given me purpose and direction." Such purpose and direction led her to James Gusella, a molecular biologist, who adopted a strategy involving restriction enzymes to cut DNA in specific places and hunt for the Huntington's disease DNA markers[13] and the gene. In 1979 Wexler led her first expedition to remote Lake Maracaibo in Venezuela, where lived the family with the largest known number of Huntington's disease members—two hundred, with two thousand more family members at risk. The pedigree consists of the blood and skin of all descendants and their relatives of a single woman with the disease who lived in the area in the early 1800s. Blood samples were obtained from both affected and unaffected individuals. This led, in 1983, to the discovery of the Huntington's disease marker on the short arm of chromosome 4, which, in turn, helped pave the way for the NIH-sponsored Human Genome Project.

Perhaps the most highly publicized example of the modern health politics of anguish occurred because Franklin Delano Roosevelt contracted polio, leading the way for the March of Dimes to mobilize against this disease, which led to the successful Salk vaccine.[14]

An advocacy group enthusiastically outlines the cost of an illness, using a methodology developed by Dorothy P. Rice, former director of the National Center for Health Statistics. However, if one added up all the cost estimates of diseases by advocates—costs of care, indirect costs such as loss of productivity, and life years lost—would the total amounts be greater than the overall health costs?[15] Not really, but, unfortunately, all advocacy groups are fighting for the same dollars. The pressure is on Congress. It is obviously impossible for an individual member of Congress, even if he or she has unusual scientific training, to always make appropriate decisions with regard to the allocation of funding. Scientists themselves are notoriously unreliable in this regard, often motivated by their own vested interests. Therefore, it would be desirable to establish some intervening group, such as the Office of Technology Assessment (killed, unfortunately, by Congress in 1995) or the Institute of Medicine of the National Academy of Sciences, in collaboration with the NIH itself to help lay out a reasonable blueprint, subject to constant revision based on changing needs and scientific opportunities. Such a blueprint would cover the range from basic science to directed science, since basic science can get lost, especially in a market-driven economy. Considerable progress follows from serendipity and curiosity-driven research. Directed research, as in the case of polio and to some measure in the case of Alzheimer's disease, has worked. Orphan diseases are of modest incidence and of little interest to the marketplace.[16]

That the individual disease mission of the National Institutes of Health has direct Congressional budgetary lines helps account for the success of American medical research. Fresh voices will continue to be heard in the fight for funds. Former First Lady Nancy Reagan and the late Christopher Reeve became ardent advocates for the treatment of Alzheimer's disease and spinal cord injuries, respectively, and supported the use of embryonic stem cell technology. And the Juvenile Diabetes Research Foundation (JDRF) has become the world's leading nonprofit, nongovernmental funding source for medical science, having contributed over $600 million by 2003.

In 1979 two Wisconsin mothers of mentally ill children started the National Alliance for the Mentally Ill. Since that time 1,100 chapters have been created in fifty states with 140,000 members, helping grow the research budget of the National Institute of Mental Health, which doubled over a ten-year period.

Perhaps we will see an American Patients' Association. International movements of patients suffering from AIDS or other diseases could lead to the development of an international coalition of patients' associations.[17] One wonders how the various disease-advocacy organizations might work together in common purposes such as fundraising and support of research rather than being subject to "divide and conquer" fights over priorities and money.

Private dollars alone, however, aren't likely to conquer any disease. As philanthropists Albert and Mary Lasker have pointed out, the big dollars are federal and it is the central government (and its power of taxation) where we must seek support for research as we seek to promote private philanthropy.[18]

NATIONAL INSTITUTE ON AGING

Although a relatively new institute, the National Institute on Aging has already supported exciting research and sponsored major discoveries, such as apolipoprotein in Alzheimer's disease, telomeres and the work of Nobelist Robert Fogel in economics, as well as the prion studies of the Nobelist in medicine, Stanley Prusiner.

Today less than 1 percent of the entire federal budget is spent on medical research. Both to improve health and control costs, *I propose that 3 percent of the nation's overall health bill ($1.8 trillion projected as of 2005) or $54 billion be available to NIH for medical research from federal revenues.* I also propose that of Medicare expenditures, 1 percent (or $3 billion) be devoted to the National Institute on Aging. This investment should increase in line with the Medicare budget, but scientific research, in fact, could lower Medicare expenditures.

To give some perspective, the FY2006 NIA budget approached $1 billion. The national Alzheimer's Disease Association argued in 2005 that the NIH as a whole should devote at least $1 billion annually for research on Alzheimer's disease. While the numbers I am suggesting may seem extraordinary, I believe the level of scientific progress in the field since the 1950s justifies such a program, which could be dubbed the Apollo Program for Aging and Longevity Science.

How would $3 billion (1 percent of the 2005 Medicare budget) be spent? It would be phased in over five years:

- $1 billion for basic biology (year 1: $200 to $400 million; year 2: to $500 million; year 3: to $650 million; year 4: to $800 million; year 5: to $1 billion) including genomics and regenerative medicine *relevant to longevity science.*

- $1 billion for age-related diseases, in collaboration with the disease-mission NIH institutes with coordination through a trans-NIH decision-making committee[19] (with similar phase-in funding).

- $500 million for clinical trials with proportionate representation of older persons (65+) including head-to-head studies of drugs including lifestyle comparisons, cost-effective studies, and a national system for post-marketing drug surveillance (with phase-in funding).

- $500 million for a national preventive medicine research initiative *through the lifespan* including studies of safety and health in the home and workplace and of physical inactivity and obesity as well as genetic and other early-life pathological influences. This includes studies of social-behavioral and economic means to effect positive changes in health behavior (since the nation is confronting a health crisis that could lower life expectancy), studies to evaluate and improve health services delivery (via the Health and Human Services Agency for Healthcare Research and Quality), and studies to evaluate the overall impact of population aging and longevity on society.

These amounts would also be used for research training in the four categories and for necessary infrastructure development (e.g., human longitudinal study programs, laboratories and equipment, bioinformatics, animal resources, etc.).

LABORATORIES 2030

It is not just the traditional scientific laboratories that need to be funded. The social sciences also have much to contribute to the challenges posed by the revolution in longevity; research in these disciplines should also be supported.

One such program is Laboratories 2030, an experimental paradigm I envision to conduct empirical studies. Just as Supreme Court Justice Louis Brandeis saw our states as great laboratories for social experiments, Laboratories 2030 would seek communities to assist states and nations in planning for the twenty-first-century extension of longevity in the context of worldwide urbanization and possible re-ruralification.

Communities in the United States and elsewhere whose older populations are at or above a country's demographic expectations for the year
2030 (excluding artificial retirement communities) could be studied,
including naturally occurring retirement communities, clusters of
apartment houses or suburban areas with large proportions of older
persons referred to as NORCs.[20] Laboratories 2030 is designed to explore, in advance of the year 2030, the conditions and consequences of
population aging. These communities would be the laboratories for
practical studies that provide an opportunity to observe relationships
across the generations. Funding for this endeavor would come from the
program I propose.

Selected communities would begin with descriptive research to establish a baseline profile of needs, problems, and existing services. Then,
interventions would be designed and implemented, based upon ideas
drawn from the project leadership and a national grants program. The
central focus of the different interventions would be to enhance quality
of life for mid- and late-life older persons and their families, in recognition of mutual transgenerational interdependence and involvement,
while also addressing the needs of children and younger adults.

With national funding and community consent, Laboratories 2030
could conduct practical, action-oriented studies, focusing on such qual-
itative and quantitative issues as the methods or processes by which
these communities now respond to population aging; the results of
these community adaptations to population aging in terms of the impact on families, service providers, businesses, employment practices,
philanthropic and religious organizations, political groups, and the
other various elements of a community; and the implementation of
practical interventions based on what has been learned about social and
environmental adaptations.

Laboratories 2030 would look for investigations whose focus tran-
scends a specific generation in order to benefit all age groups and their
quality of life. For example, changes in volunteer programs (meal delivery, respite services, etc.), adaptations of traffic systems and design
of highways and intersections to accommodate slower moving individ-

uals as well as improvements in public transit, living arrangements, and many other programs would be of enormous benefit to a vast number of people who are not old. Core studies might include the implementation of family service centers in order to reduce costs associated with home care or nursing home care. Such centers, which might be established at transformed senior centers, would provide physical therapy, nutritional services, flu shots, and other health care services and information to frail older persons while also providing an opportunity for socialization, education, and caregiver respite to the family.

Another core study might involve the establishment of a limited transportation system, such as a continuous jitney (or use of school buses outside of school hours) that operates on an established and regular circuit between a medical center, shopping center, library, and other sites to be determined by the individual community. This could include volunteers to assist older persons who are frail or suffer dementia.

Perhaps the best way of counteracting bias toward older people in the workforce would be through the study of exemplary programs and policies for productive aging in the workplace (see Chapter 13) and of the impact of these programs on the psychosocial and financial experiences of older persons, their families, and the community in general.

Finally, Laboratories 2030 could confront the myths and prejudices regarding the nature of aging and older people that have prevented communities and nations from utilizing the productivity, creativity, and resourcefulness of its older populations.

Strong historical precedence supports the use of communities as laboratories for subsequent policy and program development. Examples include the classic work of Robert and Helen Lynd in two examinations of Muncie, Indiana, reported in *Middletown: A Study in American Culture* in 1929 and *Middletown in Transition* in 1937.[21] The important epidemiological study of the residents of Framingham, Massachusetts, that began in 1948 led to the establishment of the coronary risk profile (sedentary lifestyle, tobacco use, high-fat diet) and subsequent reduction of deaths from heart disease and stroke through changing health behaviors. Finally, historical community studies that use existing data from the census or other state-collected information to reconstruct historical profiles of a society and behavior are frequently conducted.

URBANIZATION

Thomas Jefferson believed that cities were "sores on the body politic," but with the advent of the Industrial Revolution, urbanization proceeded rapidly. In Jefferson's United States at the turn of the nineteenth century, less than 1.7 percent of Americans lived in cities; in the 1940s the urban population had risen to less than a third, and by the 1990s to fully one-third. Even more dramatically, by 2000 about half the world's population lived in cities; by 2025 it is predicted that two-thirds—five billion people—probably will.

Massive urban conglomerations have grown in many nations, with eighty-six million people added to world cities each year—well over a million a week. Nearly a third of the population lives in the largest cities of some nations, including Mexico City, Tokyo, Athens, and Buenos Aires. The shantytowns and squalid hill slums of Rio de Janeiro are grim reminders of the bleak conditions that exist throughout the world. Moreover, huge numbers of able-bodied young people leave their older family members to care for the youngest in the countryside and come to the cities searching for better lives.

While urban culture brings excitement and more choices, it also brings anomie, pollution, and crime. How will megacities ever be able to supply the basic public services of sanitation,[22] safety, security, privacy, transportation, health services, clean air, and water?

Consider water and sanitation alone: By 2011, according to the World Bank, 1.4 billion people will lack adequate water supply and sanitation. Cairo, Beijing, and São Paulo will probably experience these acute water shortages even sooner.

As with many other aspects of the move away from rural economies, urbanization has both enriched and destroyed the lives of older persons. For example, the old in Africa are often left behind in the villages as the young throng to cities, hoping for a better life. On the other hand, cities often offer more services and amenities for older persons who live there.

In America, suburbs grew after WWII, but now they have grown gray and do not possess the environment and resources needed for their aging residents.[23]

Communities that become part of Laboratories 2030 might include entire towns, portions of cities or large suburban areas, and naturally occurring retirement communities. In the United States, sites might be in Utah (in order to examine the unusually healthy Mormon population and to have access to the genealogical archives of the Mormon church), Olmsted County, Minnesota (to have access to excellent available epidemiological data and the Mayo Clinic and School of Medicine), or Arkansas (whose university medical school has a center on aging and a department of geriatrics).[24]

Most people want to age in place, and housing built to last a lifetime of personal and social change would be ideal. Since aging in place is likely to become the norm in the future, universal, or life-cycle housing needs to be built to serve people of all ages and conditions. Still, a wide range of choices, such as assisted living, a hybrid of housing and a continuum of services, and opportunities for learning, are necessary. (Such universal housing is advantageous to all, from a young person recovering from a skiing accident, to a pregnant woman, to the older person living alone.)

Continuing care retirement communities have been developed near or in coordination with academic institutions such as Oberlin, Cornell, and Dartmouth. These are very attractive to retired faculty and to other retirees interested in lifelong learning. Kendall-Crosslands was an early creator of such communities. Barney Osher, a successful banker, provided over $1 million each to 115 institutions that set up lifetime learning centers by 2007. Elderhostel is another pioneering popular source of education for older persons that involves travel. These various innovations should be evaluated for their effectiveness.

When all is said and done, we need direct community experiments and computer experts, architects, builders, social scientists, and gerontologists—and even philosophers such as a Lewis Mumford—to understand the needs of *all* people of *all* ages and to emphasize quality of life in our communities. Whether the world is divided by nationalism or religion, affected by the growing power of multinational companies, or influenced by the global culture and economy, ultimately it is in the local community where people usually live out their lives.

What we do not need is imperious planning from the top, a lesson learned painfully through urban renewal destructiveness. Jane Jacobs has written that "designing a dream city is easy. Rebuilding a living one takes imagination."[25]

In AARP[26] surveys conducted over the last decade, 77 percent of respondents, on average, said they wanted to live in mixed-age communities, while only 13 percent said they wanted to live among people of their age group. But children are not welcome in some retirement communities. The Federal Fair Housing Act bars discrimination on the grounds of race, religion, sex, or family status, but it does allow discrimination against children.[27]

Other studies might address:

- Adaptation of job risks, work-site training programs, and workplace experiments to make the workplace more accessible and safer for older adults and to perpetuate working life, which I call productive aging or productive engagement.

- Creation of new forms of *functional* assessment for persons *of all ages* with respect to driving, use of machinery, etc.

- Medical prevention strategies for common problems of older persons, such as cognitive function, pneumonia, falls, and hip fractures, etc.

- Innovative partnerships with corporations to generate products and services that are of particular use to an older or disabled clientele, such as telemedicine, home safety mechanisms, home robots, and other assistive devices, and cars that are more user friendly for older adults.

- Development or expansion of networks of voluntary and paid drivers of cars,[28] vans, and school buses in urban, suburban, and rural areas, especially outside of rush hours.

- Special smart housing and prosthetic interventions such as kneeling buses to reduce the effects of frailty and to maintain mental and physical functioning, and the testing of relevant technologies.

- The high suicide rate among older men and the testing of different interventions to reduce depression, despair, and anomie.

- Alteration of the curriculum in local health-services schools to emphasize additional community- and home-based geriatric practices among the various health professions.

- The special problems of older women, who continue to outlive men by a significant margin, including examination of social and cultural aspects of widowhood, financial difficulties experienced by women, abuse, long-term care needs, and special health problems.

- Use of older persons as volunteers to assist in primary schools and to provide role models, historical reminiscences, and a sense of career choices.

- Integration of acute and long-term care into more hospitable, home-like arrangements such as Green House, a residence for six to eleven older persons who require skilled nursing care but want to enjoy a life of high quality.[29]

The World Cities Project

From the paradigm of Laboratories 2030 came the concept of comparing specific urban communities in the United States, Europe, Japan, and the developing world. With international foundation support,[30] the International Longevity Centers in the United States, France, and Japan, in collaboration with New York University, developed the World Cities Project (WCP).[31] Begun in 1997, the WCP is currently studying urban health and the quality of life of young children at risk and older persons in four world cities (New York, London, Paris, and Tokyo), providing a laboratory in which to test interventions to improve the lives of these vulnerable populations.

WCP could be useful anywhere in the world but particularly in the developed world, which is aging the fastest. Its studies include comparisons of infant mortality; aging and long-term care; vulnerability; coronary artery disease in London, Paris, and New York; avoidable hospitalizations in Paris and New York neighborhoods; social interaction; and the health of older people. While this work is largely descriptive at present, the ultimate goal is to test interventions to improve the lives of older persons as originally conceptualized in Laboratories 2030.

A Department of Education, Science, and Technology

Aristotle, a physician as well as philosopher, established the first great zoological garden. He is said to have received eight hundred talents (an estimated $4 million in 1961 dollars) from the government for research. According to Will Durant, this was the first example in European history of "large scale financing of science by public wealth."[32] Do modern states support research on proportionately as large a scale?

In the United States today, about half the $74 billion the government spends annually on all research and development does *not* expand fundamental knowledge or create new technologies. It goes instead to the evaluation and testing of military weapons, aircraft, and other systems. An orbital jump in financing of science is required to advance longevity and health as well as national wealth. Investigator-initiated research is essential, of course, but must be distinguished from ordinary targeting of specific research problems. Interdisciplinary research—breaking the boundaries of disciplines—is more valuable than ever. A large science initiative such as the aforementioned Apollo Program for Aging and Longevity Research should focus upon all science including bioscience, in particular basic biology, as well as applied research to justify the hope that the twenty-first century will be the *century of the life sciences* and to advance human health.

I propose the development of a department of education, science, and technology of the kind that other nations have, such as the Ministry of Science in the Netherlands, which could even be a virtual organization, bringing together the National Science Foundation, the NIH, DARPA,[33] the Department of Energy, the Department of Education, and other entities, coordinating activities but avoiding stultifying bureaucracy. The Department of Energy and the NIH enjoyed the success of the Human Genome Project, which also had an international component. Since national productivity and prosperity are linked to education, science, and technology, this would be a valuable step—certainly of the significant level of other departments that have helped our economy, such as those of agriculture, labor, and commerce.

As an example, a new Department of Education, Science, and Technology would necessarily place alternative energy as a high priority, given present dependence upon petroleum, ultimately limited in supply and a trigger for war.

Such a department should consider the aging factor in health and disease, which has not yet been extracted from the complexities of human biology. To make such research a priority will require passionate and effective public advocacy, not trivial pseudo science, of the kind that only a government commitment can accomplish. Market forces cannot do it. It should be disappointing to all who are interested in long, healthy lives that money made available for biomedical research sponsored principally by the National Institutes of Health has flattened out, and that aging research itself has enjoyed only modest support.

Financially, the historical gains from increased longevity have been enormous, on the order of $2.8 trillion annually from 1970 to 1990. The reduction in mortality from heart disease alone has increased the value of human life by about $1.5 trillion per year over the 1970 to 1990 period. The potential gains from future innovations in health care are also extremely large. Eliminating deaths from heart disease would generate approximately $48 trillion in economic value. A cure for cancer would be worth $47 trillion.

CHAPTER 8

ALZHEIMER'S:
THE EPIDEMIC OF THE TWENTY-FIRST CENTURY

Few diseases have had as great a social, economic, and personal impact as Alzheimer's, the most common form of dementia. Destroying the mind, devastating the family, and making considerable demands upon the health- and social-care systems, it also robs society of important contributors. In recent years the costs of nursing home care have risen dramatically, in large part due to the increasing number of dementia patients who require total care. In 2005, about 1.5 million persons were in nursing homes, about 50 percent of them suffering from dementia. As we celebrate the triumph of a century marked by dramatic increases in longevity, we see a parallel rise in the number of persons suffering from this devastating disease.[1] Lewis Thomas has called it the "disease of the century." Unless we find ways to prevent or cure Alzheimer's and other severe dementing diseases, the world will shortly be confronted with epidemic proportions of these diseases. People over eighty-five, the fastest-growing population, also manifest the highest rates of Alzheimer's disease, approaching a conservatively estimated 20 percent.

There is no evidence that Alzheimer's disease is increasing in frequency.[2] Rather, it is more common because older people are more numerous, and the diagnosis is more commonly made. In addition, breakthroughs in the treatment of diseases associated with old age, such as cancer and heart disease, will raise the number of the so-called oldest old and further propel dementias into prominence.

Despite the generally bleak picture, longstanding despair about dementia is giving way to hope. Recent advances in neuroscience are providing leads to its causes and suggesting means of prevention and treatment. In the past, the hopelessness of patients, doctors, and scientists stemmed from their association of senility[3] with the false belief that it was an inevitable result of the aging process—if you lived long enough, you would surely become senile. In addition, senility was often presumed to be due to cerebral arteriosclerosis (popularly referred to as "hardening of the arteries"), for which preventive measures were not recognized until the 1970s. Indeed, until recently, senility was felt to be not only inevitable, but unpreventable, untreatable, and incurable as well.

Contrary to traditional belief, brain cells do not simply die off as people grow older, although dendrites, the branches of brain cells, do thin out. Thus, senility, the popular term for the dementias, is not inevitable in old age, although it is highly age-associated. The conditions listed under the label of senility can and should be subjected to vigorous scientific investigation, like all other diseases. However, negative stereotypes about old age based on the inevitability of senility have hindered medical progress for a long time. This has translated into hesitancy on the part of social and scientific policymakers to invest in research aimed at the health problems of the later years. The rationale presented is that since money spent on the aged is money spent ultimately on the decrepit and incompetent, it represents an investment with little or no payback.

The idea of a second childhood, the alleged signs of supposedly childish decline in old age, haunts people to this day. Only recently have we come to realize that this is not a true image of aging, thanks in part to modern bioscience and in part to gadflies of the aging movement. Perhaps the most momentous result of successful dementia research will be to confirm our view of ourselves in our later years as disease-free and productive almost to the very end.

Opposition to aging research has lessened due in part to our growing, vigorous older population and to the dissipation of the myth of senility in the medical profession—a process that was aided by the 1978 consensus conference at the National Institute on Aging on the many reversible confusional states associated with old age.[4] Indeed, real progress toward the understanding of Alzheimer's awaited the definition of what it was *not* almost as much as what it was. Significantly, the meeting distinguished between those disorders that can be successfully treated and reversed (perhaps 5 to 15 percent) and the so-called irreversible dementias, such as Alzheimer's disease and multi-infarct dementia (MID) or vascular dementia (referred to as cerebral arteriosclerosis in older terminology). This consensus conference presented a long list of disorders that are detrimental to normal brain function in older persons. Included on this list were adverse drug reactions to therapeutic medications, infectious disease, heart disease, stroke, metabolic disorders such as kidney failure and thyroid disease, anemia, and a number of other physical and emotional problems, especially underlying depression.

WHAT, THEN, IS ALZHEIMER'S DISEASE?

The etymology of senile is "old," and of dementia, "deprived of mind." Historically, the Anglo-Saxons, in their word *dotage*, recognized the dementias of old age. However, it was not until the French psychiatrist Dominique Esquirol (1772–1840) that senile dementia was given a name and distinguished from some other psychoses. The German psychiatrist Emil Kraepelin (1856–1926) contributed to the classic definition of organic brain disease. At the turn of the twentieth century, general paralysis of the insane (paresis or syphilitic brain disease) was also differentiated. In the 1950s, Sir Martin Roth of Great Britain used a series of prognostic studies to separate the various forms of dementia from each other.

In 1906, Alois Alzheimer (1864–1915), an Austrian-born neurologist, studied a fifty-one-year-old woman, Auguste D., who suffered from dementia and agitation and died at fifty-five. Upon postmortem examination of the brain, Alzheimer described lesions that are now considered characteristic of the disease bearing his name. The woman's brain showed severe atrophy, and the outer layer (cerebral cortex) was marred by scar-like tissue called neuritic or senile plaques and by a clumping and distortion of fibers in the nerve cells. Alzheimer called these jumbles of filaments neurofibrillary tangles (NFTs). Because of the age of Alzheimer's patients and the other reports that followed, Alzheimer's disease was generally considered to be a presenile form of dementia and a rare neurologic disorder.

In 1948, R.D. Newton and Meta Neuman, working separately, concluded that Alzheimer's disease and what had been labeled senile dementia were one and the same. With the availability of electron microscopy and advances in histochemistry in the 1950s and 1960s, all investigators agreed that the lesions were identical in both.[5]

The brain of an Alzheimer victim is characterized by shrinkage or atrophy, especially of the frontal and temporal lobes, and enlargement of the ventricles or cavities with an average 8 percent loss in brain weight. The disease primarily affects the posterior hippocampus,[6] where memories are first registered but not stored, as well as the posterior cerebral cortex.

When the disease is studied, the changes most commonly observed are in the proteins of the nerve cells in the cerebral cortex. Under the ordinary compound microscope these changes appear as NFTs. The tangles consist largely of abnormal forms of cellular proteins called *tau* in twisted fibers that are strikingly insoluble. In healthy neurons, which are impulse-conducting cells, tau helps stabilize the neuronal skeletal framework. Two classes of enzymes, called kinases and phosphotases, chemically modify tau through a process called hyperphosphorylation, causing NFT formation. These enzymes are therapeutic targets for Alzheimer's. Tau protein is not mutated in Alzheimer's disease.

The second commonly seen structure, the aforementioned plaques, occurs outside nerve cells. These plaques have a dense central core made up of a substance called amyloid, which is an abnormal protein not usually found in the brain. Amyloid is produced by a precursor protein called beta amyloid precursor protein (APP), which occurs in normal brains (as well as in all human cells) and is harmless when broken down. But in Alzheimer's disease, beta amyloid is formed from the precursor protein for reasons yet to be established and apparently damages cells and affects synapses, the bridges through which nerve impulses pass. Proteoglycans, which are substances that occur outside of nerve cells in supportive cells called glia, appear to play a role in the development of amyloid deposits in Alzheimer's disease. Alzheimer's disease may be unique to humans since NFTs have never been found in any other animal, although senile plaques can be found in cats, dogs, and monkeys (and other nonhuman primates). On the other hand, tau inclusions are found in other brain conditions in humans such as Pick's disease, another dementia.

Dennis J. Selkoe of Harvard calls Alzheimer's disease "a synaptic failure."[7] Nerve impulses pass across the junctions called synapses that join nerve cells with each other, with muscle cells (to produce movement), and with gland cells (to produce hormonal secretions). The best way to determine the extent of mental impairment is by measuring synaptic density, which is the extent of communications among neurons and was first observed by the neuropathologist Robert Terry. The more advanced the disease, the fewer the nerve impulses.

Diagnosis

Alzheimer's disease is usually diagnosed by excluding other conditions, as mentioned, and by the patient's history, mental status, and a neuropsychological examination. To complicate matters further, different diagnostic manuals in the world list different criteria for the disease. Among most doctors the rate of misdiagnosis is between 10 and 20 percent. The disease is only confirmed definitively at autopsy or biopsy. In an average group of one hundred cases of so-called senility or dementia, Alzheimer's disease probably accounts for some 55 to 65 percent.

Alzheimer's disease must be separated from multi-infarct dementia (MID, or vascular dementia). About 20 percent of the cases of so-called senility are the result of MID, which is characterized by the occurrence of many repeated ministrokes in the brain and can often be prevented by control of hypertension. Another 10 percent have both Alzheimer's disease and MID. As noted earlier, the remaining five or so percent of so-called senility cases are caused by other dementias or by the dozens of underlying physical conditions arising inside and outside the brain that may affect cerebral circulation and metabolism.

Alzheimer's disease must also be differentiated from the mild forgetfulness that is sometimes mistakenly attributed to aging. Canadian clinician V.A. Kral first applied the term "benign senescent forgetfulness" to this condition.[8] He described this phenomenon as the "inability of a subject to recall unimportant data and parts of an experience [e.g., a name, a place or a date], whereas the experience of which the forgotten data form a part can be recalled." While Kral distinguished this from the "malignant" syndrome found in Alzheimer's disease, he stated that this does not necessarily mean that they are separate neuropathological processes. In time, benign senescent forgetfulness may become malignant.

The condition of benign senescent forgetfulness may be what has now been labeled *age-associated memory impairment* (AAMI). By 1998, the term *mild cognitive impairment* (MCI) was also introduced. It more sharply identifies a population with memory complaint and memory impairment but with normal general cognitive function and normal daily function.[9] Forgetfulness can also be due to stress, anxiety, and depression, as well as physical disorders.

Alzheimer's disease must especially be distinguished from depression. On the one hand, Alzheimer's disease may provoke depression. An estimated 25 percent of individuals with the disease respond by becoming depressed, which results for some in suicide. On the other hand, there is up to 25 percent coexistence of Alzheimer's disease and depression that may be due to the underlying physical pathology. Moreover, depression can mimic dementia, a condition called pseudodementia.

It is essential that the potentially reversible forms of memory impairment be promptly and accurately diagnosed and treated. Indeed, whether the patient is suffering from dementia or a delirium, such conditions are medical/psychiatric emergencies. The underlying physical disease may be fatal, and brain recovery is related to the duration of the disease's cerebral effects. Comprehensive medical, psychiatric, and neurological histories and examinations are mandatory.

WHAT CAUSES ALZHEIMER'S DISEASE?

Several theories have been advanced regarding the causes of Alzheimer's disease. These include aluminum toxicity and other environmental factors that might affect the brain, genetic factors, slow or latent viral infection, bacterial infection, prions, autoimmunity, and changes in brain chemistry. Significantly missing from this list is the theory that age or aging processes *cause* Alzheimer's disease. Aging processes *set the stage* for the development of Alzheimer's disease. The passage of time is integrally involved in many changes in the nervous system related to the disease. There is a doubling of the prevalence of Alzheimer's disease each decade after age sixty-five. *Aging is a risk factor,* and Alzheimer's disease is a disorder of longevity. Indeed, genetic bases and other factors may have been operative from the beginning of the patient's life but are only revealed when the biological changes of aging reach certain levels. In addition, many of the neuropathological and neurochemical changes found in Alzheimer's disease are also found in normal aging. Perhaps ultimately, a measurable threshold will be found,[10] distinguishing normality from pathology.

In addition to age, the risk factors for Alzheimer's disease include the presence of apolipoprotein E-4 (Apo E-4), a gene found on chromosome 19,[11] family history, Down syndrome (triplication of chromosome 21), and, apparently, female gender. Women's longevity may account for this possibly greater susceptibility. Some research suggests that estrogen deficiency may make postmenopausal women more vulnerable to Alzheimer's than men of the same age since neurons affected have estrogen receptors. (But estrogen used as a preventive or therapy has not proven successful.) There is also some evidence that head injury is a risk factor, perhaps due to damage to the brain's small blood vessels. WWII veterans and NFL football players who sustained head injuries have a higher incidence of the disease. Depression has even been suggested as a risk factor. There is a higher incidence of Alzheimer's disease among African Americans. This may be secondary to the fact that African Americans have more hypertension and heart disease since vascular dementia may be involved in Alzheimer's disease. People with high blood pressure and high cholesterol levels are more apt to get Alzheimer's disease. People with high blood levels of homocysteine, an amino acid, have twice the risk of developing Alzheimer's disease. This is based on the study of healthy participants (ages sixty-eight to ninety-seven) in the Framingham, Massachusetts, heart study.

In addition, higher education (including bilingualism) may protect individuals from Alzheimer's disease. On the other hand, it may simply disguise losses rather than protect one from them. It may also be a proxy for risks associated with poverty, such as poor nutrition, high blood pressure, poor health care, low income, exposure to toxins, head trauma, and occupational hazards. In general, well-educated persons have higher socioeconomic status. It is hypothesized that early education may stimulate synaptic activity (and establish what is called a cognitive reserve) and that a lack of stimulation may reduce brain development. It is not easy to reconcile this view with the occurrence of Alzheimer's in highly creative individuals.

Genetic Aspects

Identical twins share identical genetic makeup and also show a higher (but not universal) incidence of Alzheimer's disease than do fraternal twins, pointing to the possibility of an inherent chromosomal abnormality. Molecular genetic studies have identified gene markers for Alzheimer's disease on chromosomes 1, 14, 19, and 21. (The gene containing the code for APP was found on chromosome 21.[12])

People worry understandably about the inheritance of Alzheimer's. Three familial Alzheimer's disease (FAD) genes have been located on chromosomes 1, 14, and 21.[13] When mutated, these genes apparently account for all the cases of rare early-onset Alzheimer's disease. The disease has been observed in only 120 families worldwide (as of 1999). Late-onset Alzheimer's disease, associated with chromosome 19, is much more prevalent than the early-onset form. About 98 percent of all cases are late onset.

Scientists focus upon the molecular processes associated with the four identified genes since they may help explain the underlying pathogenesis of both nongenetic *and* genetic-based Alzheimer's disease.[14]

About 60 percent of cases are labeled sporadic Alzheimer's because only one person in the family is affected. Yet genetic factors probably play some role in the majority of cases since Apo E-4 is a gene that indicates susceptibility to the disease. About 30 percent are called familial because several members of the family are affected but in a random pattern. Those remaining, less than 10 percent, are autosomal dominants, whereby the disease is passed from parent to offspring for at least three generations. We see this in the American descendants of Volga Germans who develop early-onset disease; the established linkage is to chromosome 1 (the gene named Presenilin 2). These German families settled along the Volga River in Russia in the eighteenth century at the behest of Catherine the Great and migrated to the United States during World War I. This finding suggests a common ancestor. The sporadic form more often affects older, rarely younger, persons and could involve a genetic abnormality, environmental factors, or a combination. The risk of developing the disease is slightly increased if the individual has a first-degree relative with the disease, such as a sibling or parent.

The study of inheritance is confounded by numerous issues, including the fact that the onset of the disease typically occurs late in life and competing causes of death may conceal the true extent of hereditary factors.

The Cholinergic System

An attractive early hypothesis regarding Alzheimer's disease attributes the disease to a selective deficit in the chemistry of the brain's cholinergic system, which is associated with memory. Transmission of impulses across synapses depends on release of the neurotransmitter acetylcholine, a form of the vitamin choline, from nerve endings. Acetylcholine is synthesized by the enzyme choline acetyl transferase (CAT) and is broken down by another enzyme, acetylcholinesterase (ACE). In 1976, Peter Davies and A.J.F. Maloney assayed brains obtained after death from three patients with Alzheimer's disease and from ten persons who had no evidence of neurological or psychiatric disorder. The groups differed in ACE and CAT values.

Might Alzheimer's disease be treated or prevented chemically? In Parkinson's disease, the drug L-dopa is given to make up for the loss of natural production of the neurotransmitter dopamine to reduce the characteristic tremor and poverty of movement of the disease. Will drug therapy produce symptomatic improvement of Alzheimer's and other dementias? For the moment, attempts have been limited. Three drugs—donepezil (Aricept®), rivastigmine (Exelon®), and galanthamine (Reminyl®)—are in use and offer some limited effectiveness. They slightly and temporarily improve memory, though only for some patients, but do seem to reduce caregiver burden somewhat. Called cholinomimetics, these agents neither reverse the basic neurodegenerative changes nor prevent them.

Memantine is a different chemical approach to Alzheimer's disease; it acts on NMDA (N-methyl—D-aspartate) receptors in the brain, which, in turn, regulate levels of the excitatory neurotransmitter glutamate, thought to contribute to learning and memory.

ALZHEIMER'S DISEASE AND AGING PROCESSES

Do aging processes set the stage for Alzheimer's disease or exacerbate its consequences?

The cells of the intestine are replaced nearly daily and blood cells almost weekly, but it was believed until recently that the number of adult brain cells are not replaced; the dogma has been that, like muscle cells, brain cells do not usually divide after birth. Thus, they may be vulnerable to the passage of time and to various concomitant events ("hits") not yet understood. Free radicals and environmental neurotoxins may damage brain cells.

In fact, the brain does change after birth and is affected by internal factors and outer experiences (e.g., a mother's cuddling and love). Trophic factors promote interconnections of nerve cells, and these factors decline in later childhood. Newborns have fewer synapses. In short, the brain in infancy and childhood is plastic. But critical windows of development may be missed (e.g., most people cannot easily learn a second language without an accent after ten or so years of age). It may be possible to "catch up" regarding certain aspects of character development through psychotherapeutic intervention.[15]

A variety of age-related changes in the central nervous system may contribute to Alzheimer's disease in some way. Many older neurons show a loss of dendrites as well as reduced neurotransmitter synthesis and release. While one might hypothesize that the Alzheimer's disease brain is at one end of a spectrum of neurochemical losses and the healthy aged brain at another point, the facts present a mixed picture. For example, the rising incidence of Alzheimer's disease with aging falls after ninety.

When functional features of old and young adult brains are compared, the general trend is toward decline but hardly in a manner easily identifiable as pathological. To differentiate human aging from diseases, the NIMH conducted a study of forty-seven healthy or "super healthy" older men, beginning in 1955, and evaluated those still alive eleven years later. A slight slowing of reaction time was noted, especially in complex behaviors. Total brain or cerebral blood flow and oxygen consumption were normal, but cerebral glucose consumption was slightly reduced. The study's findings have been replicated by positron emission tomography. Modern research technology and neuroscience advances may help us determine whether the observed brain changes are primary or secondary to some other changes within the aging brain. A critical question remains: Does brain pathology follow reduced blood flow or does it result in less need for blood flow?[16] Is there a critical vascular element in Alzheimer's disease? Many believe so, as do I.

DO NEW BRAIN CELLS GROW IN OLD AGE?

Dendrites continue to grow in at least one part of the brain. This finding suggests a potential for growth in old age. Some older persons may have telltale signs of Alzheimer's disease—tangles and plaques— but reveal no clinical symptoms. This and the reverse have been reported as long ago as the 1930s by David Rothschild, a psychiatrist.

A judicious overview of brain aging may be found in a comment by scientists Henry Wisniewski and Robert Terry: "There are many changes found in the aged brain [that] may be present in pathological situations in some young and middle-aged brains. The situation is somewhat analogous to that in oncology, in which the majority of cancers occur in people over 50 but obviously cannot be considered entirely as the effect of aging. However, genetically programmed, time-dependent changes in the cell plus exposure to a harmful environment (hormonal, immunological, infectious, etc.) may open the way to an increased incidence of pathological conditions."[17]

Alzheimer's, Parkinson's disease, and amyotrophic lateral sclerosis (Lou Gehrig's disease) are all diseases of longevity and result from the death of a specific set of nerve cells. In fact, these diseases may be related. Sadly, a subset of patients suffering from Parkinson's disease also have dementia. Why are certain subsets of central nervous system cells vulnerable?

Repairing the Brain

Some scientists believe that neural circuits start to organize in the brain before birth, and change and growth continue throughout life under the influence of environmental factors, life experience, and culture. They are in agreement that new neurons are detectable in the olfactory bulb that conducts the sense of smell and the hippocampus where new memories are processed.

But does adult neurogenesis, the making of new brain cells, actually occur naturally in humans? Can brain cells ever be artificially replaced in older people? Can scientists manipulate immature cells to repair parts of the brain that wouldn't naturally be renewed?

Santiago Ramón y Cajal, a Nobelist and the preeminent figure in the genesis of neuroscience, writing at the beginning of the twentieth century, was first to publish his observation of the inherent capacity of the nervous system to show limited repair by axonal and dendritic collateral sprouting.[18]

Fernando Nottebohm, a Rockefeller University neuroscientist, found that parts of the canary brain are reborn seasonally, enabling the birds to learn new songs each year. Cells in parts of the male canary brain die off in the spring as the breeding season begins. They are replenished in the fall, so that birds learn fresh musical material for courtship the next year. Nottebohm believes the old cells must die to make room for the new ones. He also found that the birds that exercised their new brain cells by singing retained those cells longer. Birds whose songs were interrupted produced less of a chemical that advances cell growth and survival.

At the Salk Institute in La Jolla, California, Fred Gage found that mice that ran grew more brain cells and retained them longer. Gage also found that brain cells survived better in mice given an enriched environment consisting of more cage space, a running wheel, and a tunnel.

In 1998, Gage and Peter S. Eriksson reported the generation of brain cells in old persons. This surprising work countered the prevailing contrary dogma. They studied the brains of five Swedish cancer patients, ages fifty-seven to seventy-two, three weeks to two years before their deaths. The patients were injected with a chemical, bromodeoxy-uridine (BrdU), which marks dividing cells. Later, the researchers found BrdU markers that indicated immature brain cells were prolifer-ating and producing new neurons in the dentate gyrus, a portion of the hippocampus involved in learning and short-term memory. The find-ing of cell replication in adult humans and of neuroplasticity (the ca-pacity for repair) suggests the possibility for therapeutic replacement interventions.

Other findings in 1999 reinforced this possibility. The neural stem cell, the founding or originating cell from which presumably the entire brain develops, was noted by two neuroscientists to be in different lo-cations in laboratory rats.[19] Stimulating environmental conditions ap-parently promote survival of uncommitted stem cells. Identification of the genes that participate in neural regeneration would open the door to treatment. Elizabeth Gould and her colleagues at Princeton reported "a naturally regenerative mechanism" in the brains of mature nonhu-man primates. They injected monkeys with BrdU, which is taken up by cells in the process of making new ones. The new neurons migrated into the neocortex, which is the center of the brain's ability to think and reason. It raises a question: Could Alzheimer's be linked to a de-cline or failure of such a regenerative process? It must be noted, how-ever, that, to date, scientists have not been able to confirm Gould's work.[20]

There is still hope for embryonic stem cells, which are more versatile than fetal cells and can be grown in the laboratory. However, in 1999, embryonic stem cells became controversial. Such work had been held back by the federal moratorium on the use of fetal tissue on the un-founded grounds that such research would encourage abortion. Yet "regenerative medicine" or cell replacement therapy, based on embry-onic stem cells, could conceivably treat this dreaded disease as well as others. *However, replacing cells does not replace memories.*

RESEARCH LOGISTICS

The pace of brain research related to Alzheimer's disease and the processes of aging is quickening. Research efforts include studies of neurotransmitter deficits and clinical trials of replacement therapy; the role of brain peptides and hormones in regulating complex physical and psychological functions; neuroplasticity or the reestablishment of neural connections; neural stem cells; the effects of growth factors; immunologic interventions to strengthen the brain against infection and autoimmune complexes; inheritance of Alzheimer's disease; the role of proteoglycans; the character and functions of amyloid and its precursor; the nature and functions of tau protein; the development of animal models for experiments to define the mechanisms of Alzheimer's disease; and more.

In the 1950s, there was very little research or clinical activity concerning senility. There were no significant biochemical or neurotransmitter leads; indeed, little knowledge of neurotransmitters or of neurotransmitter chemical activity existed. We had available the Kety-Schmidt method (since 1948) for measuring *total* cerebral blood flow and glucose and oxygen consumption, but we did not have techniques for precise quantitative, regional measurements of cerebral metabolism as we do today with PET scanning. MRI was not even a gleam in anyone's eye. Our diagnostic formulations were extremely weak, even downright misleading.

Back then, Alzheimer's was considered a rare neurological disease of the middle years. Until the introduction of Medicaid and Medicare in 1965, older supposedly senile patients were largely in public mental hospitals, where they received minimal custodial treatment. Very little research activity was devoted to organic brain diseases of old age. As recently as 1976, when the National Institute on Aging became operational, there was little research on Alzheimer's disease. Two years later, the NIA became the lead agency in supporting Alzheimer's disease research and launched the field.

Today in the United States, older patients with significant brain disease who are not cared for at home are in the sixteen thousand or so nursing homes throughout the country. In part because of the failure of public mental hospitals to show sufficient interest in this common disease, and due to limited financial reimbursements, nursing homes have stepped into the vacuum and are taking over the care of patients with dementias. For example, for-profit and other nursing homes now develop special settings for the care and treatment of people suffering from Alzheimer's disease. According to studies, it is *not* clear how useful these special care units are or whether they are simply attractive marketing devices for the nursing home industry.

We have seen some improvement in diagnosis. Distinctions between multi-infarct, or vascular dementia, and Alzheimer's disease[21] can now be diagnosed more accurately. Moreover, there are exciting developments in the technology of imaging, based on PET scanning,[22] and MRI, which promise early detection. Nonetheless, we do not yet have clear-cut, direct indications of the disease in the blood, cerebral spinal fluid, or elsewhere.

One of the impediments to progress is that we do not have the data, epidemiological or otherwise, to support most of our presently held notions, including exactly how many persons are suffering from Alzheimer's,[23] the degree to which the disease ends life, and transcultural comparisons. Screening methods are uncertain and tend to vary from study to study. Postmortems, the final peer review, are rarely performed in nursing homes, where most dementia patients die and where death certificates are woefully inadequate. Indeed, the accuracy of death certificates is generally questionable and Alzheimer's disease is underrecorded. We lack the final peer review since, as mentioned, Medicare does not pay for autopsies. We have little neuropathological data on deceased subjects of the few longitudinal studies, such as the NIA's Baltimore Longitudinal Study on Aging in which extensive psychometric testing was done.

So, too, the degree to which NFTs may be specific to dementia of the Alzheimer's type remains uncertain. After all, a number of conditions present NFTs, such as postencephalitic Parkinsonism, dementia pugilistica, Down syndrome, and aging itself. While NFTs in humans may prove important, we may be dealing with changes associated with certain kinds of causes of dying cells. We know very little about the death of nerve cells, indeed about cell death in general. The death of nerve cells is exceptionally important. We still know too little about the biochemistry of NFTs and senile plaques, including the chemical character and biologic functions of tau and of amyloid.

What about the currently prevalent hypotheses concerning all the factors that may be involved in the genesis and pathogenesis of Alzheimer's? Some are of uncertain significance.

As we gain greater knowledge of the central nervous system as a whole, discovery of the cause of any disease of the brain will be advanced. Many excellent researchers in the dementias know their own discipline (virology, molecular biology, etc.) but may not possess a working knowledge of the central nervous system. Dr. Kamil Urgubil, director of the Center for Magnetic Resonance Research at the University of Minnesota School of Medicine, has said that "most neuroscience research is conducted at the cellular level. But if you are interested in the human brain, you have to study patterns at the organizational level, what groups of neurons are activated and how they interact with each other during the performance of any complex task."[24]

The brain has an estimated 100 billion neurons, 100 trillion synapses and one trillion supporting cells, yet it only requires sugar and oxygen. It is responsible for more than learning and memory, as it regulates the homeostasis of the entire body. There is no organ that is not innervated by the nervous system. It is the premier integrative system, the governor of all. More genes are expressed in the brain than in any other organ. With aging there is a gradual decline in neural function that accelerates under certain environmental and pathological conditions. Thus, neural aging plays a paramount role in aging. Science is poised to make great progress in the study of neural aging and aging in general, but an orbital leap in funding is required. Dementia costs nearly $110 billion a year in direct and indirect costs,[25] and in the twenty-first century the cost could be unsustainable. This figure includes the costs of medical care, nursing home care, and social services, as well as early death and lost productivity. It is estimated that 70 percent of patients are cared for at home by families with an average annual cost of twelve thousand dollars each, with the average duration of the disease being about nine years. For fiscal year 2007, federal funding for Alzheimer's disease at the NIH was increased to $643 million—a step in the right direction.

The 1990s were labeled the Decade of the Brain by the NIH's Neurological Institute, but funding was not remarkably increased. We really need a full-scale Century of the Brain to better understand the complexities of human behavior.

CARE AND TREATMENT OF ALZHEIMER'S DISEASE PATIENTS

The impact of dementias on the American family is made all the worse by reimbursement limitations that dramatically affect the care of the patient at home, in the community, and in nursing homes. The private insurance industry has made only a minimal contribution to covering people suffering from Alzheimer's dementia. Indeed, the private insurance industry has been cautious overall about long-term care insurance, and managed-care organizations have sidestepped coverage of long-term care. Unquestionably, the behavior of doctors, hospitals, and nursing homes is influenced by reimbursement arrangements, and an inadequate level of reimbursement exists for the care and treatment necessary for people who suffer from Alzheimer's disease.

Fortunately, many pharmaceutical and biotech companies perceive a huge market and are committing resources to discover and test medications for Alzheimer's disease, in addition to the presently available cholinomimetics to increase acetylcholine, the memory chemical. One major target is to find methods and drugs such as protease inhibitors to reduce amyloid deposition. Encouraging reports in 1999 pointed to an enzyme—beta-secretase, similar to the HIV enzyme protease—that may have a role in the production of amyloid. It is one of three enzymes that snips the amyloid fragments from the harmless precursor protein. The other so-called scissors are alpha and gamma secretase. Drug companies are seeking to develop medications to inhibit the enzymes, allaying or even reversing effects.

Indeed, there is a continuing debate among scientists over the origins of Alzheimer's disease: the two principal camps are those who consider amyloid and those who consider tau the locus of pathology. For example, tangles, not plaques, form first. There is no quantitative correlation between the amount of amyloid deposition and clinical impairment.

Given the complexities of the brain and of Alzheimer's disease, there may never be a single magic bullet to prevent or treat it, at least in the near future. Rather a brew of agents may need to be assembled, with dropouts and additions as new knowledge dictates.

In addition, there may be ways to help preserve intellectual function, even perhaps in the presence of dementia. Continuing education and intellectual stimulation,[26] good nutrition, social interactions, and physical exercise could "complete" or "supplement" the experimental cocktail, though it must be stressed that these activities have not been proven to prevent Alzheimer's disease.

The Neuroimaging Initiative

It would be desirable to have means of assaying the effectiveness of interventions, but there are no simple tests or indicators (comparable to blood pressure or cholesterol levels) available for Alzheimer's disease.

William E. Klunk and Chester A. Mathis have developed PET imaging agents for beta amyloid that are useful to diagnose, manage, and evaluate putative anti-amyloid interventions.

The Neuroimaging Initiative, sponsored by the NIA, intends to track mental deterioration more extensively than ever before, but the project costs at least $50 million. Magnetic resonance and other diagnostic images track changes in brain structure, matching them to scores on a battery of cognitive tests.

The study could benefit the drug industry, and so the NIH requested that it provide more than half of the funds. Public-private cooperation is modeled on an NIH osteoarthritis study, launched in 2001, for which four drug companies pledged to pay $22 million of the $60 million cost.

One unanswered question is whether progressive shrinkage of the memory storage and processing area in the brain known as the hippocampus—as measured by MRI—is a "reliable surrogate" for the progression of Alzheimer's. For example, one's weight gain or loss may change the volume of the hippocampus; so can kidney dialysis. It is also possible that a drug could swell the hippocampus without improving memory.

Until more is known, the goal is to preserve and enhance existing cognitive functioning whenever possible. Even if a disease is incurable, it may not be untreatable. Under the direction or counsel of the geriatrician, psychiatrist, neurologist, psychologist, nurse, and/or social worker, or preferably a team, much can be done when Alzheimer's disease develops:

• Direct treatment of common primary symptoms of disorientation and confusion through the provision of memory aids and orientation procedures, such as cues, color-coding, calendars, and clocks, is of only modest benefit. So-called memory training, as that employed in the ancient Roman method of *loci et res* (i.e., using visual cues, such as statuary, to remind one of events or activities that required attention), has limited value. Social interaction, intellectual stimulation, and physical activity may delay cognitive decline but not necessarily Alzheimer's disease.

• The treatment of the common secondary symptoms (sometimes called excess disability) which derive from the patients' self-incapacity and their reactions to this incapacity. Such symptoms and treatments include: Painful self-awareness, such as agitation, anxiety, and depression, which are patient reactions to psychological and motor symptoms. Treatment includes counseling the patient and family to provide emotional support and teaching specific skills with which to handle the patient's hostility, anxiety, agitation, and depression. The judicious use of medications (mild to major tranquilizers and antidepressants) may be in order.[27] Agitation and restlessness can often be lessened through physical exercise.

- Self-incapacitation—symptoms directly arising from physical and mental incapacitation, including malnutrition, dehydration, falling, leaving the stove on, making mistakes in taking medications, and confused wandering. The provision of proper nutrition, exercise, and a secure environment is essential.

- Stimulation-threshold problems—symptoms reflecting overstimulation by the environment; that is, stimulation beyond the capacity of the patient. (This is called the Goldstein catastrophic reaction with anxiety and irritability; it was originally described by the neurologist Kurt Goldstein, who found that brain-damaged soldiers in World War I became agitated when confronted with a challenge they could not master.) I have found understanding of this phenomenon very useful in my clinical work. Clinicians can regulate the intensity, multiplicity, complexity, and timing of stimuli to which the patient should be exposed. Here the development of simple routines and techniques for stress moderation is helpful. For example, four grandchildren do not have to spend an entire Sunday afternoon with the patient; one child at a time could visit for a limited time. The dangers are isolation on the one hand and overstimulation on the other. The capacities for pleasure and love may remain intact, as Peter V. Rabins and Marylin Albert, experts on Alzheimer's, have stressed, even when the condition is severe. As we learn more about the treatment of sleep disturbances, such treatment may improve the life of the Alzheimer patient.

From the perspective of science it is important to be able to make the diagnosis early; but from the perspective of the patient it is often best made *known* as late as possible.

WHERE DO WE GO FROM HERE?

Dementia is an extraordinarily rich research topic to pursue because it draws together such disciplines as epidemiology, neurology, immunology, genetics, molecular biology, the neurosciences, chemistry, neuropsychology, neuroanatomy, neuropathology, gerontology, and virology. I question how truly effective academe is in creating multi- and, especially, interdisciplinary programs, yet these are just the programs we must plan and execute to solve Alzheimer's disease. This is why organized centers of excellence are necessary. There should be an international consortium to undertake epidemiological studies and clinical trials sensitive to possible cultural and ethnic variations in the incidence and manifestations of the disease.

Some estimate that about twenty-six million people in the world already suffer from dementia, a number that will grow as the world ages. Between 1990 and 2000, the world's population increased by 17 percent, but those over age sixty-five rose by 30 percent. By one estimate, there could be fourteen million Americans with Alzheimer's disease by the middle of this century. Then the surviving boomers will be between 86 and 104 years old.

Efforts are under way to delay the onset and slow the occurrence of the disease, a useful strategy but not good enough to deal with the huge challenge ahead. We don't yet know for certain the causes and cannot significantly slow progression of Alzheimer's disease.

Dementia both shortens life and impairs it; surely, people wish most of all to enjoy *longevity of the mind,* a key factor in quality of life. Moreover, maintenance of cognition appears to be an independent, positive factor in gaining longevity.[28]

Alzheimer's does more damage than most diseases. It has powerful cultural connotations that go beyond the humorous reference to a senior moment. It destroys a lifetime of memories and one's core identity. In full form, it is a personal and cultural disaster.

It is essential to meet this scientific and cultural challenge.[29] Neuroscience is making great advances, propelled by cell biology, chemistry, genomics, proteomics, and imaging technology. Environmental possibilities require concurrent investigation. We need to understand the causes and relationships between gene function, brain wiring, the environment, and behavior.

Understanding the pathway(s) to cell death could result in prevention and treatment of many diseases, including Alzheimer's disease and AIDS.

Progress against the family of varied Alzheimer's diseases and other dementias, as well as other forms of neurodegeneration, would markedly diminish pessimism over longevity, curb ageism, and open up new possibilities for both a longer and a happier life. While science is far from a fundamental solution, I oppose declaring war on Alzheimer's disease and declaring a timed goal. Rather, we must realistically accept the fact that we have a long way to go and must undertake major steps, such as building a truly major public-private research initiative I call the Century of the Brain.

Old age without dementia would be the medical conquest par excellence of the twenty-first century. To accomplish this will require an extensive, generously financed initiative to understand the brain *and* the dementias and the other neurodegenerative conditions that would not be any less than efforts against AIDS, malaria, and tuberculosis. How optimistic dare we be?

CHAPTER 9

THE "BIOLOGY OF EXTENDED TIME": EVOLUTION AND LONGEVITY

This chapter and the next concern the biology of longevity and aging. I will consider evolutionary theory before I review the specific contemporary theories of aging. How does evolution relate to longevity? Does the study of different species aid our understanding of the basis of longevity? Can we ever expect to influence evolution? What genes are involved?

Caleb ("Tuck") Finch, a biologist at the University of Southern California, explores what he calls the "biology of extended time,"[1] or longevity, and shows how necessary it is to study a wide range of different species. He concludes there is no single nor universal mechanism of senescence, not even among all cells of a specific organism. He uses the term *senescence* in a special way, considering it distinct from aging and defines it as "deteriorative changes that caused increased mortality." Principally because of financial considerations, the biologists of aging and longevity focus primarily upon flies, worms, yeast, and mice. Other short-lived species are necessary in this work in order to expedite studies of aging within reasonable time frames.

Finch argues that there may be different genomic influences in the entire evolutionary history of a particular organism. One striking conclusion of his is that "most somatic cells of young adults are genetically totipotent [and] may be provisionally extended for later ages." Totipotency is the ability of a cell, such as an egg, to give rise to cells that are *unlike* it and that develop into a new part or organ. If his arguments are valid, many aspects of senescence should be strongly modifiable by interventions at the level of gene expression. This would be accomplished by "modifications of senescence through manipulations of the external or the internal (physiological) environments."[2]

The cloning of the six-year-old sheep Dolly (and the subsequent cloning of cows, pigs, mice, deer, and mules) supports this view of totipotency because the Scottish experiment counters the dogma that differentiated cells are forever "committed" and cannot recover totipotency. Cloning promises considerable insight into development, cellular aging, and longevity (while also challenging our thinking about reproduction and individuality). Moreover, the discovery of embryonic stem cells that have the power to grow into any cell or tissue of the body, and of cloning by which adult cells can revert to stem cells, in effect reflects the biology of extended time.

THE ROLE OF GENES IN LONGEVITY

Some species die within a predictable time after mating or reproducing, such as Pacific salmon,[3] spiders, octopuses, and marsupial mice. When the act of reproduction is followed by death, the species is referred to as being *semelparous (once-only)*. In contrast, an *iteroparous* organism is capable of repeated reproduction. Immediate death prevents parental care, of course. As if to compensate, semelparous life is usually associated with a large number of progeny to offset high infant mortality. If semelparous organisms are prevented from breeding, for example, by castration in a laboratory experiment or by chance in the wild, they may live through periods in excess of a normal life span, which suggests that death (and longevity) is *directly related to reproduction and not to independent aging processes.*[4]

There is also evidence that certain genes influence human longevity by promoting immunity. Why do some people avoid having diseases such as cancer despite disease-promoting habits like smoking, drinking, and an unhealthy diet? One theory is that they may possess genes that make enzymes that detoxify carcinogens. These genes are involved with the major histocompatibility system (HLA), which is a master genetic system located on a single chromosome that controls and regulates a number of immune functions. According to Roy Walford, pathologist and author of the immunological theory of aging, the major histocompatibility complex affects the rate of aging through influences upon DNA repair and the production of free-radical scavengers. Louis Kunkel, the geneticist, found suggestive evidence of longevity genes on chromosome 4[5] in humans, which contains one hundred to five hundred genes.

According to Sir Peter Medawar and others, natural selection declines with age in multicellular organisms. As noted before, Medawar saw aging as an expression of deleterious genes that are delayed from developing by evolutionary selection in order to protect reproduction and the perpetuation of the species. In his view, genes with early-acting effects are more strongly selected than late-acting genes. So we see the accumulation of genes with harmful effects late in life. In his view, aging is due to genetic molecular effects, reinforced by environmental factors. Aging is not programmed, it is random.

Finch accepts Medawar's theory and believes the reproduction schedule is coordinated by neural and endocrine mechanisms in multicellular organisms. Therefore, genes determining the lifespan could be expressed in neural and endocrine cells in diverse animals.[6]

Regarding the question of the genetically fixed lifespan, James Vaupel and colleagues studied Danish twins (fifty-five thousand, identical and fraternal, born since 1870) and concluded that while genetic factors play a significant role in longevity, there is a 3 or 4 percent variance in the actual *age at death* or length of life of humans.[7] Identical twins do not die at the *exact same time*. Thus, environmental factors play a role, as do genetic factors, in the determination of the precise end of life.

The human genetic material is nearly identical (nearly 99 percent) to that of the chimpanzee. However, half a million generations, or about five million years, of evolutionary time separate chimpanzees and humans. We live about twice as long as a chimpanzee that lives in the zoo. Other species that are genetically related have very different life spans, suggesting that a few genes or regulators are in control.

What of the relationship of genes to environmental factors? Consider the honeybee. Queen bees and worker bees (sterile females) have identical sets of genes, but the queen bee can live five years and a worker bee no more than one. Queen bees get special attention, including a special diet. Moreover, the role of the environment rather than genes is obvious, for workers are more apt to experience wing breakage. The queen flies but once.

Longevity runs in some families. Working at Columbia University, Raymond Pearl and, later, Lissy Jarvik and Franz Kallman undertook a twenty-year study of 134 pairs of twins and found a greater concordance of life expectancy between identical twins compared to fraternal twins, suggesting familial or genetically determined longevity. Yet, as noted, genes have been shown to account for only some 25 percent of one's longevity, a finding uncovered in the Danish twin study.

Life expectancy may not always decrease at the same rate, as organisms grow older; in fact, it may actually increase toward the end of life, as noted in mass studies of fruit flies, medflies, and nematodes. Thus, the Gompertz equation[8] may not hold at the very end of life. There may not be a decisive midnight hour when the organism falls apart.[9] In simple terms, the longer one has lived, the longer one may live.

The enthusiasts who seek an expanded world of longevity are encouraged by possibilities like longevity genes. But experiments with worms the size of a comma, fruit flies, and yeast that support this concept may prove a far cry from the human situation, despite the hopes of interested biologists that longevity genes in lower forms may have complementary genes in our own species.

THE ELEGANT WORM

The free-living, soil-dwelling nematode, no larger than a printed comma, has the wonderful name *Caenorhabditis elegans* and has thirteen thousand genes and 959 somatic cells. On the fourth day it reproduces; by the middle of the eleventh it dies. Male nematodes live much shorter lives than hermaphrodite males, presumably because of sperm production—8.1 days compared to 11.1 days. (Making sperm is apparently a complex and debilitating process.)

Manipulating the mutant Age-1 gene involved in oxygen metabolism, Thomas Johnson caused his nematodes to produce one-fifth as many offspring. These offspring were healthier, albeit less fecund, and lived 75 percent longer. Johnson says that nearly every gene in the worm has a complement in the human genome (perhaps an article of faith among students of this creature); therefore, he speculates that the human species may have an Age-1 gene as well.

A daf-2 gene was identified in the nematode by Cynthia Kenyon in 1993. It controls an alternative state in the worm's development, called the dauer formation. This state occurs during times of stress, such as heat or food scarcity, when the worm responds by hibernating. The worm's life is extended under this condition. Daf-2 encodes an insulin-like growth factor receptor that regulates expression of a number of target genes that slow the rate of aging by changing cellular metabolism.

Reduced insulin-like signaling might be linked to longevity because fasting reduces the need for insulin secretion required for glucose homeostasis and because reduced insulin-like signaling promotes the expansion of antioxidant enzymes believed to be associated with longevity. How all this will relate to humans is unclear since reduced insulin-like signaling can be associated with small stature, metabolic disease, and diabetes. But this work is suggestive enough to warrant special attention.

The longevity genes in *C. elegans* are similar to those found in fruit flies, mice, and possibly humans. S. Michal Jazwinski of Louisiana State

University has identified two longevity assurance genes (LAG1 and LAG2) in yeast. And Richard Miller created a long-lived mouse he named Yoda.

The extent to which genes survive over evolutionary time is called conservation. This is an important consideration in evaluating potential longevity genes. Traditional genetics depends upon the production of mutants in order to study genetic alterations. The goal, of course, is a successful search for disease-causing genes or longevity genes, and even death genes. (The latter have never been found.) As Finch has noted, transgenic experiments (inserting genes from one species to another) could determine whether genes that influence life span in one species could modify the life spans of others.[10]

Although longevity genes have not been *definitively* demonstrated in humans, Francois Schachter and his group in France have presented evidence that the apolipoprotein E-2 gene on chromosome 19 may have a protective role in promoting longevity. There are other examples of genetic polymorphisms (different external manifestations of genes in the same species)[11] associated with exceptional longevity in humans. Moreover, as already noted, the HLA system may be relevant to human longevity. Only scientific knowledge, not speculations, will ultimately clarify the applicability of such work to the human species. The Human Genome Project will probably facilitate the identification of human longevity genes.

Genetic Mutations

Mutations offer insights into understanding longevity. When is a mutation a genetic strength and when is it a genetic liability? There are no known mutations that extend the human life span but many that shorten it. However, if mutations were purely negative, they would disappear. So it is theorized that there must be a selective advantage when there is only one copy from one parent of the mutation. Cystic fibrosis and sickle cell disease illustrate the complexity of genetics. Individuals who possess only one copy of the sickle cell gene are more resistant to malaria than those with no sickle cell gene. The sickle cell trait first evolved in areas where malaria is endemic. However, those who inherit two copies of the recessive sickle cell gene end up with a painful and fatal disease. Cystic fibrosis alters water and salt balance in the body against dehydration and may protect against cholera. Thus, cystic fibrosis mutation may confer an evolutionary advantage upon individuals who inherit only one copy of the mutated gene.

Natural Selection

Longevity presumably evolved through natural selection. As already noted, the driving force of evolution is reproductive success and the predictable life span of various species makes genetic determinants obvious. But it is certainly difficult to calculate these life spans under conditions in the wild. Indeed, wild animals do not usually live long enough to grow old. Predation (victims of nature's food chain), diseases, and accidents are their usual fate.

As biologist Leonard Hayflick says, "Aging is an artifact of civilization." We can calculate life spans in humans and their protected pets and to some degree in animals captive in zoos. We know, for example, a fly that lives for only one day; the annual fish and the shrew, one year. Reptiles such as the Galapagos turtle live in excess of a hundred years.

The short-lived, hyperactive little shrew has a heart rate that can approach 780 beats per minute. It uses a relatively large amount of oxygen per gram of its tissue. The long-lived elephant (about seventy years) has a metabolic rate of about 1/130 that of the shrew. These differences could be due to genomic-based differences and/or protection from various forms of environmental damage, such as from free radicals by antioxidants. The apparent inverse relationship between longevity and rate of metabolism has been called the rate-of-living theory.[12]

The Brain and Longevity

The bigger the brain relative to the size of the body, the longer the life span. This relationship between brain and body size among mammals is called the index of cephalization. It was first reported in 1910 and independently rediscovered by gerontologist George Sacher in 1955. John Allman of the University of California, San Diego, conducted comparative studies of human and nonhuman primates and offered suggestive evidence that certain brain structures (primarily the cerebellum but also the amygdala and the hypothalamus) may regulate life span.

According to the study of fossil remains, the brain of our human ancestors (hominids) apparently doubled in size, possibly from five hundred cubic centimeters to a thousand cubic centimeters. The increase in brain size and intelligence has been associated with an increase in longevity, beginning somewhere between two million and 1.5 million years ago. Because the increase was relatively rapid, some gerontologists, including George Sacher and Richard Cutler, have suggested that relatively few genes, perhaps twenty or forty, are involved. The life span (not life expectancy) apparently doubled from about fifty-seven years to 115 and then stabilized, *if it has,* about a hundred thousand years ago at the time of the Neanderthal. It is possible, of course, that the human life span and brain size have not stabilized at all but could increase further and may be doing so now. It would be virtually impossible to pick up changes at the end of life over a short duration.[13] For example, until the 1930s the United States did not have reliable birth and death records. However, anthropologists Donald Johanson and Blake Edgar believe that the conditions for significant evolutionary change in human species "have greatly diminished, perhaps [have been] eliminated."[14]

Cell Therapy

Could cell therapy extend lives?

In 1891 the great biologist August Weismann suggested the distinction between the somatic line (the mortal body) and the germ line (the potentially immortal egg and sperm).[15] The germ line must be immortal in order to perpetuate the species, but the somatic line is not. They are the same in a unicellular organism but separate in multicellular organisms. Thus, in humans the egg and sperm that contain the germ plasma continue to exist in any progeny, but the individual soma or body, exclusive of the germ cells, dies.

Somatic cell therapy does not affect the germ line or hereditary transmission. It involves gene insertion limited to somatic cells of the body precisely in order to avoid the introduction of exogenous genes into germ lines and, therefore, potentially modifying the gene pool. Transgenic approaches, of course, do just that; they introduce genes from another species and create animal and plant models in which germ-line cells are subjected to genetic manipulation. Such manipulation has not been applied to studies in humans.

Immortality

There do appear to be some supposedly immortal species such as coelenterates, sea anemones, and flatworms which have such remarkable regenerative power that they escape senescent or aging change. Various fish, amphibians, and reptiles keep growing larger but do not age. The term *negligible senescence* applies here, meaning that for all practical purposes aging does not occur.[16] Telomerase, central to immortal cells, has been reported at high levels in negligibly aging animals, such as the American lobster. Rockfish and sturgeon may reach 150 years of age. They die of disease, accidents, or predation.

So why did the evolution of human (and mammalian) aging occur as it did? Why was this particular evolutionary path taken that did not allow indefinite survival? One theory is that it is adaptive to the species rather than to the individual. Without aging, our planet could be overcrowded, to the detriment of the young. Aging, as noted by H.W. Woolhouse, "serves to promote evolutionary progress by accelerating the turnover of species generations."

According to Finch, senescence, or aging, was not the original state of living things on the planet. The mammalian line, one of the latest and certainly the most sophisticated of the chordate classes (which include vertebrates), can be considered an "island of senescence or aging among most of its evolutionary relatives." What then are the adaptive benefits of senescence? In general, aging species die off sooner but reproduce earlier in life and more successfully than nonsenescent organisms. This trade off is called negative pleiotropy because genes have evolved that confer marked advantages on an organism at certain stages in life at a later cost. (Pleiotropy refers to the control by a single gene of several distinct and seemingly unrelated manifest or phenotypic effects.[17]) This hypothesis, proposed in 1975 by George Williams, evolutionary biologist, specified that genes might be selected on the basis of advantages to fitness early in life but then be harmful later. For instance, some genes help humans store fat so they can survive enough cold winters to reproduce. But high fat levels promote heart disease in older adults. Other genes may give young persons the adrenalin they need to escape danger. The same process promotes hypertension in older persons.

Many do not accept the adaptive explanation of the evolution of aging. For example, violent deaths in wild populations are so common that individuals seldom survive long enough to age. Since aging is scarcely seen in the wild, how could an adaptive aging mechanism evolve? As already noted, Medawar relates aging to reproduction. If the goal of reproduction is met, then aging is simply a byproduct of this beneficial trait. Genes, therefore, would influence reproductive success, and there would be "only loose genetic control" over the last portions of the life span. Only bad genes get weeded out before reproduction, but problems linked to genes after reproduction tend to accumulate.

There are two arguments against Medawar's hypothesis. One is that in many of the longer-lived species and certainly in humans, time involved in the rearing of the offspring creates a selective advantage for longer life. An offspring whose parents (or grandparents) live long enough to nurture and protect it has a better opportunity of living to adulthood and continuing the gene pool than when orphaned, because of a lack of longevous genes.

The second view contrary to Medawar is the belief that good health is programmed by genes. The more successful the genes in regulating the developmental phase up to reproduction, the longer will be the period afterwards, since the genes that confer robustness and good health would have selective advantages beyond reproduction.

TRADE-OFF BETWEEN REPRODUCTION AND MAINTENANCE

Thomas B.L. Kirkwood, British biologist, looks at aging in the following way: The organism must have enough energy allocated for reproduction and growth (to the germ cells) on the one hand and for growth, defense, and the maintenance and repair of the body or soma on the other. Therefore, the greater the energy allocated to one particular activity, the less available for others. If there is too little investment in body maintenance and repair, for example, the organism might disintegrate before reproduction. If there is too much of an investment, the organism might reproduce only very slowly. Kirkwood's theory is called the disposable soma theory; in it, reproduction takes precedence over maintenance.

In a fascinating argument, Kirkwood also suggests that menopause is a means of protecting older women from the risk of death through childbirth. After menopause, they can no longer become pregnant and risk death in childbirth. He theorizes that menopause is not a primary feature of the aging process but a secondary adaptation allowing women to come to terms with aging and to delay death.

Why Do Women Outlive Men?

As discussed, women decisively outlive men.[18] This is the cumulative result of excessive male mortality *at all* ages beyond conception, beginning with fetal deaths. There is an excess of male fetal mortality which still yields a ratio in live births of about 106 males to 100 females. There are also more male deaths than female deaths during childhood. This is primarily due to infection, suggesting differences in the immune system. By early adolescence, numerical parity is achieved, but in early adulthood there are more male deaths due to accidents and violence. And, happily, maternal death rates associated with childbearing have become rare in the industrialized world due to medical advances, cleanliness, contraceptive use, and family planning.

The sex ratio of persons over eighteen in the United States is 52 percent female to 48 percent male. The ratio grows for women as the years pass. We know that women outlive men because of both lifestyle and biological factors. For example, it has been estimated that up to two-thirds of the difference in life expectancy is due to greater tobacco use and alcohol abuse by men. Although women have entered the workforce in increasing numbers, there has been no decline in their life expectancy. This contradicts the idea that men die earlier than women because they work outside the home. According to William R. Hazzard, a geriatrician, the sex differential in coronary heart disease alone contributes fully 40 percent to the differential in longevity in the United States and is due to biological, genetic, and behavioral factors. Biological factors include hormonal protection against heart disease (at least until after menopause) and perhaps immune protection. It has been theorized that the female immune system is especially adapted to retain and *not* reject a foreign body, namely the embryo and fetus. In addition, women have two sets of X-linked genes, which may be protective. Men have an X and a Y chromosome. Many of the defective genes that cause diseases such as hemophilia and color blindness are on the X chromosome. Women can compensate for a defective X chromosome gene if the counterpart gene on their second X chromosome is functioning properly. Thus, women's second X chromosome may be a longevity factor in itself.

Women survive the worst of conditions better than men perhaps because their biology equips them to endure environmental challenges like cold and famine. Women have less body mass on average than men and therefore need less food to survive. They have twice the percentage of body fat right beneath the skin so they have better insulation.

Women were disproportionately the survivors from the famous Donner expedition of 1846, in which the pioneers were caught in a blizzard and stranded for six months.

The role of the female in human evolution is emphasized by the finding that mitochondrial DNA outside the cell nucleus is inherited only from the mother. The mother's ovum is an entire cell, cytoplasm and all, while the father's sperm contains only DNA in the cell nucleus. Thus, the mitochondria in our bodies are all inherited from the mother and exist in all cells. Mitochondrial DNA has the same double-helical structure as chromosomal DNA, but unlike chromosomal DNA, it is not reshuffled when a sperm unites with an egg. Mitochondrial DNA is passed down from the mother, and only the mother, intact and unaltered, except for occasional chance mutations. It is a kind of evolutionary clock, a reliable witness to past events.

Researchers Rudi Westendorp and Thomas Kirkwood examined the detailed historical family records of 19,380 male and 13,667 female British aristocrats, born between 740 and 1876. Because these families were relatively well-off, there was minimal possibility that poverty might shorten life span. This study suggests that women who have fewer children and have them later in life may survive longer.

ACCELERATED AGING

Study of diseases that appear to accelerate aging or aspects of it contribute to our understanding of the genetics of aging and longevity. Insights gained thereby may offer opportunities to delay aging and age-related diseases in man.

Progeria, from the Greek word for early old age, is a very rare congenital disorder of childhood (fewer than fifty children worldwide are affected) characterized by rapid onset of some, but not all, of the physical changes typical of old age and resulting in early death. The condition is characterized by baldness and prominent blood vessels (because of reduced subcutaneous tissue), a big head and a small face, and an average to high intelligence. Coronary thrombosis or congestive heart failure is the common cause of death in these individuals, who have an average life span of thirteen years (a range of seven to twenty). Progeria is believed to be caused by a mutation in one of the nuclear lamin genes that leads to the production of a truncated form of lamin A.[19] Lamin A provides structure to the membrane of the nucleus.

Patients with the more common Werner's syndrome, another progeroid syndrome, live much longer. Their disease begins in puberty and they die in their forties. This syndrome occurs in one in a million births. It is a rare, recessive, single gene disorder; a mutation has been localized to the short arm of chromosome 8. Patients develop atherosclerosis and a high-pitched voice as well as cataracts, Type II diabetes, wrinkled skin, and osteoporosis. It may be a better model of aging than progeria.

In 1995 an international team led by Arno Motulsky, Gerald Schellenberg, and George Martin of the University of Washington discovered for the first time a gene that presumably affects aging in humans among families in Japan and Syria[20] with Werner's syndrome. Do others carry different versions of the gene that might affect life span? The Werner's gene appears to carry instructions for the production of helicase. This is an enzyme necessary to unwind the coiled strands of DNA whenever it is replicated, repaired, or transcribed for gene expression. However, Werner's patients do not usually develop Alzheimer's disease or hypertension. Both Werner's and progeria patients produce excessive hyaluronic acid whose effects are not yet understood. Hyaluronic acid is widely distributed throughout the body in joints, skin, and neural tissues and has been called a goo molecule, holding cells together.

The most common progeroid syndrome with a reduced life expectancy is Down syndrome. It is due to a trisomy, an extra copy of chromosome 21. No specific gene is mutated. It is not hereditary. As more and more Down syndrome patients have survived into middle age, they develop dementia that appears similar to the Alzheimer's type. Those with either Down or Alzheimer's have the same pathological brain findings and clinical signs. Both show similar brain amyloid deposition. The gene for beta amyloid protein is located on chromosome 21 near the locus for earlier onset Alzheimer's disease. Down syndrome patients manifest accelerated deterioration of various organ systems, a greater likelihood of developing cancer, and die earlier than their age-peers. Down syndrome and diabetes are useful human examples of possible accelerated aging that should be further studied to acquire fresh insights into the biology and physiology of aging.

EXCEPTIONAL LONGEVITY: CENTENARIANS

There is approximately one centenarian for every ten thousand people in the United States. One projection from the Census Bureau said there could be as many as 840,000 centenarians by 2050, when the oldest baby boomers have turned one hundred. Centenarians are ideal subjects for the discovery of polymorphisms (or lack of polymorphisms)—different ramifications of the same gene associated with a survival advantage. The importance of the absence of the wrong polymorphisms has been shown by Tomita-Mitchell and colleagues, who used American mortality records to calculate the expected decrease in single nucleotide polymorphisms (SNP) that code for premature mortality in newborns compared with centenarians. (SNP is a small genetic change, or variation, that can occur within a person's DNA sequence.) Frisoni and colleagues propose that the increase in apolipoprotein E-2 with extreme age suggests it is a longevity assurance gene; it is also possible that the lack of the apolipoprotein E-4 allele is what matters. Thus, the terms *longevity enabling* or *predisposing* gene are probably more appropriate than longevity gene.

In a number of major projects exceptional longevity is being studied, beginning to explore its occurrence.

Nir Barzilai directs the Longevity Genes Project at the Albert Einstein School of Medicine. It includes more than three hundred Ashkenazi Jews[21] with an average age of ninety-eight as well as their offspring and age-matched controls to the offspring from Ashkenazi families with average life expectancies. The blood of the centenarians contains large lipoprotein particles, normally seen only in young, exercising adults. These aged Jews and their offspring tend to have a mutation in a gene which raises levels of the so-called good high-density lipoprotein (HDL) cholesterol and also increases particle size of both HDL and low-density lipoprotein (LDL).[22] This is a gene variant that is involved in a lipid pathway connected to long life in *Caenorhabditis elegans*, the nematode.

The New England Centenarian Study began in 1994, under the direction of Thomas Perls.[23] It is the largest genetic study of centenarians in the world, with about one thousand subjects, including centenarians, their siblings, children, and control subjects. Few centenarians are obese and a substantial smoking history is rare. Many centenarian women have a history of having children after thirty-five or forty, suggesting slower aging. Exceptional longevity has been found to run in families. Ninety percent of the centenarians functioned independently until they were ninety-two on average and 75 percent did so until they were ninety-five. Men tend to be better off than women, both in terms of physical and cognitive function, but fewer become centenarians.

Dr. Alan Shuldiner is studying longevity in the Old Order Amish, a population that is relatively homogeneous, genetically and environmentally, and therefore is desirable for identifying genetic predictors of longevity and varied manifestations. The focus of this study is on cardiovascular disease, since this is the most common cause of death.

For every population segment of one hundred thousand in Okinawa, more than thirty-three are aged one hundred or older, making them the longest-living people in the world. Okinawa's population of 1.3 million includes more than four hundred centenarians (85.7 percent are female), many times the rate in Western countries. Bradley Willcox at Harvard Medical School is coinvestigator of the Okinawan Centenarian Study, begun in 1976, which concentrated on the genetics and lifestyle of over six hundred centenarians and elders in their seventies, eighties, and nineties. Coronary heart disease, strokes, and cancer, the three leading killers in the West, are the least likely to afflict people living in Okinawa. They suffer less from ovarian and prostate cancer than Westerners, and have half the rate of colon cancer.

The Okinawan centenarians shared certain lifestyle factors, including consumption of low-calorie, cold-water varieties of fish such as tuna, mackerel, and salmon, which contain high concentrations of omega–3 fatty acids, believed to reduce the risk of heart disease and breast cancer; an abundance of soy foods, vegetables, fiber, and fruits; and regular physical activity, moderate alcohol intake, avoidance of smoking,` and strong belief systems and social networks. They keep physically active by dancing and engaging in soft martial arts, walking, and gardening.

The Okinawans' rate of dementia is low. Older females were found to have the lowest suicide rate in East Asia, an area notable for suicide among older women. Women in the family are responsible for keeping a spiritual connection with ancestors, which engenders a sense of usefulness and self-worth.

Okinawans incorporate quiet time into their day. A cultural habit, called *hara hachi bu*, involves ceasing to eat when feeling about 80 percent full. Unfortunately, the younger generations eat fast food near U.S. military bases. Okinawans under the age of fifty have Japan's highest rates of obesity, heart disease, and premature death!

THE MIND AS A LIMITING FACTOR

In many ways the mind becomes the limit of our scientific endeavors to extend life. If we were able to maintain the machinery of the body, what would we do with our mind, which represents the essence of our personality, unless we were able to directly intervene to extend the life of the brain? Nerve cells of the brain seem to have the capacity to last more than a hundred years. There is some evidence of plasticity. Among the goals that should be central to our proposal for a century of the brain should be expansion of knowledge of neuroplasticity and its application.

Eugenics

Gene hunters seek to isolate genes related to enhancement as well as specific diseases. With the prospect of "longevity genes,"[24] already identified in primitive life forms and in mice, the possibility of eugenics to *promote* longevity comes into view. But dangers accompany humankind's hubris, perhaps especially in Western society, when we imagine ourselves able to alter both the environment and our biological destiny.

There are perils in efforts to improve the human race by changing its gene pool, whether for the elimination of disease, alteration of traits, the curbing of violent aggression, or the promotion of longevity. Unanticipated consequences destructive to the human gene pool could be set in motion.

Eugenics could serve special groups or promote certain causes if controlled by ideologies. The prospect that the wealthy and the powerful, already likely to be the healthier, would corner the market on longevity is a frightening possibility. They would have the capacity to purchase life-extending drugs and even genes through genetic therapy, resulting in a powerful class that would overwhelm and easily control those in society with shorter lives marked by illness and disability.

The future of eugenics depends upon governance and the attractions of the marketplace, and totalitarianism is always the great danger. The medical profession has been the willing servant of the totalitarian state in the past. It could happen again.

CHAPTER 10

CYCLES, CLOCKS, AND POWER PLANTS: LONGEVITY SCIENCE AND AGING RESEARCH

Over the last seventy-five years there has been considerable progress toward understanding the basic biology of aging and longevity, first in 1934 with the discovery of the role of caloric restriction in longevity, then in the 1950s with the work at the National Institute of Mental Health and at Duke University on healthy aging, and after 1975 and the founding of the National Institute on Aging, which allowed for a more predictable and expanded flow of financial support for biogerontologists. The results have been so encouraging that it may soon be possible to delay significantly both aging and age-related disease in humans.

Along with other research, this work has led to the proposal discussed in Chapter 7 to adopt a new research paradigm and to provide major new infusions of money to be devoted to the basic biology of aging and longevity to complement funds allocated to the study of specific diseases. I believe it is possible to further expand productive healthy aging by the combination of improved lifestyles (e.g., smoking cessation), continued disease-targeted research, and commitment to basic biological studies directed at how and why we age and live as long as we do.

Most contemporary theories have contributed to our understanding, and none is mutually exclusive. For example, reductions in both free radicals and glycosylation appear to occur in caloric-restricted animals. That aging has many causes was concluded at a major published symposium, *Biological Theories of Aging*,[1] in 1987. This remains true today. This chapter will provide an overview of some of these theories, beginning with that concerning the telomere.

A MOLECULAR CLOCK OF LONGEVITY?

In 1992 the scientific community began to focus on the telomere, a structure found on the tip of each chromosome that shortens as a person ages. Telomeres are made up of long chains of repeated DNA letters and are believed to be protective of chromosomes, very much as the plastic tips of shoelaces keep them from fraying.

With each cell division the telomeres shorten, and some biologists have suggested that when the telomeres fall below a certain length, death follows. Telomeres don't shorten in cells that do not divide, such as most neurons of the brain and heart muscle cells, nor in human sperm and egg cells. Sperm cells do divide, as do other body cells, but their telomeres retain their length because of the activation of telomerase, an enzyme that is present in all cells but inactive in most.

Telomerase has been called an immortalizing enzyme because it resets the clock counter. The telomere does not shorten while the cells divide. Thus, the germ line is protected from the process of telomere shortening.[2]

In cancer cells, telomerase also comes to life and telomeres are rebuilt. Greider and Harley have noted[3] that since cancer cells are capable of an infinite number of cell divisions, telomeres should shorten like those other cells, but because of telomerase they do not. Harley and others reported in 1994[4] that telomerase is found in all cancer cells studied but is absent in all normal cells, with the exception of germ cells. (It is also found in single-cell organisms that reproduce by cell division, which is a form of immortality.)

We do not know how cancer cells activate the dormant gene (or genes), but scientists hypothesize that a mutation may be responsible. More information related to telomeres or telomerase *might* lead to diagnostic tests and/or therapies for cancer. Since cancer cells produce telomerase, could a telomerase blocker be a cancer therapy? With chemotherapy often killing healthy cells as well as malignant cancer cells, this would mark a different and valuable approach in cancer therapy.

We certainly have more questions than answers. Are changes in telomeres simply markers (like the graying of hair) rather than among the many causes of aging and death? Can we diminish the effects of aging by increasing the length of our telomeres? Can we take cells from certain cell types and rejuvenate their telomeres in order to grow new skin for burn victims and retinal cells to overcome age-related macular degeneration and to rejuvenate blood vessel linings? Finally, how does work on telomeres and telomerase link to that of Hayflick?

THE HAYFLICK LIMIT

Leonard Hayflick is a biologist at the University of California, San Francisco. In 1961 he and biologist Paul S. Moorhead of the Wister Institute in Philadelphia published their classic paper which reported that cultured, normal human fetal fibroblasts, obtained from the living tissue of human embryos, underwent a finite number (namely fifty) of population doublings in glass (in vitro) over a period between seven and nine months, followed by cell death. A fibroblast is a type of cell that makes and maintains the structural framework of tissues. It is the most common cell found in connective tissue.

With a proper nutritional medium, fibroblasts are grown until they reach confluence; that is, they have filled all the space. At that point they are divided in half and placed on fresh culture vessels. Again they divide until confluence is reached. This process is called doubling. When fifty doublings have occurred, they reach the Hayflick limit. Hayflick interpreted this as aging at the cellular level.[5] He, in effect, proposes the location of a biological clock of aging in the cell nucleus. His work upset a dogma of cell biology that had been entrenched for more than fifty years: once cells were cultured in glass, they could remain immortal, short of a technical failure by researchers or lack of nourishment.[6]

Hayflick also observed that normal human cells can be frozen for indefinite periods of time. One of his cell strains, WI–38 (named after the Wister Institute where he had worked), was frozen at subzero temperatures in liquid nitrogen for more than 28 years. Upon reconstitution in 1990, these cells proceeded to divide again but never exceeded the fifty population doublings. In short, they appeared to have remembered. This "memory" constitutes the biological clock at the intracellular level.

In 1965 Hayflick first reported that cultured normal human cells derived from older donors underwent significantly fewer than fifty doublings. He found that the in vitro life span of cells from Werner's disease patients is shorter than those of progeria, and both are shorter than normal cells. Both can be considered accelerated aging. Similarly, a species' maximum life span is correlated with the number of population doublings[7] that its cultured normal fibroblasts underwent. From his work Hayflick concluded that animals and people do not age because their cells stop dividing; rather, they age before the cells lose their capacity to replicate.

It is possible to study the physiological and biochemical changes that precede the Hayflick limit. Thus, Hayflick sees population doublings as an expression of the longevity of each species and the physiological increments and decrements preceding this limit as an expression of age changes. These changes occur before the cells lose their ability to replicate and include changes in DNA, RNA, enzymes, lipids, and carbohydrates.[8]

Hayflick theorizes that his in vitro findings and those of others support the view that longevity is indirectly determined by natural selection and that only those cellular changes that occur from sexual maturation to the end of the life span should be considered aging. Thus, he carefully differentiates between how and why we live as long as we do from how and why we age. Genes, along with environmental events and chance, determine longevity; with regard to aging, he supports Sir Peter Medawar's theory that the deleterious genes in a species, through the process of natural selection, postpone their manifestations to the postreproductive period of life.

Hayflick's work is widely accepted. I will now review other important contemporary theories of aging. These theories could help account for the Hayflick limit.

THE FREE RADICAL THEORY OF AGING[9]

Earth is estimated to have begun up to fifteen billion years ago without free oxygen. It is believed that life began 3.5 billion years ago when, through photosynthesis, blue-green algae used energy from the sun and produced oxygen, thus beginning the oxygenation of the biosphere, which put evolutionary pressure on the many organisms that had lived and evolved in an anaerobic (i.e., oxygen-free) world.

Nearly 95 percent of the oxygen we breathe is used by our cells to produce energy for life processes. The remaining 5 percent has an unpaired electron and becomes a type of free radical.[10] Free radicals are chemically unstable; they seek stability by searching for electrons wherever they can find them. For example, a free radical can attach itself to another electron and be changed in the cell into hydrogen peroxide, which, in turn, may form still other more potent free radicals. They are highly and indiscriminately reactive and can tear holes in the cell membrane by combining with fats. They can react with DNA in the nucleus and mitochondria to cause mutations, damage proteins and lipids, and cross link with collagen to alter its strength. Free radicals can mutate genes and set the stage for cancer. They can kill cells.

The generation of free radicals is part of life itself; it follows from living and breathing. There are thousands of daily oxidative "hits" but, fortunately, also daily defense and repair, though this is never complete defense and repair; hence, aging, according to the free radical theory of it.

That cigarette smoking exacerbates many of the diseases of aging is in itself indirect evidence in support of the free radical theory. Other environmental risk factors that result in free radical damage include the sun, toxic chemicals, alcohol ingestion, car exhaust, and ozone.

Free radicals *may* play a primary role in age-associated diseases that include cancer, emphysema, atherosclerosis, cataracts, arthritis, Parkinson's disease, Alzheimer's disease, and osteoporosis, as well as in immunity and infection.

Our bodies defend against free radicals by producing antioxidants, so-called free radical scavengers that give electrons to free radicals to end the harmful cycle. But they have a limited effect because oxidative damage continues to accumulate at chronic but low levels.

While free radical damage appears to occur, and accumulate, with aging and in age-associated diseases, it has not been definitively proven to be causally linked to aging, nor have antioxidants proven to be useful in humans.[11] Free radical scavenging is not an established medical treatment to prevent or modify aging or aging-related diseases, but the possibility cannot be dismissed and research evidence is suggestive.

Dietary antioxidants from fruits and vegetables, such as selenium (concentrated in seafood, sunflower seeds, and nuts), ascorbic acid or vitamin C, vitamin E, and beta-carotene,[12] may possibly help. The chemist Bruce Ames, quoting Gladys Block, an epidemiologist, emphasizes that the vast majority of some 150 epidemiological studies support the advantage of eating fruits and vegetables.[13] Of course, there may be chemicals in fruits and vegetables other than antioxidants and vitamins that may have healthful effects. Moreover, fruits and vegetables might have favorable effects against cancer simply because people eat more fats when they do not eat fruits and vegetables. Green tea[14] is said to have especially potent antioxidant activity.

DNA DAMAGE AND REPAIR

Richard Setlow of the Brookhaven National Laboratory and Ronald Hart of the National Center for Toxicological Research were leaders in developing the DNA damage and repair theory of aging, proposing that aging results from the accumulation of random breaks or mutations in DNA. There is some evidence of a relationship between DNA repair and species longevity and some data suggesting that aging is associated with DNA damage and failure to repair, consequent to various deleterious environmental agents such as ionizing radiation, ultraviolet light, various chemicals, and free radicals. The capacity for DNA repair is limited.

Immunological Theory of Aging

One critical aspect of human aging is the decline in immune function. Some see it as one key pacemaker of aging. The thymus is a small glandular organ in the neck containing lymphatic tissue. It is the site of the differentiation of cells called T-cells, which play a central role in cell-mediated immunity. Starting with thymic involution beginning at puberty, there is a decrease in T-cell function with age and a concomitant decrease in control of antibody production. As the immune system declines with age to perhaps 20 percent of its peak function in adolescence, there is the prospect for intervention, should means of restoring or maintaining immune function be discovered. On the other hand, the decreased ability of older persons to react to an antigen, a substance that stimulates an immune response, may be protective. Autoimmune diseases like rheumatoid arthritis may be diminished by senescence or aging of the immune system. Clearly, the pros and cons of interfering with immune senescence must be carefully evaluated.

Stress and the Adrenal Gland

The glucocorticoid (corticosteroid) cascade hypothesis of aging developed by Robert Sapolsky suggests the emergence of hyperadrenocorticoidism, hormonal excess, largely caused by stress encountered during life that may be responsible for a host of age-associated problems. He used a rodent model. Specific groups of neurons in the hippocampus (the seat of learning and memory) of aged rat brains are found deficient in corticosteroid receptors. Thus, stress is seen as a pacemaker of senescent neuronal degeneration. Additional support for the hypothesis is found in noting age-related diseases in humans that have a significant component of hyperadrenocorticism, i.e., immuno-suppression, muscle atrophy, arteriosclerosis, osteoporosis, and steroid diabetes. Glucocorticoids, such as hydrocortisone (cortisol) and dex-amethasone, damage hippocampal neurons, apparently by impairing their ability to store energy. Sapolsky's work suggests that it might be wise to restrict the use of steroids in medicine. Research has explained that the death of Pacific salmon (not Atlantic salmon) a few weeks after spawning is due to an enormous secretion of glucocorticoids. When the adrenal gland is removed, death does not occur.

Bruce McEwen's concept of allostatic load greatly adds to under-standing of the role of stress in aging, and Epel and Blackburn's finding of stress-induced shortened telomeres adds further to the belief that stress hastens aging. The latter study found that chronic psychological stress effectively took the equivalent of nine to seventeen years off the length of telomeres.[15] Clinical psychologist Elissa S. Epel and molecular biologist Elisabeth H. Blackburn studied thirty-nine healthy mothers who were caring for a chronically ill child and compared them with nineteen age-matched mothers of healthy children. Women who were more stressed, as measured both objectively (years spent caregiving) and subjectively (self-perceived stress levels), had shorter telomeres, lower levels of telomerase, and higher oxidative load.

The Heat Shock Response

Aging is considered by many to be due to molecular damage by a va-riety of toxic factors, either internal or environmental. Genetic systems have evolved to detect damage and activate genes to repair the damage. One such response is the production of heat shock proteins. The re-duced ability of older persons to respond to acute stress may be con-nected with the difficulty of mounting a heat shock response. Heat shock proteins (HSPs) are the products of a family of highly con-served[16] genes that are found from microorganisms to mammalian cells when exposed to various stresses. Thus, normal homeostasis is ad-versely affected. Perhaps a pharmacological or gene therapy approach could be found to maintain or rejuvenate this important homeostatic response and thus delay late-life dysfunction and perhaps aging per se.

Glycosylation

Just as the normal necessary life process of oxygenation may be a principal cause of aging, so, too, another normal life process, the body's utilization of sugar, may play a role. There is an age-related increase in blood sugar after eating in humans, and clinicians have observed that diabetics age faster. Diabetes has been described as accelerated aging because those who suffer from this disease exhibit certain signs of old age, such as cataracts. Glycosylation results in the accumulations of disabled proteins called advanced glycosylation end products (AGEs). This process involves glucose chemical action on free amino groups of proteins and DNA. This complex nonenzymatic chemical process, called a browning or Maillard reaction, is seen when a Thanksgiving turkey roasts, or when bread is toasted for breakfast. It also occurs in our bodies. Proteins that are cross-linked through the interaction of sugar and proteins include the membrane proteins of the kidney, con-nective tissue, blood vessels, and the crystallins of the eye lens. AGEs are believed to toughen or stiffen tissues.

The author of the theory, chemist Anthony Cerami, hoped to reduce the formation or enhance the removal of AGEs by use of chemicals such as aminoguanadine that would prevent conditions common among diabetics and older persons, notably complications such as kid-ney disease, atherosclerosis, nerve damage, retinopathy, and cataracts. So far, success has proved elusive.

IS AGING A MITOCHONDRIAL DISEASE?

Common experience suggests that a loss of energy, even in the absence of disease, is basic in human aging. It is not difficult to imagine that reduced energy levels and cell activity could be due to drops in the supply of blood circulation or nutrition. This thinking leads one to study mitochondria, which are the power plants in cells. Mitochondria are the sites of a biochemical process called oxidative phosphorylation and the main site where adenosine triphosphate (ATP) is generated in the cell. ATP provides the energy for most cellular processes, such as pumping ions across cell membranes, contracting muscle fibers, and making proteins. When ATP, present in all cells, is split by enzyme action, energy is produced. Oxidative phosphorylation is progressively reduced as the severity of certain diseases increases with age.[17] Changed ability to utilize food substances (fuel) or changes in production by the mitochondria could slow down vital functions, causing them not to work properly.

In fact, studies show that deletions in mitochondrial DNA increase markedly with advancing age. These are most common in brain and muscle, tissues generally without cell division, and could occur because the DNA in the mitochondria might be damaged, as predicted by random or stochastic theories of aging (such as by toxins, free radical damage, mutations, etc.) or by theories such as the playing out of the biological clock, which could lead to diminishing cell energy. Or, of course, some combination of the two. In mammals the major source of free radicals is oxygen consumption by the mitochondria.[18]

Mitochondrial DNA is more exposed and vulnerable than DNA within the cell nucleus and suffers more oxidative damage. The work of repair enzymes may be insufficient. These changes in the DNA decrease the cell's ability to synthesize protein so that the heart, for example, is less able to repair itself. Damaged mitochondria, the cellular energy source, may also decrease heart function. These events might be subject to moderation through scavenging of free radicals and/or other means. But eventually there is probably *less coordination* or an incoordination of our varied biological cycles and clocks as well, perhaps in part due to reduced functioning of our power plants.

THE MCCAY EFFECT: CALORIC RESTRICTION THEORY

In 1934 Clive McCay built upon preliminary observations by others that caloric restriction or "undernutrition without malnutrition" can lead to increased survival of rodents. This is one of the few experimental manipulations that *appears* to increase the length of life. For example, in 1917, Jacques Loeb and John H. Northrop reported that cold temperature increased the life span of fruit flies. But the McCay experimental paradigm remains the most popular. In fact the McCay effect may be more of a contribution to the understanding of longevity than of aging per se and to demonstrating the effect of diet upon health in a compelling way.

When caloric restriction is initiated before sexual maturation, sexual development is delayed. In addition to rodents (rats and mice), the McCay observation has been extended to species lower on the phylogenetic ladder, including hamsters, fish, and protozoa.

Mice that ordinarily live thirty-six months have lived fifty-five months—Methuselah mice! With sixty million evolutionary years separating rodents from humans, dare we hope that it might apply to us?

The mechanism of the McCay effect remains a mystery. We do not know its basis, but it is such a compelling and repeated finding that it is important to undertake major longitudinal dietary trials in primates. These are now in process in monkeys and in human beings, where the preliminary results are encouraging. Does caloric restriction really pay off in humans? Whether it is an increase of life expectancy or a true extension of life span, it is nonetheless important to understand the underlying mechanism(s).

The McCay effect could simply reflect the evolutionary history of short-lived species and not be universal. Or it could be a strategy, formed early in evolution and perhaps built into all animals, which allows an organism to live longer and postpone reproduction when food is scarce.

Caloric restriction[19] leads to a reduction of disease through nutrition and moves the curve of survival to the right. It delays the onset of tumors specific to a particular species. Some have written that one drawback is that the McCay effect reduces fertility. But this could well be a societal advantage. Caloric restriction may simply reflect improved nutrition rather than a fundamental increase in life span. Killer diets that mice and rats have been fed, courtesy of laboratory husbandry, are improved with the McCay experiment, very much as the Pritikin diet improves human health. At this time it must be considered that vigorous restriction of caloric intake has not been established to increase life expectancy *or* span, nor reduce or retard disease *in humans*, and in any case it is presently impractical for widespread human application. Consumption is restricted to about sixteen hundred to twelve hundred calories per day. If we understood the underlying mechanisms, however, we might secure effective interventions, such as a palatable pill that mimics caloric restriction.

No one knows for certain why a calorically restricted diet prolongs life in a variety of animal species. But the gene sir2, first found in yeast cells (*Saccharomyces cerevisiae*), a common subject of laboratory study, is known to govern the rate of aging in these cells and has also been found to be active in mice.[20] When endowed with an extra copy of the sir2 gene, the yeast cells lived longer because the gene suppresses production of waste genetic material. Sir2 is an acronym for "silent information regulator no. 2." (Sirt1 is the mammalian ortholog[21] or counterpart of Sir2.)[22]

Sir2 silences other genes and quiets various physiological processes (stabilizing blood sugar and reducing free radicals) so that yeast and the nematode[23] can survive in a nonreproductive state at times of famine. Lenny Guarente, who discovered the sirtuin genes, thinks this single gene may be a universal regulator of aging.[24]

Wine and Resveratrol: A CR-Mimetic

Resveratrol exists in the skin of both red and white grapes but is found in amounts ten times as high in red wine as in white because of the different manufacturing process. It is evident especially in pinot noir, which is grown in cooler climates, such as in New York and Oregon. It is also found in blueberries and peanuts. It promotes or activates Sir2 enzymatic activity[25] and helps cells create more mitochondria. It has prolonged the life span of yeast by 70 percent, in mimicking the effect of a very low-calorie diet. Resveratrol is synthesized by plants in response to stress, such as the lack of nutrients and fungal infection. Besides resveratrol, another class of chemicals found to mimic caloric restriction is that of the flavones, found abundantly in olive oil. Were a CR-Mimetic, that is, a calorie restriction mimetic, ever developed, it would be the most cost-effective drug ever.

Body Size

Why do Chihuahuas and miniature poodles live longer than Great Danes and standard poodles? Within species, the association of small body size to extended life span applies to dogs and mice, and shorter people are relatively resistant to most forms of cancer (aside from smoking-dependent cancers) compared to taller people. Low levels of the hormone IGF-I (insulin-like growth factor I), a mediator of growth hormone action, may be responsible for the link between small body size and exceptional longevity in mammals. Data suggest that mutations that lead to dwarf body size and dramatic increases in longevity may act by increasing cellular stress resistance.

Although not firmly established, there is some evidence that animals that consume a lot of oxygen in proportion to their body size have shorter life spans. Mice, for example, live only 2.5 to three years. Larger animals such as horses, elephants, and humans live longer.[26] Animals that periodically hibernate, during which they have lower oxygen intakes, live longer than similar nonhibernating animals.

Short people may be relatively long-lived, or at least relatively resistant to certain major classes of disease. For example, on the Adriatic island of Krk, some humans are short (four feet, five inches) because of a mutation in the same gene responsible for the Ames dwarf mutation in mice. Krks have a long life expectancy.

Biomarkers of Aging

How should we measure aging and the effects of interventions in aging? There are no established comprehensive measures of biological aging.[27] Such biomarkers would measure the rate of aging and would help in the conduct of intervention studies, just as the development of the blood pressure cuff facilitated studies of anti-hypertension drugs. The success of efforts to measure biological aging has proven elusive, and calendar age is only a fair measure of aging, but still the best we have. We do have indicators or measures of disease and indirect measures of longevity, such as cholesterol. Forced vital capacity, the largest volume of air that a person can exhale after taking as deep a breath as possible, and VO^2 Max (maximal rate of oxygen consumption) are our best physiological measures of predictable decline with age. Increased reaction time (central nervous system in origin) is also associated with aging. Other markers include decreased insulin and lower core body temperature. Molecular markers of age could help quantify the rate of aging and evaluate putative anti-aging interventions. These include deletions in mitochondrial DNA, the shortening of telomeres, and oxidative damage to proteins. Of course, aging not only involves damage to molecules, cells, tissue, organs, and systems but also concerns exhausted processes of defense and repair.

AGING AND THE HUMAN GENETIC BLUEPRINT

In his sixties, Nobelist James Watson undertook leadership within the National Institutes of Health of the so-called Apollo Project of biology, the monumental Human Genome Project.[28] This initiative of Big Science cost up to $3 billion. DNA, the repository of all hereditary information and instructions, was comprehensively examined. That meant mapping the location of genes (short segments of DNA) on the twenty-three pairs of human chromosomes present in each cell nucleus and then sequencing or defining every single one of the three billion or so chemicals (called base pairs) that constitute the genetic endowment or code or genome. The human genetic blueprint was reconstructed. This would help us understand what directs the development of the egg; how it grows, differentiates, and ages; how health is maintained and diseases develop; and finally, how death occurs. Genes direct the production of body proteins. They are implicated in some five thousand disorders,[29] including Alzheimer's disease and other maladies of late life.

It is not enough to plot the position of the gene because that does not explain how a disease is initiated. Sequencing the bases tells us what vital substance is or is not created by the gene. In so doing we enter the era of proteomics, with the identification and characterization of each protein and its structure, and of every protein-to-protein interaction. It is the proteins the genes code for that actually do all the work.

From one angle, the great questions of biological gerontology are the great questions of biology (i.e., the study of life) itself. How and why do cells develop, live, age, and die? How do two cells become a human organism of myriad, specialized cells? How do genes cause cells to develop, live, age, and die? What turns a gene on or off? What turns a gene on or off at a particular time of life? Is it possible to alter or fix genes? These questions are the common domain of geneticists, cell and molecular biologists, embryologists, gerontologists, and cancer biologists, among others, all of whom exploit molecular genetics and molecular and cell biology, with their newest concepts and technical virtuosity. It is believed that all organisms have a common legacy, or conservation, and are subject to the dynamics of evolution and the environment.

New methods of study are available. Placing thousands of bits of genetic material on gene chips makes it possible to study the simultaneous expression of thousands of genes, resulting in insights into cancer, aging, and the immune system. Genetic markers called single-nucleotide polymorphisms, or SNPs, differ from one person to the next and will prove useful in identifying disease genes and assessing an individual's susceptibility to certain diseases.

OTHER COORDINATES OF LONGEVITY

As observed in an earlier chapter, longevity is certainly associated with socioeconomic status. For example, members of the *Who's Who* tend to live longer than others (except some groups such as journalists who tend to die earlier). Personality factors such as a positive engagement in life may be a factor in longevity. Symphony conductors, especially, enjoy superior longevity. A twenty-year follow-up study of 437 active and former conductors of orchestras in the United States revealed exceptional longevity, which suggests that fulfillment, pleasure, and control may be vital.[30]

A prospective cause-and-effect correlation between social relationships, activities, and longevity has been found in the Tecumseh Community Health Study after nine to twelve years following 2,754 adult men and women (ages thirty-five to sixty-nine as of 1967–1968).[31] Married people outlive unmarried single[32] people perhaps because healthier people self-select in marriage (e.g., the unmarried include persons with mental illness and disability, but, of course, for various reasons, many healthy people do not marry). Married couples can help one another when sick. A person living alone is more vulnerable.

No group of people bound by culture, environment, race, or ethnicity is any longer lived than another. The beliefs and even the presumed scientific reports of the mid-1960s and early 1970s of genetically isolated groups, such as the Hunza in Asia (the storied source of James Hilton's famous novel *Lost Horizon* about Shangri-la) reaching very long life spans, have proved false.

The Gender Gap

As already discussed, there is a difference in life expectancy and probable life span between the sexes throughout most of nature. Females are favored, but there are unexplained exceptions. What gender differences are there among diseases and reactions to them? How are drugs handled differently metabolically? What are the underlying biological mechanisms that account for various gender differences? What are the psychological and social issues that differentiate women from men? What really is a hot flash? Answers to such questions would be invaluable. To date some studies have been undertaken to obtain such information, most notably the NIH's Women's Health Initiative. It includes a randomized clinical trial involving fifty-seven thousand postmenopausal women to test the effects of low-fat diets, hormone therapy, vitamin D and calcium on heart disease, cancer, and osteoporosis, as well as an observational study of a hundred thousand women to ascertain disease markers or predictors. The studies have been criticized for various reasons, including concerns about the role of estrogen.

We also need a well-conceived men's health research initiative. This will help to raise national awareness, especially among men, of the importance of health promotion and disease prevention. For them the medical issues are clear—heart disease, stroke, colon cancer, prostate cancer, high cholesterol levels. The behavioral issues are just as important—smoking, alcoholism, men's fear and avoidance of doctors, and greater suicidal tendency, for example.

A successful men's health research initiative would help men live as long as women. Greater equality in life expectancy would also have an emotional and social impact upon the lives of women, for widowhood often brings not only grief but economic destitution.

AGING—DISEASE INTERACTIONS

Happily, there are collaborative programs involving the National Institute on Aging with nearly all of the other National Institutes of Health. This has been true since the NIA began. For example, there are NIA's efforts with the arthritis and dental institutes to study osteoporosis (which also helps us to understand related periodontal disease), the painful, debilitating arthritides that so often plague old age, and the extraordinarily important interrelationship of cancer and aging, e.g., the possibility that cancer is, in effect, failed aging, an escape from senescence at the cellular and molecular level. If we understood the biochemistry of aging, we might be able to age (and kill) cancer cells.[33] The activity of the p53 tumor-suppressor protein has a key role in controlling both cancer and aging: underactivity encourages the growth of cancer, and overactivity can accelerate the aging process. Cells need four to six mutations to become malignant.

Such studies illustrate how a major new trans-NIH coordinating effort as proposed earlier in these pages could be devoted to both the basic biology of aging and longevity and to clinical studies of age-related diseases. One task of such a trans-NIH efforts should be to raise the number of older persons participating in studies at the NIH Clinical Center at the NIH research hospital and to push for including more older persons in extramural studies including clinical trials supported by NIH at universities. Exclusion has been explained on the basis of protecting frail old persons or that multiple, complex illness confounds studies, but protection of older persons from research also protects them from the fruits of research.

It would be very useful for the NIA to develop a stronger collaboration with the National Institute of Environmental Health Sciences, as suggested in a report of a committee I cochaired at the National Academy of Sciences that evaluated various chemical and physical effects upon aging. It is difficult to disentangle environmental toxic effects from intrinsic aging processes. Observe the similarities between sun damage (called photoaging of the skin and cataracts) and aging.[34] Moreover, the role of environmental factors, such as industrial particle pollution, in promoting aging and age-associated disease must be evaluated in the era of environmentalism. (What we call skin aging today is largely due to excessive sun exposure, which can be reduced.)

There should also be active cooperation with the Occupational Safety and Health Administration. Largely unexplored is work history as a constituent element in causing the disorders of later life. The addition of occupational and environmental data within the NIA-Baltimore Longitudinal Study on Aging (BLSA) as well as other longitudinal studies would allow us to better understand the disorders of industrialization and their impact on later life. This knowledge would help reduce workmen's illnesses, compensation costs, and health costs. There should be increased emphasis upon studies of human performance—the maintenance and restoration of various physical and psychological factors conducive to productive aging and engagement.

We need national human population laboratories,[35] with widespread racial, ethnic, age, and social class representation and with families to explore the interrelationships among genetics, disease, occupation, environment, and lifestyle. The Department of Defense now has a DNA bank on all soldiers. With its large stable of families and the vast genealogical archives of the Mormon Church, Utah is an exceptionally valuable location for genetic studies. Epidemiological and longitudinal studies could be conducted and national health promotion and disease prevention programs built. Study of potentially valuable interventions through clinical trials conducted in representative samples would be another spin-off. "Outcomes" research to test old as well as new diagnostic and treatment methods could be accomplished.

Longitudinal studies examine biological and behavioral functions over time. Such studies are essential to identifying and measuring changes, imperative to the successful pursuit of human gerontology, indeed of the life course in general. We would learn which developmental changes are universal and whether any individual differences represent simple variability or different developmental pathways. It is difficult to sustain the interest of outstanding scientists (and volunteer subjects) in conducting such studies because they are naturally impatient to obtain results. This obstacle can be avoided by using the longitudinal population for specific cross-sectional and short-term studies. The resulting data sets should be made widely available. There have been relatively few longitudinal studies dating back to the nineteenth century.[36]

Through longitudinal and other studies we learn about the long reach between childhood traits and events and late-life outcome, including longevity. One very significant result of the Baltimore Longitudinal Study on Aging has been the extent of *individual differences* in aging due to varied lifestyles, diets, diseases, *and* genes. There is no single, simple pattern of aging, a fact that helps explain why one person at age sixty can look years younger than someone the same age. The very fact of the individuality or individualism of aging suggests that its rate and the incidence of its concomitant dysfunctions can be reduced. The BLSA is the longest-running scientific study of human aging and is a successful partnership between volunteer participants and researchers.[37] Among other objectives, it presently focuses upon functional capacities and how to maintain them against frailty.

The Inner Life

We need to understand human psychological development and the inner life from conception to death, a full portrait of the stages of life. James Birren, Paul Baltes, David Guttman, Laura Carstensen, George Valliant, and Barbara Haight are among the few endeavoring to build a psychology of late life. "A complete and adequate notion of life can never be attained by anyone who does not reach old age; for it is only the old man who sees life whole and knows its natural course; it is only he who is acquainted—and this is most important—not only with its entrance, like the rest of mankind, but with its exit too,"[38] wrote Arthur Schopenhauer.

THE NEW GERONTOLOGY

This brings me to a concept which is interventionist in character. Its beginnings can be traced to the longitudinal studies of healthy, community-residing older people that were started in 1955 by the National Institute of Mental Health (NIMH) and Duke University. First came the debunking of myths and stereotypes about aging, such as assumptions about the inevitability of senility. This work helped forge the promising theory that senility was a potentially treatable and preventable disease that could be studied. It demonstrated that age-associated changes were not inevitable, but were due to disease, personality, and social circumstances. More broadly, this new gerontology emphasizes the evolution of positive personal growth, not just the inevitable decrements.

Next came a *better understanding of the underlying mechanisms of aging* such as immune senescence and osteopenia and the downward curves of physiological functions. The latter, a litany of decrements in conduction velocity, basal metabolism rate, cardiac index, vital capacity, standard renal plasma flow, and so forth, have dominated. The profile of losses describes not only how a body ages but also how it deconditions *and* the effects of disease. One-third of the BLSA sample revealed no decline of kidney function with age, and even the variable rate of decline in the other two-thirds probably followed from disease, some of which was undetected. There are older persons whose functions are no different from those of a twenty-year-old.

The original curves were obtained from studies of volunteers in generally good health when they entered the BLSA, of course, but who did not necessarily remain so. Nathan Shock, pioneering gerontologist and founder of the BLSA, understood this limitation of the study. He said, "As we learn more and more about diseases and come to be able to control them, there won't be much left to old age except aging."

Hearing is another example of a need to revise our understanding of the effects of aging. Hearing problems may be a consequence of noise pollution, vascular disease, and untoward drug effects (e.g., by large amounts of aspirin). Better environmental control of noise and increasing effectiveness in the prevention of atherosclerosis could do much to reduce hearing impairments. What will be the hearing capacity of the baby boomers exposed to loud rock music often magnified with earphones? Greater hearing loss in men may result from noise at the workplace, military service, or from leisure activities such as hunting and woodworking, as well as listening to loud music. Even some age-associated problems of vision, such as sun-related and traumatic cataracts, can be prevented.

There is no denying that aging occurs, of course, and the changes are profound and inexorable, but it is necessary to strip away the concomitants of aging and get to the core of what Ewald Busse, who led the Duke University studies, called "primary aging." But it is not easy to separate aging from disease. How do they interact? Threshold remains a key concept. For example, when does rising systolic blood pressure become pathological systolic hypertension in late life? When does osteopenia or bone thinning become osteoporosis?

It is important to stress that the effect of the environment on aging is not uniformly deleterious. There is biological adaptation, for example, with regard to the immune system, which has a memory of past and still potential adversaries. We see the development of bone density through exercise and time. Through education and memory, people learn how to react to adverse events or stress and avoid them. One can bank immunity, bone, and memories. Thus, in addition to the deleterious effects of aging, survivors show adaptability.

Interventions

The third and most important aspect of the new gerontology is the introduction of preventive, therapeutic, and rehabilitative interventions, which include social and behavioral strategies, such as widow-to-widow (to deal with bereavement) and reminiscence programs (to conduct one's life review). Both have been used by the AARP, and biomedical interventions are often borrowed from various fields (such as dietary, pharmacologic, and endocrine therapy). Soon, genetic strategies will be introduced as the race to identify genes accelerates. Furthermore, regenerative medicine and genomics may prove to offer powerful treatments. Some interventions in gerontology focus on delay of occurrence and slowing of progression of diseases, a reasonable approach in late life. Short of a cure, if we could delay the onset of Alzheimer's disease by five years, for example, its incidence would be reduced by 50 percent. Moreover, if dependency could be postponed for one month among all persons over sixty-five, we would save $5 billion per year (last calculated in 1993 dollars). *Research is both the ultimate cost containment and the ultimate human service.*

Some interventions *claiming* to retard aging are questionable. For example, a 1990 study of the human growth hormone by Daniel Rudman and his associates at the Medical College of Wisconsin[39] seemed promising, but the results were temporary and there are certain possible risks, including cancer. In addition, the cost of this therapy is great—some twelve thousand dollars or more per year. The use of human growth hormone "off-label" is possibly illegal.[40]

It is worrisome that businesses, medical practices, and clinics have sprung up advertising human growth hormone (hGH) replacement therapy, as part of a growing industry called anti-aging medicine. The average adult loses muscle at the rate of six or seven pounds per decade until age forty-five. The losses are even faster thereafter. By age seventy there are one-third fewer muscle cells than at twenty. We know that muscle strength declines an average of up to 30 to 40 percent over an adult's life span and that to some degree this reflects our sedentary lifestyle. Muscle weakness, particularly in the lower part of the body, contributes to falls, injuries, and disabilities.

The NIA followed up on the Rudman work and encouraged research on various factors in promoting growth and maintenance of tissues, and, therefore, in preventing physical frailty.[41] Growth factors, as well as anabolic hormones[42] such as growth hormone, insulin-like growth factor labeled IGF–I, testosterone, and estrogen, are under study regarding impaired strength, mobility, balance, and endurance. Success could result in preventing osteoporotic fractures, improving the healing of fractures, leg ulcers, and bed or pressure sores. Side effects and complications are also under study and remain worrisome. Testosterone can enlarge the prostate, for example.[43]

Reductions in trophic factors and general deconditioning or disuse atrophy (and they undoubtedly reinforce one another) lead to frailty, a key element impairing quality of life and being remarkably costly to society. Thus, the ideal approach to optimal conditioning may prove to be a combined program of exercise and eventually some kind of growth factor administration along with a healthy diet.

Why not accomplish the same goal claimed of growth hormone therapy by diet and exercise, which are far less expensive? Here we turn to the work of Harvard geriatrician Maria Fiatarone (Singh)[44] conducted at the Tufts Human Nutrition Center in Boston. Her studies, characterized in the press as "pumping iron by older persons," demonstrated the capacity for strength training in individuals in their eighties and nineties. Fiatarone's volunteers, who were about ninety, completed an eight-week strength training program involving their legs. They worked out on a high intensity resistance machine doing leg lifts against increasingly heavy weights three times per week. The first phase of this work was done in 1990. William Evans, with whom Fiatarone worked, writes of sarcopenia[45] and observes the extent to which muscle loss is key to much of the decline in late life—to a degree preventable and even, as we see, treatable.

A NEW PARADIGM[46]

The historic biomedical research approach has been largely devoted to the study of one disease at a time, as if each was totally independent of the other, whereas underlying all diseases is the reality of aging. It is therefore time to supplement the disease-by-disease research approach by providing new intellectual and material resources dedicated to the underlying biology that predisposes us all to disease, disability, and ultimately death.

That the limits on life span are not immutable is suggested by evidence that demonstrates a decline in mortality at older ages in fruit flies, medflies, nematodes, mice, and perhaps humankind.

Some of the hormones and cellular pathways that influence the rate of aging in the lower organisms also contribute to many of the manifestations of aging that we see in humans, such as cancers, cataracts, heart disease, arthritis, and cognitive decline. Experiments have demonstrated that by manipulating certain genes, altering reproduction, reducing caloric intake, and changing the signaling pathways of specific physiological mechanisms, the duration of life of both invertebrates and mammals can be extended.

Further investigation may yield important clues about intervening pharmacologically in humans. A similar set of hormonal signals, related in sequence and action to human insulin, insulin-like growth factor (IGF-I), or both, are involved in aging, life span, and protection against injury in worms and flies and extend life span in those animals. Extension of disease-free life span of approximately 40 percent has already been achieved repeatedly in caloric restriction experiments with mice and rats.

Some people, including a proportion of centenarians, live most of their lives free from frailty and disability. Genetics plays a critical role in their healthy survival. As noted earlier, chromosome 4 has been identified as a longevity modifier in a sample of centenarians.

"In Pursuit of the Longevity Dividend"[47] proposes a goal we believe is realistically achievable: a modest deceleration in the rate of aging sufficient to delay all aging-related diseases and disorders by about seven years. This target was chosen because the risk of death and most other negative attributes of aging tends to rise exponentially throughout the adult life span with a doubling time of approximately seven years. If we succeed in slowing aging by seven years, the age-specific risk of death, frailty, and disability will be reduced by approximately half at every age. People who reach fifty in the future would have the health profile and disease risk of today's forty-three-year-old; those aged sixty would resemble current fifty-three-year-olds, and so on.

Because a healthier, longer-lived population will add significant wealth to the economy, an investment in the longevity dividend would pay for itself.

Imagine an intervention, such as a pill, that could significantly reduce the risk of cancer. Another that could reduce the risk of stroke or dementia. Now, imagine one intervention that does all these and more. Aging interventions already do this in animal models. Such an intervention is a realistically achievable goal for people. A concerted effort to slow aging should begin immediately.

The present level of development of aging and longevity research justifies an Apollo-type effort to control aging, extend the healthy life, and equalize life expectancies adversely affected by socioeconomic class, gender, ethnicity, and race. Now we have both past work as a foundation and new scientific tools offering hope that we may soon have a more prolonged, vigorous, and productive life and added longevity. During the twenty-first century, the century of the life sciences, longevity science should truly come of age.

PART IV: SOLUTIONS

CHAPTER 11

TOWARD A PRESCRIPTION FOR LONGEVITY AND QUALITY OF LIFE: HEALTH PROMOTION AND DISEASE PREVENTION

". . . old age begins too soon, . . . it is not what it ought to be under normal conditions, and human life itself does not last nearly so long as it ought to do in ideal conditions. We may predict that when science occupies the preponderating place in human society that it ought to have, and when knowledge of hygiene is more advanced, human life will become much longer and the part of old people will become much more important than it is today."[1]

Elie Metchnikoff, 1903

Healthy old age makes longevity more affordable for individuals and societies. If we are to truly enjoy long life, we must have good health that naturally supports independence and vitality and, by extension, facilitates the contributions that older persons make to society. This requires more than good genes, money, and fine medical care. It requires that individuals take responsibility for their own well-being.[2] It also requires effective public health measures that include a healthy environment and workplace. Attention must be paid to the powerful role that cultural forces play in keeping people healthy, as seen in the varied attitudes of racial, ethnic, and religious groups to nutrition, physical activity, and health.

Because many of the diseases of longevity originate in utero and in early childhood, we need a life-span perspective to guide all aspects of health care. With genes accounting for only some 25 percent of one's health and longevity,[3] our environment and personal behaviors account for the rest. Exercising informed choice, we have many opportunities to control our future health and longevity. Freedom of choice is important and the desire for short-term gratification may lead one to consciously decide to enjoy whatever life sends his way. Others, however, may lack the knowledge to make informed choices.

In this chapter I address general strategies and goals for health promotion and disease prevention while being deeply aware of the political and economic factors that interfere with both. More than talking about *what to do*—which could be a separate book itself—I will emphasize concepts and skills; in short, *how* to advance our healthy behaviors. I will touch upon some topics that are not ordinarily highlighted in disease prevention and health promotion writings and campaigns and offer advice and opinions that may seem contradictory, but which, in fact, reflect the realities of today's complicated and evolving body of knowledge.

Inasmuch as the life-span perspective is essential to understanding the genesis of health and disease, we must begin our considerations by addressing the care of a human being before, during, and after birth, and the early detection of harmful conditions.

It seems obvious that one would want early detection; however, it must be noted that early detection has it downside, especially when it involves an invasive procedure or when it may lead to decisions that inadvertently inflict harm, both physical and emotional. For example, early personal testing for evidence of genetic factors for Alzheimer's disease could result in chronic anxiety throughout life; the presence of genes related to breast cancer (BRCA1 and 2) might lead to unnecessary mastectomies.

Genetics tests using saliva or blood samples are already available to consumers to find if they are at higher risk for breast cancer or other medical problems. However, these tests are not well regulated and we cannot be sure of their accuracy, their interpretation, or emotional impact.

Currently the United States spends only about 1 cent of every health care dollar on prevention, but in order to reduce the high costs of illness and accomplish personal and social goals, serious investments should be made in relevant research and scientifically based health promotion and disease prevention.

It is seldom emphasized that medical studies are almost always works in progress.[4] As recent events illustrate, the results of ongoing research can be disconcerting and cause confusion. One example can be seen in the results of the Boston nurses' study, which led to the conclusion that estrogen reduced the risk of heart disease and stroke in menopausal women. The NIH Women's Health Initiative study contradicted these findings, but the issue is still not settled. There are frequent reports that taking a particular substance—a vitamin, an herb, or a drug—promotes health, and that the public may interpret that as a call to action. But an association or correlation does not prove a cause-and-effect relationship. Epidemiological observational studies can help point the way to more sophisticated studies, but they should be interpreted with caution. Once again, the Boston nurses' study is illustrative of the problem of attributing success by observing a limited group. Involving nurses who are probably healthier and more likely to engage in favorable lifestyles than women in the general population, this study could not confirm the effect of hormone replacement therapy on the majority of women. Sometimes meta-analysis which combines the results of relevant existing studies helps.[5] But the gold standard is the one that is randomized, blinded, crossover,[6] and placebo controlled.

PUBLIC HEALTH

Good health care depends upon scientifically tested diagnostic and treatment interventions applied to both individuals and to the population as a whole; in other words, population medicine. To date, the individual doctor-patient relationship has received the most attention and funding. However, although this relationship is central to the ideal practice of medicine, health maintenance of the society at large must be organized into a system of public and environmental health and health education that is devoted to the health of the entire population.

Achieving broad public health goals involves health promotion and disease prevention as well as maintenance of the quality of the air, water, and food supply.

Food safety, quality, and security are difficult to ensure due to importation, technology limitations, the realities of the free market, cost, and politics. The Centers for Disease Control and Prevention report that contaminated food causes roughly seventy-six million illnesses (mostly from severe diarrhea accompanied by fever) and five thousand deaths annually in the United States. The Food and Drug Administration, responsible for the safety of most foods, is underfunded and understaffed.

For those who may be pessimistic, there have been significant public health successes, including reductions in smoking, seatbelt use, breast-feeding, and the designated driver rule.

Medicare

Medicare should strongly focus on prevention. It is never too late to improve one's lifestyle and always too soon to stop. Moreover, it is easier to test the preventive value of measures in the shorter horizon of the later years than over a lifetime. An annual physical remains controversial, since it is expensive and generally yields little information during most of life, but I believe it is valuable after the age of sixty. Doctors should take advantage of the time to counsel patients to improve their lifestyle.

And with the population of older people growing, caregiving responsibilities will increase and take their toll. This also requires counseling.

Ideally, of course, we should have a Medicare health insurance program for all ages to assure lifelong efforts at health promotion and disease prevention, as well as the immediate treatment of diseases.

Adult Immunization

Vaccinations are not just for children. While most adults have some immunity, people over fifty should have booster shots for diphtheria and tetanus every ten years, as well as pertussis. Seventy percent of tetanus (lockjaw) infections occur in people over fifty, especially among gardeners. Hepatitis A and B, pneumococcal pneumonia, and flu vaccines should become routine. Vaccines against vaccinia (to avoid painful and debilitating shingles) and human papillomavirus (to prevent cervical cancer) are available and should be taken.

Foods as Medicine

Many of the disorders of longevity are due to inappropriate nutrition and over-nutrition. Basic is a balanced diet in the caloric range of eighteen hundred to two thousand[7] with the standard intake of vitamins and minerals such as vitamin D and calcium.

We have very little conclusive evidence regarding the healing properties of a plant-based diet. Vegetables such as bok choy, broccoli, brussels sprouts, and cabbage contain phytochemicals that may be effective in fighting cancer, and eating a garlic clove or two each day can very modestly reduce high blood pressure and cholesterol levels.

Neutriceuticals[8] are also of interest. These chemical or natural food products, such as fish oils, oat bran, or hyperimmune milk and eggs that contain human antibodies, may have medical benefits.

Yet another aspect of food as medicine involves edible plants, including bananas and potatoes, which could be genetically altered to provide inexpensive edible vaccines and other medicines.

Nutrition and Longevity

It is uncertain whether homo sapiens were intended to be intensely carnivorous and, especially, to eat as much fat as is consumed in the Western diet. Most of the world's peoples are vegetarian or semivegetarian. It would help greatly if we better understood the role of nutrition in biological evolution and the extent to which the modern diet is out of step with our nature as defined by evolution.

Like medicines, foods have side effects and can be pathogenic. The natural basis of foods and the healthy and natural selection of a diet are as important as the development of appropriate, naturally based medications. For example, the usefulness of milk in osteoporosis in the African American population is compromised by the occurrence of genetic-related lactase deficiency but countered by physical activity, special milk, or drinking regular milk with meals.

Redesigning the government's food pyramid is a formidable task, profoundly affected by politics.

U.S. agriculture policy subsidizes and promotes sugar and cheese; it offers little or no assistance to fruit and vegetable growers. The goals of the departments of Agriculture and Health and Human Services are at cross-purposes. In passing, it should be noted that the Department of Agriculture now calls hunger "food insecurity." The government's rhetoric about obesity emphasizes exercise and personal responsibility—messages that the food industry favors—instead of reducing consumption.

OBESITY

A genuine crisis. The Centers for Disease Control and Prevention believes obesity was responsible for 112,000 premature deaths in 2002 and for $75 billion in medical costs in 2003. Unfortunately, the American marketplace does not support good health. By promoting unhealthy eating habits, special interests such as the meat, dairy, sugar, salt, alcohol, soft drinks, and tobacco industries encourage obesity, diabetes, cancer, and high blood pressure, and contribute to the nation's poor health.

People seek several means of overcoming obesity, which is exceedingly difficult to do by diet and exercise. More than 140,000 Americans had weight-loss surgery in 2004, mostly gastric bypass, which staples the stomach shut. A tiny pouch is left that can hold a few tablespoons of food. Patients usually lose about a third of their body weight and keep most of it off. In 2005, Medicare agreed to cover weight-loss, or bariatric, surgery. And Medtronic Corporation made available an implantable stomach pacemaker that allows the stomach to signal the brain when it is full.

Various drugs purport to help people lose weight. The vast majority, however, regain the weight as soon as they go off the drugs. And medications can have severe or even fatal side effects. Over time, in the absence of diet and lifestyle change, medications are not effective.[9]

Some believe that having more muscle speeds metabolism and increases the calories a person burns while at rest, but there is little evidence that this causes weight loss. It is nearly impossible to increase muscle while cutting calories because muscle mass is lost along with the weight. However, among the many benefits of regular resistance training, it can prevent some of the muscle loss, as well as lowering body fat levels and helping to preserve bone mass.

Childhood obesity. According to an Institute of Medicine report, 31 percent of American children (including teenagers) in 2005 were overweight and 16 percent obese (whereas 5 percent were in the 1960s).[10] In 1946, the National School Lunch Program was set up to deal with childhood malnutrition by providing low-cost and free lunches to schoolchildren. But federal meal subsidies are not enough to buy good food. In the 1980s, the advent of corn sweeteners (fructose) marked the beginning of the rise in obesity and Type II diabetes, and today we are struggling to overcome the obesity crisis in children. For the first time in American history, children may not live as long as their parents.[11]

Nutrition should be brought into schools, with a required educational class during lunch that teaches nutrition and health.[12] Unfortunately, junk food, with its advertisements targeted to children, continues to be highly lucrative. And the American Dairy Association has fought the decisions of Los Angeles and New York City authorities to eliminate milk from their school lunch programs. Fast food, laden with fat that is understandably attractive and filling, especially for those with low incomes, has yet to be dealt with.

It must be acknowledged that some companies have devised programs to both counter negative publicity and to promote health. Coca-Cola[13] has a program called Live It. It will not sell its products in elementary schools or market to children under twelve. It gives pedometers to students and rewards those who take the most steps in a week. Kraft Foods, the maker of Oreo cookies, also pledged to stop advertising to children under twelve. In 2006, the American Beverage Association announced a new policy curbing sales of soda in schools to address the problem of childhood obesity. However, some critics see these moves as clever marketing rather than as protecting children.

EXERCISE

The body was adapted during evolution to a physically demanding environment. Aerobic exercise is an effective preventative and a treatment for many conditions, including high blood pressure, diabetes, and heart disease, as well as obesity. Exercise can protect against the development of cognitive dysfunction; it helps sleep, reduces depression, and even prevents constipation. At the same time, the environments in which most of us live do not facilitate healthy habits such as walking and biking.

We cannot absorb a limitless amount of continual physical punishment; what we consider healthy behavior may take a toll on our bodies. For example, in the long term, brisk walking is better on the knees than are jogging and running or step-and-stop sports like tennis and squash. Osteoarthritis, which twenty-one million Americans suffer from, is the leading cause of disability in the United States. Weekend athletes are particularly at risk, and too few people have received good training that helps them prevent injury. Wear and tear is especially hard on our musculoskeletal system because it breaks down cartilage at the ends of bones, resulting in joint pain. Hips, spines, fingers, and knees are affected.

On the other hand, regular, moderate exercise is a good way to reduce stress and the progress of disease. It helps retain the mobility of arthritic as well as healthy joints and builds the strength of supporting muscles. In addition, minimal weight loss can take pressure off the knees. For example, losing eleven pounds takes twenty-two to thirty-three pounds of force off the knee. (Obesity is the leading risk factor for osteoarthritis.)

People should routinely practice balance and maintain their strength, flexibility, and posture. They should engage in some aerobic activity to burn roughly in the range of 2,000 to 3,500 calories per week.[14] An optimal workout is forty to sixty minutes, four to six times each week, but at least thirty minutes four days per week is suggested.

Keeping joints and muscles moving is important for older people. Tai chi, which offers moderate range-of-motion exercises for the neck, shoulders, trunk, hips, knees, and ankles, also improves balance. It can be done without causing pain and stiffness because of its slow circular movements without external impact. Stressful activities to be avoided include prolonged standing, kneeling, squatting, and stair climbing. Going down stairs is often more painful and debilitating than going up. Any excessive changes in the body, i.e., in muscles, joints, tendons, ligaments, cartilage, and knees, need time to heal. The strength of the quadriceps, the thigh muscle, is a valuable predictor of frailty. Its strength can be tested (Table 11.1). Squats to exercise the quadriceps can make a huge difference. Frailty, along with dementia, destroys the quality of life and can shorten it. It results, among other things, in falls, the twelfth-leading cause of death in persons sixty-five and over.

Exercise in our schools. American third-graders exercise nowhere near the thirty to sixty minutes a day that experts recommend for children.

They log an average of only twenty-five minutes per week of vigorous physical activity in school physical education classes.[15] Ironically, college and professional sports have become a distraction from the country's adoption of a health-oriented culture.[16] People become spectators instead of active participants in physical activities. No one should graduate from high school and college without a professional, tested, personal fitness program for use throughout all stages of life. Health education and training should be well under way by first grade.

Cholesterol

We owe the Framingham Heart Study, the longitudinal study sponsored by the National Heart, Lung and Blood Institute, a debt of gratitude for creating the concept of risk factors and specifically identifying the coronary risk profile. This invaluable Public Health Service study was initiated in 1948 and has been followed by the Framingham Off-Spring study and the Third Generation Study.

The Framingham Heart Study showed that at least half of heart attack victims had cholesterol levels of 240 or more. (The goal is total cholesterol of 200 or less, preferably 180.) Cholesterol levels can be reduced by a low fat diet and exercise and, if needed, a statin, a cholesterol-lowering drug.

Trans Fat. Hydrogenated oil contains trans fats that reduce the good cholesterol. Beginning in 2006, companies had to disclose trans fat amounts on food labels. Unfortunately, companies, searching for trans fat alternatives, are using unhealthful fats, such as palm oil, a saturated fat believed to promote heart disease.

Tobacco

People often believe that it is difficult to change behavior, but public health and education as well as economic, legal, and legislative efforts to curb tobacco use, especially taxation, resulted in a 50 percent reduction in the number of people smoking since the surgeon general's report, *Smoking and Health: Report of the Advisory Committee to the Surgeon General of the Public Health Service,* was issued in 1964. About 40 percent of Americans now live in areas that limit smoking in all workplaces, restaurants, and bars. Cigarette consumption is down to World War II levels.

In 2005, the Centers for Medicare and Medicaid Services (CMS) added coverage for tobacco use cessation counseling to help certain Medicare beneficiaries overcome the habit. However, Medicare coverage for smoking cessation counseling will be available for smokers whose illnesses were either caused or are complicated by tobacco use. This includes cerebrovascular disease, heart disease, lung disease, osteoporosis, blood clots, and cataracts—the diseases, in fact, which account for most current Medicare spending. An estimated four million people will be eligible for the new coverage.[17] Moreover, the Medicare Part D drug benefit begun in 2006 includes smoking cessation medications.

It is never too late to change habits. Even after age sixty-five, health risks from tobacco begin to decline within a few months of quitting, and the risk of dying decreases.[18] Immediate and long-term benefits of quitting are lowered blood pressure, improved lung function, and decreased risk for heart disease. Approximately 10 percent of smokers over sixty-five quit each year; of these, only about 1 percent relapse. A fifty-year study showed that smoking shortens life by an average of ten years and that almost all the risk can be avoided if one quits by age thirty. Stopping smoking at ages sixty, fifty, forty, or thirty gains three, six, nine, or the full ten years of life expectancy. Unfortunately, despite a decline, teenagers, especially girls, are smoking, in part perhaps because smoking is an appetite suppressant and it is thought to be cool.

ENVIRONMENTAL REGULATION

What explains the rising rates of some cancers and childhood brain disorders? Perhaps these are due to the industrial chemicals found at extremely low levels in the environment. Studies of low-dose exposure challenge current ideas about toxic substances and environmental law. Modern pollution restrictions aim to limit exposures to levels studies have found safe. For example, after studying the effects of lead on children, the Environmental Protection Agency (EPA) came to the conclusion that some common substances are unsafe at any level. Given that toxologists test chemicals individually, they are challenged by the greater toxicology exhibited by some chemicals in combination with others.

Radiation. Safe levels of radiation are in dispute. A 2005 report by the National Academy of Sciences on the health effects of small doses of radiation, a report incorporating nearly fifteen years of new data on atomic bomb survivors in Japan, reinforces the idea, opposed by some experts, that even tiny doses may add slightly to risk. Women are more likely than men to get cancer, given equal doses. *There is no truly established safe radiation level.* For typical Americans, 82 percent of exposure stems from natural sources such as radon gas seeping from the earth; the rest is manmade, coming mostly from medical procedures such as from diagnostic and therapeutic X-ray exposure.[19]

Soot. Sooty pollutants account for thousands of premature deaths and illnesses each year. Such particulate matter comes from factories, cars, and forest fires. Some particles are so small they penetrate deep into the lungs and are especially serious in children with asthma, frail older people, and others who are vulnerable. The Clean Air Act requires scientific reviews every five years of the standards for particulate matter and other pollutants to ensure that public health is fully protected, although the reviews are often delayed by litigation. Proposed standards include one for annual exposure based on a daily average and another for exposure within twenty-four hours. Sulfur dioxide and nitrogen oxides are other significant pollutants.

DRUGS

The definition of a modern drug is a chemical compound that inter-acts with a biological target such as an enzyme, receptor, or other pro-tein to enhance or suppress the function of the target. The target might be H2 receptors, which are antagonists, to treat ulcers; enzymes such as ACE inhibitors, which treat hypertension and heart failure; or an ion channel, as in calcium ion channels, blockers of which treat angina and hypertension. Advances in genomics research will lead to the identifi-cation of genes implicated in human disease and to the proliferation of new biological targets.

Individuals receiving many medications (polypharmacy), especially if older, may experience fatigue, intellectual confusion, memory loss, impaired sexuality, or urinary incontinence. In addition, longevity it-self poses a danger, for medications that might have been safe for use over a short time may have untoward delayed effects. Drugs should, when possible, be used sparingly and at low dosage.

Notwithstanding the enormous value of medicines in keeping peo-ple healthy, they often fail to work in individual patients and constitute one of the leading causes of illness and death.[20] One study found that in 1994 there were 2,216,000 hospitalized patients with various adverse drug reactions (ADRs) and 106,000 deaths from them.[21] ADRs would rank from the fourth to the sixth leading cause of death! These figures "excluded errors in drug administration, noncompliance, overdose, drug abuse, therapeutic failures and possible ADRs," which makes the findings even more shocking. The author acknowledged the results of his meta-analysis "must be viewed with circumspection."

Understanding the role of genes, along with age, gender, disease, and interaction with other drugs and nutrients in drug responses, will re-sult in personalized medicine, such as customized drugs that will en-able a physician to match a patient's ability to metabolize a particular drug based on genetic information. Customization, however, will be costly. The pharmaceutical industry will change. Boutique, niche com-panies may compete with, or possibly even replace or become divisions within, "big Pharma" firms presently dedicated to the search for one-size-fits-all blockbusters.

Post-Marketing Research

Phase 3 studies are expanded controlled and uncontrolled trials. The FDA conducts Phase 3 studies after Phase 2 studies suggest the drug is effective. A Phase 3 study usually involves only about three thousand patients, yet idiosyncratic reactions occur in less than one in ten thousand. Thus, it would be best to test at least thirty thousand.

The Cox–2 inhibitor Vioxx, believed to increase serious coronary heart disease, was removed from the market voluntarily by Merck in 2004. This case illustrates why post-marketing surveillance is necessary. At present it is carried out through the FDA's passive, reactive Med-Watch program. It is believed that MedWatch receives reports on fewer than 10 percent of adverse drug reactions. There needs to be a national system for phase 4 post-marketing evaluations run independently of both the FDA and companies, which may also reveal unanticipated drug benefits[22] as well as untoward side effects and complications. Such a safety system[23] would also reduce liability issues confronting drug companies and doctors. It should be paid for jointly by taxpayers and drug company contributions. Such a program is all the more necessary as efforts to accelerate drug approval are encouraged. It is difficult to accelerate the evaluation and use of medications because our biological nature itself is the limiting factor.

Utilizing the NIH-type centers mechanism and a peer review through competition, a medical center would be selected in each of the ten regions of the Department of Health and Human Services. This would assure that all racial and ethnic groups as well as rural and urban populations would be covered. Each winning center would have a special team of clinicians, pharmacologists, pharmacists, statisticians, toxicologists, epidemiologists, nurses, and computer and other experts who would track *all* medicines used in the outpatient and inpatient populations. The teams' understanding of the mechanisms of action of varied drugs would be essential. The use of information technology would be critical.

According to one analysis, the number of off-label prescriptions doubled between 1998 and 2004, another reason for post-marketing surveillance. Indirectly, head-to-head comparisons of drugs could be made from post-marketing surveillance data. Long-term surveillance is especially important given the frequent use of drugs such as Ritalin and SSRIs that affect the central nervous system but whose long-term effects are unknown, such as the rewiring of the brain.

Before such an ideal system is put into place, presently available electronic health records from managed care organizations such as Kaiser Permanente, the Veterans Administration, and Medicare/Medicaid data could be used as an interim measure. Some doctors will not prescribe or use a new drug themselves until it has been on the market for at least five to seven years.

The Role of the FDA

Uncontrolled, unsupervised experiments are going on while risking possible danger and high cost to hopeful, at times unsuspecting, American adults who take chemicals purported to extend longevity. We have no way to gauge effects on the basic biology of aging, since we do not yet have acceptable biomarkers of aging.

Some 40 percent of Americans take at least one prescription drug. According to the National Center for Health Statistics, there are some fourteen thousand approved prescription drugs available in the United States. Pharmaceuticals have advanced the health of people but, as noted, with surprising perils.

What should be done?

Without a doubt the FDA should be strengthened. It should have a larger budget, more senior and other scientists (it has about ten thousand employees), and subpoena power. Along with the NIH, it needs a major infusion of funds for large clinical trials that would be conducted independent of the drug industry.

Relationships

Linda Waite of the University of Chicago analyzed data from more than 8,600 people ages fifty-one to sixty-one to examine the connection of relationships to physical health. Among her findings: any post-marital state, such as divorce or widowhood, creates stress, which has been linked to chronic health problems. Deaths and losses, the nearly constant companions of old age, create something more than stress. There is a huge emptiness that requires finding new relationships.

Baby Aspirin

Deciding whether to take aspirin for the prevention of a stroke or heart attack depends on one's risk for these conditions compared with the risk of such serious side effects as bleeding. The risk of major complications from aspirin may be reduced with lower doses of aspirin (i.e., 81 vs. 325 mg) the preventive effects of which remain the same.

Silent Conditions

Hypertension, diabetes, and glaucoma may go unnoticed. For example, half of the more than three million Americans who suffer from glaucoma don't know they do. Glaucoma results from damage to the optic nerve usually caused by increased pressure within the eye. People over age sixty should have their eyes examined every two years, those under this age every two to four years. But anyone of African-American descent, with a family history of glaucoma or with a history of high pressure in the eyes, is at higher than average risk. Medications generally work by decreasing the production of fluid in the eye.

Moreover, seventy percent of people with hypertension do not have their blood pressure under optimal control. Uncontrolled hypertension is a leading cause of preventable heart attacks, stroke, and death. Hypertension occurs in one of three American adults and about 30 percent are undiagnosed. Yet, exercise, a low-salt diet, weight loss, and, if necessary, anti-hypertensive medicines are effective, beginning with thiazide diuretics. It is questionable whether someone with prehypertension (with 130 to 139 systolic blood pressure or 85 to 89 diastolic blood pressure or both) should be treated with drugs.[24]

Diabetes can also be a silent condition, undiagnosed in about half of those who suffer from it.

OTHER ISSUES

The automobile. Congress repealed the national maximum speed limits (fifty-five in urban areas and sixty-five on rural roads) and some states raised their speed limits to seventy or seventy-five. There followed a 15 percent increase in vehicular occupant deaths in twenty-four states that had raised their limits. The estimated total cost of motor vehicle crashes in 2000 was $230.6 billion. Total Medicare expenditures that year were $221.8 billion. Obviously, we need stringent speed limits, effectively enforced laws against drinking and driving, and national, state, and local safe-driving campaigns. Alcohol-involved automobile crashes resulted in an estimated 16,792 deaths in 2000 as well as 513,000 nonfatal injuries, and $50.9 billion in costs, 22 percent of all crash costs. Altogether, motor vehicle crashes accounted for 2.3 percent of the gross domestic product and included some $80 billion in lost productivity, $59 billion in property damage, and $32.6 billion in medical costs.[25] In 2005, the number of highway deaths also rose for the first time in twenty years.

Genes. We need comprehensive studies of the interaction of genes, the environment, lifestyle, age, and disease. In the next several decades, genomics will play a direct role in medical care, offering the ability to predict certain diseases and prevent them.[26] Until that time comes, families should gather data on conditions that affect them, such as miscarriages and stillbirths, infant deaths, psychotic disorders, hypertrophic cardiomyopathy, and cancer.

It is useful to have a detailed family and an occupational history as a baseline. Every person should have a list of their continuing health status, including records of vaccinations, medicines, surgeries, mental illnesses, hospitalizations, overseas travel, etc.

Populations at risk should receive focused and frequent examinations; if at risk of colon cancer, frequent gastrointestinal evaluations should be conducted.

Life itself is a risk factor. The tempo of the environment, notably the speed, multiplicity, and intensity of stimuli, such as noises, that surround us affect our mental and emotional life as well as our body's circadian rhythms. Modern life requires multitasking that adds further pressures. Chronic noise exposure is associated with a higher risk of high blood pressure, heart disease, and hearing loss. Difficulty balancing work and home life is common.

Weather and other hazards. Families, doctors, community agencies, and health clinics should have lists of names and addresses of older persons most at risk—people older than eighty who live alone and are frail and demented and who have had repeated recent hospitalizations. TV and radio advertisements should remind people to drink lots of water and stay inside during the hottest hours of summer. A 2003 heat wave killed thousands of older people in Italy, France, and Spain. In the winter, older people are especially susceptible to cold and can also die from exposure. Both air conditioning and central heating are life saving. The catastrophes of 9/11 and Katrina revealed how ill prepared we are in general and how inadequate our efforts are on behalf of older disabled and vulnerable people.

Negative emotions. Anxiety and depression are natural to the human condition. Depressive feelings are especially unavoidable as one witnesses the realities that confront humanity, from human actions to natural calamities. Myrna Weisman, an epidemiologist, reported increased depression in the twentieth century. Dan G. Blazer, a psychiatrist, wrote *The Age of Melancholy: Major Depressions and its Social Origins*, which looks beyond psychiatry to understand the role of social forces.[27]

People use alcohol and other addictive substances to ease anxiety and depressive feelings of a kind felt all the more by the poor who live under high tension and frequent disappointment, as they struggle daily to survive.

It is nearly impossible to prescribe relaxation and stress relief. Arranged quiet times, habitual rituals, diversions, and pastimes can help. Meditation, massage, yoga, and baths ease life's troubles. There is an unfortunate separation between body and mind. Contemporary medicine has not succeeded in integrating what we know about the brain above the neck and the body below the neck.[28]

Sleep: A Vital Sign. As many as seventy million Americans may be affected by sleep problems at an annual cost of $15 billion in health care expenses and $50 billion in lost productivity (2005). Sleep loss may be related to obesity and diabetes.[29] Since 1960 the number of sleep hours has fallen and the levels of obesity and diabetes have risen. Eighty-three million Americans work shifts outside the typical 9–5 work schedule.

It is important to treat the underlying disorder, not simply the insomnia. In the long run, sleep hygiene, cognitive behavioral therapy, exercise, light, sleep restriction, and various forms of stress reduction may be more effective than sleeping medications. The latter are valuable for short-term use.

Taxing Addictions. Addiction results in family distress, domestic abuse, automobile accidents, and death. Alcohol abuse burdens hospitals, curbs national productivity, shortens life expectancy, and causes liver and brain damage resulting in dementia, aggression, and paranoia.

Taxes can favorably change behavior. Alcohol taxes should be raised to pay for health and social services. At the moment, taxes on alcohol do not even begin to pay the costs of alcohol abuse. Federal estimates put those costs at $185 billion, while federal, state, and local alcohol tax revenues yield only about $18 billion. Alcohol excise taxes used to be a significant part of the federal budget receipts, representing 11 percent in 1941. Because of pressure from the alcohol industry, federal liquor taxes have increased only twice since 1951, and beer and wine taxes only once.

Maintaining Cognitive Health. The Romans emphasized the concept of a sound mind in a sound body—*mens sana in corpora sano.* In our own day, there are many ways to maintain cognitive health as we grow older. There is paid work, volunteer activities, and intellectual stimulation— what some call brain jogging—such as learning a new language or a new musical instrument. Social interaction and physical activity are also believed to help one remain cognitively healthy.

Why don't men take better care of themselves? Cultural attitudes expect men to rise above pain, discomfort, and disease—to be brave and uncomplaining. Unlike women, men do not have routine contact with doctors (women are typically in contact with the obstetrician in pregnancy and the pediatrician during child care). Men are not expected to visit the doctor during the workday. They engage in more high-risk behavior—alcohol use and abuse, tobacco use, higher risk sports—which they regard as appropriate male behavior, not requiring medical intervention. Men are probably not as exposed to health information as women.

PRESCRIPTION FOR LONGEVITY

The reality is that some people begin life with key advantages that promote longevity, such as good genes and a higher level of income.

The American health crisis involves our sedentary life, fast food, and growing obesity. Indeed, projections show that life expectancy could be between two and five years shorter based on current trends in obesity. To give some context, if all cancers were conquered it is estimated that we would gain 2.4 to 3.5 years.[30] As unfortunate illustrations of global-ization, first our tobacco and then our fast food have moved around the world. America has exported serious health problems, including transferring lung cancer and our obesity crisis.

There are so many unknowns with regard to health and longevity that people are willing to experiment with little knowledge or guid-ance. An estimated $27 billion is spent on complementary and alterna-tive medicine. When conventional or allopathic medicine does not provide answers, when treatments are ineffective, understandably des-perate people look elsewhere. And why not? Some efforts are intended to strengthen the immune system, for example, to fight cancer. Spiritu-ality is promoted. The New Age! Unfortunately, doubts about science grow. Doubt is of value so long as it is displayed regarding nonconven-tional as well as conventional medicine and the world at large. For ex-ample, people often forget that just because a substance is natural does not necessarily mean it is safe. Snake venom, arsenic, cyanide, and nicotine are all natural substances. Nature's own healing properties do misfire at times with side effects or complications similar to medica-tions. Immune responses may provoke autoimmune diseases, whereby the body becomes its own enemy. So-called anti-aging medicine gives medicine, geriatrics, nature's pharmacy, and modern pharmaceuticals a bad name. It is not a recognized specialty and plays into the unwar-ranted hopes of people of all ages, especially the middle-aged and older persons who seek vigor and health. They hope longevity is only a pill away.[31]

Unregulated natural remedies. Natural remedies are available in health food stores,[32] but their manufacturers are not required to prove that they work or that they are safe. These include ginkgo biloba, ginseng, and echinacea. Though the FDA has issued warnings about the toxic effects on the kidney and liver of products containing kava, comfrey, and aris-tolochic acid, the Dietary Supplement Health and Education Act (DSHEA) of 1994 provides a dangerous loophole,[33] since it has signifi-cantly curbed the FDA's ability to regulate the industry by classifying a variety of substances, including plant extracts, enzymes, and minerals, as dietary supplements, and allowing them to be sold in some ten thousand health food stores and drugstores and on the Internet.

Rather than placing the burden on the manufacturers to show that the product is safe, pure, and effective, the law places the burden on the FDA to prove the product is unsafe before it can be taken off the market. Impurities discovered have included bacteria, glass, pesticides, and lead, and label strength can be off by 20 percent.

The DSHEA permits the manufacturer to claim that a product affects the "structure or function" of the body as long as no claim is made for the effective prevention or treatment of a specific disease. The only requirement is a disclaimer informing the consumer that the FDA has not evaluated the agent. The DSHEA should be rescinded and dietary supplements should be evaluated by the FDA.

OBSTACLES TO PREVENTION PRACTICE

Why has prevention received so little attention or, specifically, funding? While hospital executives and physicians express support, it is not in their best interest. The incentives are perverse. They benefit from disease and injury. Employers might at first be supportive, believing prevention would reduce cost, but, in fact, significant testing often reveals unexpected pathological findings and false positives. Testing is expensive and follow-up treatment, when indicated, is as well.

Physicians need to be taught prevention in medical school and compensated in professional practice for preventive measures on behalf of their patients, so they can write prescriptions for health promotion and disease prevention that go far beyond medicines. This expanded prescription should emphasize lifestyles and social, economic, cultural, and personal factors and deal with everyday stress. It would be a huge step forward.

The prescription for longevity and quality of life should include the appropriate use of medications and the development of a strong support system. A network of neighbors, friends, and relatives is especially valuable in the wake of illness, loss, and grief. Most important are intimate relationships, especially with one's spouse, partner, companion, or friend.

To have goals and purpose in life, something to get up for in the morning, especially a passion, is known to enrich and possibly extend life.

Perhaps, most of all, what is needed now is a greater emphasis upon public, environmental, and occupational health.

CHAPTER 12

REDESIGNING HEALTH CARE FOR AN OLDER AMERICA

Geriatrics is the branch of medicine concerned with the medical problems and care of older persons. Its practical application is capacious, encompassing social and emotional barometers of health and physical well-being. Older patients may present with a multiplicity of complex interacting, acute, and chronic medical and psychosocial conditions. They require that the physician be involved in health promotion and disease prevention, long-term care, including home care, community-based care, and hospice care, as well as outpatient and in-patient acute hospital care.[1]

Salient health issues in old age include congestive heart failure (the number one reason for hospital admission),[2] coronary heart disease, chronic obstructive pulmonary disease, diabetes, stroke, cancer, dementia, disability, frailty, impaired mobility, sarcopenia, sexual dysfunction, insomnia, incontinence, depression, and suicide (see Table 12.1), and, of course, pain and suffering and, ultimately, dying and death. Furthermore, many older people do not report their conditions, which they consider natural, and therefore inevitable, unpreventable, and untreatable. So, unfortunately, do many health professionals. Moreover, many conditions occur and evolve differently in older persons, throwing untrained doctors off course.

Patients of all ages face further complications in the hospital. Older persons especially are not always treated with respect and may be stripped of their identity and dignity. Doctors and nurses may not introduce themselves properly and may call older persons by their first names without permission. In 1978 Maggie Kuhn said, "The ultimate indignity is to be given a bedpan by a stranger who calls you by your first name."[3] The discomforts of the modern hospital—the revealing gowns, inadequate blankets, cold gurneys, and generally unappealing food—offer an unpleasant environment for patients of any age. Many older persons develop delirium in hospitals. Furthermore, a striking study by the Institute of Medicine found that the perils of the hospital account for an estimated forty-eight thousand to ninety-eight thousand deaths per year,[4] and thousands more acquire nonlethal infections as well as suffer from unacceptable diagnostic and procedural errors in persons of all ages. In addition, the morale and physical condition of older people descends in hospitals, where they are left largely immobile in bed.[5]

As late as the 1960s, during cataract surgery the eyes were covered with black patches, sandbags were placed along the sides of the head to restrict movement and wrists were tied to the bed. This immobilization often resulted in disorientation, bedsores, and pulmonary embolism. As many as 20 percent of patients developed vitreous hemorrhage, macular edema, eye infections, and retinal detachment. Then Charles D. Kelman invented phacoemulsion of the lens, making outpatient cataract surgery rapid, easy, safe, and effective. His work also stimulated the development of other "keyhole" surgeries, including removal of the gallbladder and lumpectomy of the breast.

Health care providers who are responsible for older persons often lack proper education. Compared with standard medical care for younger adults and children, which usually involves only one medical condition, geriatrics is much more complex and intellectually challenging, because the physician is often faced with multiple interacting physical, social, and psychological issues that require sophisticated assessment and management best accomplished by a team that includes the nurse and social worker as well as the physician. Furthermore, geriatrics involves both individual, one-on-one attention *and a specialized system* of care.

For example, geriatricians ask patients and their families to put all the medicines in their house, including old medicines, dietary supplements, vitamins, and herbals (as well as any given to them by friends and neighbors), into a brown paper bag to be brought in for review. The results can be startling. This approach helps doctors prevent adverse drug reactions based on inappropriate dosages and mixtures.

At least six imperatives should propel the development of geriatrics: the growing number of older persons, the age-associated disease burden, rising health care costs, the need for comprehensive care from health promotion to end-of-life care, the necessity for research, and, finally, the widespread ageism in the health care system.

In the United States, progress in the field has been glacial. Beginning in the 1950s in the United States, geriatric psychiatry outdistanced geriatric medicine. Landmark advances in the 1950s and 1960s included research studies of older persons living in the community at Duke University and the National Institute of Mental Health. Both helped clarify the distinctions between aging and disease, observed the individuation of aging, revised many stereotypes, and suggested therapeutic possibilities. Yet, it was not until 1982 that the first medical school department of geriatrics in the nation was founded at the Mount Sinai Medical Center in New York. Today there are still only eleven departments of geriatrics out of 144 allopathic and osteopathic schools of medicine! In Great Britain every medical school has a department of geriatrics, and geriatrics is the second-largest specialty.

It is important that there be departments (or department equivalents) of geriatrics in medical schools. Departmental status provides political power within the school, involving decisions concerning curricula, physical space, laboratory and clinical facilities, and, of course, budgets!

One of my first acts when I became NIA director in 1976 was to invite medical school deans to Bethesda to discuss needed medical school education in geriatrics. Only thirty of 125 allopathic medical schools[6] sent representatives. I concluded that the NIA should fund competitive awards to build leadership in the field. However, I was not able to fund the training of physicians in clinical geriatrics under the congressional authority by the Research on Aging Act of 1974. In contrast, the National Heart Institute, founded in 1946, was able to train some sixteen thousand cardiologists during its first twenty-two years, which helped build cardiology and undoubtedly contributed to the 60 percent reduction in deaths from heart disease and stroke since 1960.

Based on a model created at the National Heart, Lung and Blood Institute (NHLBI), I developed the Geriatric Medicine Academic Award, which resulted in the recruitment of many outstanding clinicians, teachers, and investigators in aging, among them Patricia Barry, William Hazzard, Jack Rowe, and Fred Sherman.

In 1977, I called upon David Hamburg, then president of the Institute of Medicine, to contract a study to be called Aging and Medical Education. I asked *how* geriatrics should be incorporated into the medical school curriculum. I did not ask whether it should be. Paul Beeson, a revered physician, was selected as chair, and his report published in 1978 proved to be a landmark, calling, as I had hoped, for required programs within medical schools as well as integration of geriatrics within residency training, whatever the specialty. Medicare should fund both geriatric fellowships and academic career development awards.[7]

I favor an academic specialty of geriatrics rather than a practice specialty. We must have the teachers. All doctors should have the basic training and knowledge to care for older persons.[8] Medical students should rotate through geriatrics just as they do with dermatology and pediatrics. No one should graduate from medical school or any residency program, whatever the specialty, without proper training in geriatrics to ensure outstanding patient care. But I see no advantage to creating a new and potentially expensive private practice specialty. Internists, family practitioners, and general practitioners, when well trained in geriatrics, can serve as primary care physicians to older persons. Why should people give up their primary care doctor at some arbitrary age and have to find a new one?

Of course, geriatricians outside of academia are also needed to be medical directors of the sixteen thousand nursing homes and to serve as special consultants in other locations where there are high concentrations of older persons, such as assisted-living facilities, continuing care retirement communities, and naturally occurring retirement communities. This is already happening.

Unfortunately, some doctors will identify themselves as geriatricians, even if they are not properly trained. The competent academic geriatrician should not only be a well-trained graduate of internal medicine or family medicine or psychiatry but should have been educated in basic biological gerontology, the social-economic aspects of aging, ethics,[9] public health, rehabilitation, and palliative care; in short, he should have experienced a very special curriculum. An added dose of good humor and insight is needed. For example, a seventy-six-year-old, confused hospital patient in Glasgow stubbornly insisted on wearing his cap when the nurses asked him to shower. Sir Ferguson Anderson, the great Scottish geriatrician, unraveled the mystery when he discovered that as a child the patient had been told by his mother never to go out in the rain without his cap.

Geriatrics must be integrated within all medical and surgical specialties. The Association of Directors of Geriatric Academic Programs (ADGAP) was funded by the Donald W. Reynolds Foundation to review the program requirements of ninety-one specialties and review committees of the Accreditation Council for Graduate Medical Education. It found that as of 2003 only twenty-seven of the ninety-one specialties had specific geriatric training requirements, and these had modest expectations. Geriatric medicine was finally made a subspecialty in 2006.

Of course, the proper care of older persons especially requires well-trained nurses, nurse practitioners, social workers, pharmacists, nutritionists, fitness trainers, physical therapists, and other health care providers who, at best, should work in a team, emphasizing comprehensive and continuing assessment, and coordination of care from the beginning of contact with a patient to the very end of life. These fields need extensive funding for training.

Much physical caregiving depends especially upon nurses who possess a variety of training, and there is a serious shortage in the field. Geriatrics nurse practitioners are especially important and potentially the first line of defense in the care of older persons. Technology, including robotics and sensing apparatus, should help, allowing more energy to be applied to the personal aspects of care. The health and nursing home and in-home paid aides must earn more, receive health benefits, and be given opportunities for career development; otherwise, the shortages and the huge annual turnover of aides will continue. The costs to families and national productivity are obvious.[10]

THE FUTURE OF GERIATRICS IN THE UNITED STATES

The United States requires a private-public initiative to develop geriatrics in all health fields. Government alone will not provide endowed professorships, facilities, faculties, and programs. On the other hand, the private sector alone cannot afford to pay for the fellowships necessary to establish and maintain the field.

In a bill that could become a national model, the South Carolina Legislature established a state loan repayment program in its Office of Aging. It reflects the shortage of geriatricians South Carolina is experiencing just as a strong demographic shift of older persons is emigrating there. This legislation responds to the high debt medical students have upon graduation (an estimated average of $150,000) that propels many to seek a procedural specialty and not enter primary care medicine, where there is a major shortage.

The American Medical Directors Association, which credentials physicians in long-term care, has been able to certify only nineteen hundred such doctors in the entire country; only 2 percent of physicians in training say they want to go into geriatric care. Recruitment of young physicians is difficult for various reasons. More significant obstacles, however, include few endowments, the absence of role models for teaching, low reimbursement rates under Medicare,[11] and personal discomforts people have about old age, which can be accompanied by despair, depression, dementia, deterioration, and death.

Students who experience how even modest medical and social interventions can make an enormous personal difference in an older person's life often become more interested and supportive of geriatrics. That is why medical school training in geriatrics must be required and not just an elective, which is the usual situation. They come to know that touch, the physician's laying on of hands, is therapeutic. It has been established that 90 percent of the time a diagnosis can be made without expensive diagnostic technology. Giving a patient time to provide a life history and receive a thorough physical examination also helps build the doctor-patient relationship.

It is hoped that baby boomers will be powerful political advocates on behalf of geriatrics, insisting, for example, that the field be expanded.

Longevity Medicine

On the front line, geriatrics will be able to deal with some of the complexities of older persons through advances in genomics, pharmacogenomics, regenerative medicine, and nanotechnology, as well as with social and behavioral medicine. Eventually, two academic, scientifically based specialties will emerge and possibly merge: a *longevity-oriented medicine,* concerned with preserving and promoting health and longevity, and *geriatrics,* which will deal with both the assessment and coordination of care of those who are impaired. It will be dedicated to the restoration and rehabilitation of functions, and the struggle against frailty and dementia.

It is likely that geriatrics and longevity medicine will develop major specialties and perspectives for the medicine of the future.[12] By controlling the myriad diseases that afflict humanity, and ultimately by directly intervening in the biological processes of aging, longevity medicine works to fulfill the ultimate mission of the field of medicine itself, which is to extend life of high quality within genetically determined limits. Operating from a life span perspective, longevity medicine applies to all means that would extend the healthy life, including health promotion and disease prevention far beyond the usual recommendations—diet, exercise, cessation of tobacco use—as well as advanced medical care and new discoveries that result from basic and clinical research. It also suggests the ultimate possibility of identifying and even manipulating those genetic factors that may influence the determined limits of longevity of the species.

BUILDING ACADEMIC GERIATRICS: A MODEST PROPOSAL

Seeing older patients does not a geriatrician make, any more than seeing patients who have hearts makes a cardiologist. In order to improve the health and well-being of older persons in the United States, we must develop a cadre of 1,440 or more academic geriatricians, enabling each of the 144 allopathic and osteopathic medical schools in the country to have at least ten academic geriatricians on its faculty. This is a minimum to meet teaching responsibilities based on our experience at Mount Sinai School of Medicine and elsewhere. Realistically, some of the larger schools would be able to, or would wish to, support more positions, especially to pursue research; 2,400 academic geriatricians for our nation's medical schools would approach the ideal.

In 2002, I proposed a modest but incremental investment by the federal government averaging less than $22 million per year to support the Geriatric Academic Career Award (GACA). Begun in 1998, this award is provided through the Health Resources and Services Administration. Private support could contribute toward the goal of 2,400, with the funds generated by public/private partnerships ensuring the establishment of a foundation of knowledge and leadership sufficient to improve our health care system's capacity to care for an aging population. Such a funding effort by the federal government could be modeled on the wise, bipartisan approach Congress took in 1999 with regard to funding the National Institutes of Health (NIH), although the federal funding involved in this initiative would be much more modest.

A strong commitment by the government to provide a predictable stream of financial support for geriatrics would encourage medical centers to establish or expand programs, enhance their ability to attract funds from private sources, and ultimately produce a sufficient number of academic geriatricians. Strengthened continuing medical and nursing education for doctors, nurses, and social workers already in practice should also be included. Expediting this process becomes urgent, given the impending retirement of the baby boom generation. In fact, Congress appropriated funds to begin such a process, but in 2005 the Bush administration eliminated the fiscal year 2006 budget for geriatrics training, and, under intense budgetary and political pressures, the House HELP[13] Appropriations Committee followed suit. The Senate committee restored most of the budget ($29 million versus $32 million),[14] but the House-Senate conference settled on elimination.

This was a grave setback. Young doctors who had won the competitive geriatric academic career awards were suddenly left without financial support. The overall demoralization further weakened the ability to attract doctors to enter the field.

Happily, in 2007 the budget was restored, though it remains insecure!

It is important now to gain a special set-aside of the Medicare graduate education funds to support academic geriatrics without affecting other necessary specialties dependent on such support. The cost would be modest, as already noted.

REINVENTING MEDICARE

Medicare is the *medical* insurance program that covers forty-three million Americans, thirty-four million aged sixty-five and older, of which 4.5 million are older than eighty-five (the "old-old"). Five million younger adults with permanent disabilities and all persons with end-stage renal disease who require dialysis or a transplant are also entitled to receive Medicare benefits. By 2020, there will be an estimated fifty-three million Americans older than sixty-five, 6.5 million of whom will be the old-old. Originally, Medicare had two principal parts:

- Part A for hospital insurance. Most beneficiaries get Part A automatically because they or their spouse paid Medicare taxes while working.

- Part B for medical insurance. Beneficiaries pay a premium that is generally taken from their Social Security or civil service retirement payment. Beneficiaries pay 25 percent of the program's cost.

The two parts of Medicare added since its inception include Part C, which provides managed care through a health maintenance organization (HMO) or a preferred provider organization (PPO), and Part D, which is a program of prescribed drug benefits for patients outside the hospital. It is run by private health plans subsidized by Medicare. The plans offer confusingly different drug formularies. Everyone is still not covered and all confront the "doughnut hole," which is a gap in drug prescription reimbursement.

Medicare was adopted for a variety of complex social, economic, and political reasons. Among them was the realization by middle-aged adults that it was to their collective advantage to share the costs of their parents' illnesses and at the same time provide for their own old age. State and local governments supported the program because it relieved them of the financial burden of caring for old people in municipal hospitals. The creation of Medicare involved a monumental struggle among the insurance industry, the American Hospital Association, the American Medical Association, unions, consumer groups, and legislators embracing a variety of ideologies.

Medicare established a strong government commitment to providing access to medical care for older Americans without which our current increased life expectancy and lower morbidity rates would not have been possible. Prior to Medicare's passage, medical insurance for older Americans was prohibitively expensive and was often canceled when the insured retired, became ill, or reached sixty-five. It improved the financial situation of many older people by lowering their out-of-pocket medical costs. It reduced anxiety and it worked well for older persons with medical problems such as pneumonia and congestive heart failure. Medicare promoted wider geographic distribution of medical services, especially hospitals, than had previously existed and promoted physician-based ambulatory care for many people who would otherwise have been limited to hospital-based emergency care. *Medicare has low administrative costs (2 to 3 percent compared to 14 to 16 percent in private medical insurance plans).*

Medicare is an egalitarian program that covers wealthy and poor alike. It is not a means-tested, two-tiered program, such as Medicaid, and it provides racial and ethnic minorities with access to medical care. For example, African American women die of breast cancer at a higher rate than white women with comparable tumors, but the gap disappears among women old enough to qualify for Medicare.

However, no experts in the care of older persons were included in Medicare planning to make a knowledgeable case for the specific health care needs of older people. In 1965, geriatrics and gerontology were barely beginning to take root in the United States.

Medicare's *essential* structure has not changed since its inception. Modeled after the private employer-based system of acute care medical insurance of the time, it was formulated to treat all recipients over sixty-five *as if* they were forty-year-old, employed white men.

Medicare is a reactive *medical* insurance system. It is not a proactive system of *health* care. Medicare overemphasizes acute inpatient care. Hospitals are the major recipients of reimbursement, which strains the program financially and can result in poor quality of care for the recipient. Many conditions can be treated effectively and at less cost on an outpatient basis.

As observed earlier, the hospital is a threatening environment, especially for older people; patients fall under the influence of anesthetics, pain pills, anti-emetics, and soporifics. Nosocomial infections are common. Prolonged bed rest adversely affects muscle function and morale is compromised.

Medicare carries reimbursement disincentives to caring for older, sicker patients with chronic conditions. The Physician Payment Review Commission,[15] which adopted a resource-based relative value scale (RBRVS) to determine physician reimbursements, ignored the intensity and complexity of medical conditions in older people, particularly those relating to cognitive impairment and the necessary time involved in providing decent care.

In fact, new geriatrics patients require comprehensive history and physical examinations. Medicare provides inadequate reimbursement for these essential assessments. However, in 2005 it did decide to pay for a Medicare entry examination. Since 2006, there have been constructive efforts by Medicare authorities to pay for more preventive care.

Medicare does not provide reimbursement for a team approach to patient care. Made up of a physician, a nurse, and a social worker, such a team ensures a comprehensive assessment and best practices management plan for older patients. Medicare's reimbursement schedule, especially in our litigious society, encourages physicians to order expensive imaging tests and scans. If given adequate time to take a careful history and physical examination, a physician's training and experience can often uncover important diagnostic information directly from the patient *without expensive tests.*

Medicare retains a fragmented approach to health care that is built into the payment system. Each service or procedure covered is allocated a different payment system, and some vitally needed services are left out. Fragmentation of care is not compatible with either cost containment or good quality of care. Nor is it compatible with the clinical realities of the changes older persons experience over time.

Medicare provides limited, community-based, and preventive care, and there is incomplete coverage of outpatient drugs (Part D was introduced in 2005–2006). Together with home and community-based services, these could dramatically decrease the number of hospital admissions. In general, Medicare does not fully cover services that could avert more costly interventions.

The growing diversity of the Medicare population requires that greater attention be paid to differences in gender, language, culture, religious beliefs, dietary habits, and health needs. In 2001, women were about 60 percent of those sixty-five and older. Women aged eighty-five and over outnumber men about two to one. Nearly 16.4 percent of those over sixty-five are racial and ethnic minorities. They will represent more than 30 percent of the older population by 2025.

Black and Hispanic Medicare beneficiaries tend to be both sicker and poorer than white beneficiaries. In 1999, the Kaiser Family Foundation reported that 43 percent of black and 48 percent of Hispanic Medicare beneficiaries are in poor health. Older Asian American women appear to have high rates of undiagnosed clinical depression. Extraordinarily high rates of Type II diabetes and its complications occur among older Native Americans.

Eighty-five-year-olds are more likely than younger Medicare beneficiaries to have two or more chronic conditions, at least one functional limitation, and mobility and social activity limitations. They tend to be less healthy, more socially isolated, and therefore more vulnerable than younger Medicare beneficiaries.

Efforts to promote universal or national health insurance, especially financed by taxpayers as Medicare is, have routinely and successfully been attacked as socialized medicine. However, the United States has provided one version of socialized medicine since the 1920s in the form of the fourteen hundred Veterans Administration hospitals, clinics, and nursing homes. The VA system, which has had its ups and downs, has generally been credited with doing a fine job, taking care of American veterans and their families. In 2006, the journal *Medical Care* reported that Boston University and the VA reviewed one million records from 1999 to 2004 and found that males over sixty-five who received VA care had about a 40 percent decreased risk of death compared with those enrolled in Medicare Advantage's private health plans or HMOs.

The VA's cost per patient has remained steady since 1997 at about five thousand dollars per year, while the consumer price index for private medical care has risen about 40 percent. VA prescription drug costs are also lower. Doctors, nurses, and staff are government employees, and the care is essentially free to veterans.[16]

The VA is the largest integrated health care system in the United States, with electronic records of all its patients. It cared for about 5.5 million veterans in 2007. The VA system is vehemently supported by national veterans' organizations.

It is not aging and dependency that will bankrupt the nation. Aging is not the primary factor driving health costs. New technology and medicines, the middleman health insurance industry and its administrative costs and profits, as well as inflation, are the culprits. The high cost of dying is also exaggerated. Few older people receive heroic, high tech care and end-of-life costs have remained stable at less than 1 percent of Medicare expenditures.[17]

Medicare faces significant challenges. The structure of its benefits package needs to be changed to reflect the requirements of a new era of longevity. For example, in order to prepare for an increase in the number of older persons in the workforce, it is necessary to include services to protect their health and which affect work-limiting disabilities. A proactive and comprehensive health care system that keeps older people healthy as they age will help reduce costs, with an emphasis on the services that avert serious illness and treatment errors and improve an older person's ability for self-care, independent living, and economic activity.

We need to create a proactive health care system for an older America, informed by central guiding principles.[18]

TEN GUIDING PRINCIPLES OF HEALTH CARE

Life Course Perspective

This is a conceptual framework that incorporates the continuity of a person's biological, personal, environmental, and social experience over time. It includes the cumulative effects of early life events and conditions that vary from person to person and provides a scientific basis for understanding the emergence or worsening of health problems. It can foster an appreciation of the dynamic processes of aging, which counters a stereotypic and negative image.

A life course perspective seeks to prevent during infancy, childhood, and youth the diseases of old age, such as heart disease and osteoporosis, by inculcating healthy lifestyles and disease prevention. More personal responsibility for one's health is essential.

Integrated Health Care

Effective care requires an interdisciplinary team approach, in partnership with the patient and the family. This will provide a *continuum of care* and specifically assure smooth and prompt transfer of patients from one level to another, such as rehabilitation, long-term care, or hospice. A managed care model would balance the provision of care based on need with financial incentives. Cost-effective and quality care can be made more attainable by *shifting the focus of financing from hospital-based to home and community-based health care.* Already, more procedures are conducted on an outpatient basis and hospital stays are shorter.

But prevention is still minimal. For example: Medicare pays for 350,000 hip replacements each year. It does not but should reimburse programs that teach older people simple muscle strengthening and balancing exercises to help prevent falls that result in hip fractures.

To insure a well-trained health workforce for integrated care, major reforms are necessary in medical and nursing education. Most medical education, for example, is conducted in high technology hospitals, not in the community. Needed is multisite training, including in nursing homes, in the community, and in the home.

To build community-based and home care services requires not only a well-trained workforce but an appropriate infrastructure to promote "aging in place" and noninstitutional, long-term care as required. Nursing homes and senior centers could be transformed to be family service centers to provide such services and, if necessary, to train and retrain their staffs.

Long-Term Care

Most Americans cannot afford long-term care insurance. Public or public-private means of providing such comprehensive insurance would spread the risk of becoming ill and make long-term insurance affordable. At best, long-term care should be incorporated into the continuum of care provided by the medical team.

The absence of coverage for long-term care exaggerates a caregiving crisis in the making as the baby boomers and their parents grow old and sick. Caregiving saps the nation's productivity and overburdens families.[19] Paid caregiving must be professionalized with comprehensive training opportunities for career development, decent pay, benefits, and certification. This is not the case today.[20] Nursing homes serve less than 19 percent of older persons who need care. Most is accomplished at home.

Palliative and End-of-Life Care

Studies report that a substantial proportion of dying hospitalized and nursing home patients received inadequate pain relief and suffer from a prolonged dying process because of invasive medical technology at the end of life. To ensure that each of us has control over quality of life and death, everyone should routinely complete living wills, health care proxies, and organ donation cards at maturity and discuss their wishes with their family, doctor, and lawyer. Contrary to the usual opinion, only about 3 percent of Medicare beneficiaries receive aggressive care at the end of life. Understandably, costs are high at the end of life at any age. With Medicare, about half of costs are incurred during the last sixty days.[21]

Pharmaceuticals

Drugs have played a critical role in advancing health and longevity, and there is a critical need for Medicare to cover the costs of medications for older people who are not hospital inpatients. This is good economics as well as good medical practice because many medical conditions can be effectively managed and expensive hospitalization avoided. For example, paying for drugs that control cholesterol levels may prevent the need for bypass surgery, hospitalization, and chronic care. It is hoped that the controversial Medicare Prescription Drug Improvement and Modernization Act of 2003 (Part D) will eventually provide at affordable costs the outpatient medications needed.

People sixty-five and over are woefully underrepresented or even excluded from clinical trials,[22] which evaluate the safety and efficacy of drugs and treatments. We must focus clinical research on complex patients such as the classic geriatric patient. We do not have the luxury of identifying experimental subjects with only one specific problem but must include those with traditionally regarded confounding variables—with multiple, complex, interacting, acute and chronic, physical and psychosocial pathology—because in the real world these are the individuals we must treat. This is one of the major messages of geriatrics.

In addition, clinical trials should include studies of how lifestyle alterations versus the use of a drug impact the course of a disease. For example, exercise can often be as effective against depression as selective serotonin reuptake inhibitors, and as effective as metformin against diabetes. Nutritional supplements and herbal remedies must undergo clinical trials to test for safety, efficacy, and untoward side effects.

Technology

The use of electronic medical records (EMR) can help to overcome a variety of problems associated with paper records, including illegibility, inaccuracy, and incompleteness. Patients should be able to carry their own detailed personal medical records. EMR systems can reduce the administrative burdens physicians and nurses face and have great potential for quality improvement. (The 1999 Institute of Medicine report on medical errors stressed the need for improvements in the nation's information technology infrastructure.)

Telemedicine represents one solution to the dearth of physicians and other medical professionals available to people living in rural communities. On the other hand, relying on technology too much can dehumanize medicine without affording any benefit to the patient, because clinical diagnoses are often more accurate than diagnostic technology. A major concern associated with the development and greater use of EMR systems is our ability to protect the patient's privacy.

Special Programs

Some innovative programs are already under way. The Medicare Prescription Drug Improvement and Modernization Act of 2003 Section 721 created the Chronic Care Improvement Program that may prove to be an important first step. Eighty-three percent of Medicare beneficiaries have at least one chronic condition. The cost implications are notable. Twenty-three percent of beneficiaries with five or more chronic conditions account for 68 percent of the program's spending. The objective of Section 721 is to encourage evidence-based care, reduce unnecessary hospital stays and emergency room visits, and help beneficiaries avoid costly complications.

Stroke units can decrease mortality and increase the probability of returning patients to full functional status. They include physicians, nurses, and other personnel. There is a brief window of opportunity (about two hours) to treat acute stroke that requires both patient and physician education. People need to recognize early symptoms and proceed at once to the hospital. Using emergency rooms for nonemergency situations is expensive and takes staff from genuine emergencies. Community-based urgency care centers that offer an intermediate source of care could help alleviate the misuse and expense of emergency rooms.

Research

"If you think research is expensive, try disease," said Mary Lasker, the philanthropist and health advocate. It has been estimated that thirteen dollars is returned for every dollar invested in research. The survival and health of older Americans is largely due to the investment in research by the private and public sectors. As discussed, much more is needed.

Ethical Issues

The combination of an aging population and advanced medical technology creates new ethical issues that must be addressed. These include:

- Physician-assisted suicide is considered a last resort solution by patients with terminal illnesses who feel they can no longer live with pain and suffering. Most patients, in fact, can be helped by the appropriate provision of palliative care medicine.

- Although ageism is explicit in the medical treatment given older people, it also can be a silent factor in deciding the course of medical treatment, resulting in rationing of medical services for old people.

- Older patients are often excluded from the decision-making process and are subject to neglect because they do not or cannot always speak for themselves.

- Contractual consulting arrangements of doctors with medical device makers (e.g., implantable defibrillators) and drug companies, physician-ownership of specialty hospitals, and other issues require intense scrutiny.

- There needs to be a balance between clinical autonomy on the one hand and pay for effective performance by the physician on the other.

The Rights of Patients of All Ages

Actions must be taken to ensure:

- Development of an easily accessible grievance process for patients through development of an American Patients Association to ensure that patients receive appropriate and adequate care without prejudice.

- Recognition of the needs of older, rural men and women.

- Formation of an organization to monitor social changes and keep abreast of America's changing cultural environment.

- Health disparities between minorities and white populations are erased.

- Recognition of the unique health needs of men and women.

- Study of models of care in other countries so that Medicare can benefit from the experiences of others.

HEALTH CARE FOR ALL AMERICANS

Surely it is time to introduce lifelong universal health insurance to promote a productive, healthy society through whatever prudent financial arrangements are necessary. It could help business, such as the automobile and airline industries, with huge legacy obligations for the health care of their retirees. A reformed Medicare could be the template for transformation to a high quality system of health care *for all*. Health contributes to and, indeed, is necessary to maintain and build the wealth and power of a nation and of the world as a whole, and it speaks to a positive future across generations.

The U.S. system is more bureaucratic, with higher administrative costs, than other countries. Private insurers try *not* to pay for medical care to increase profits. Is there not an economic and moral case for health care reform in America?

The challenge is not just the reform of Medicare and the design of a health care for older America, but the restructuring of the entire health care system, which is a problem for everyone—the government, taxpayers, employers, employees, doctors, and patients.

America has a fragmented, confusing, expensive system of mixed quality. Even the wealthy often do not fair well, although they often do not realize it. Yet, we spend 16 percent of GDP on health care.[23] (See Table 12.2.) Administrative costs are excessive. *Up to 15—and some say 20—percent of dollars assigned to health care insurance does not go to the delivery of health care but instead to advertising, selling, marketing, claims (and denials), paperwork administration, and profits to middlemen.*[24] *There is enormous waste, inequity, and abuse.*[25] Insurers cherry pick to avoid covering people with expensive, preexisting conditions; this undermines the risk pool.

Health insurance premiums keep rising. It becomes increasingly unaffordable for both employers and employees.

Choice in health care is an illusion for the average person. Choosing among the health insurance policies, if available, is nearly incomprehensible. The choice of a doctor is a matter of luck or word of mouth and is not based on performance. This applies to the hospital as well, for geography reduces one's choices. The right to choose is essentially almost meaningless. More accurately, people lose control in the health care system.

Is it time for the for-profit health insurance industry to be phased out?

The United States spends a higher percentage of its GDP on health care than European countries, which cover everyone and enjoy greater life expectancy. We pay for our health care out of pocket; European nations pay through the common-risk pool, taxation.

Fifteen percent or forty-seven million people (as of 2007) do not have access to health care through insurance but depend upon the emergency room, an extraordinarily expensive place to receive care of uncertain quality. America has a high infant mortality rate, greater than most of the wealthy countries that are members of the Organization for Economic Cooperation and Development (6.5 deaths per thousand). The World Health Organization ranks the United States thirty-third in quality of health care. The U.S. life expectancy has fallen from eleventh to forty-second over the last several decades. More companies are cutting back on employee and retiree health benefits. Despite modern technology, the United States is behind in the use of health information technology. Doctors and hospitals do not function under pay for performance. Hospitals are dangerous places (see Table 12.3). No one can understand hospital costs. There is little transparency and there are critical health disparities related to race, socioeconomic status, and education.

Doctors also have less clinical autonomy than those practicing in other nations, and they spend much less time with their patients. Under managed care, doctors may have less than ten minutes to give to each patient, and they face huge administrative costs as well as liability pressures resulting in unnecessary but defensive tests and treatments. True prevention and health promotion barely exist, and the number of primary care doctors and nurses in the United States is declining.

A study by the Institute of Medicine of the National Academy of Sciences looked at the potential economic value to be gained by health insurance coverage of all Americans and estimated a range of between $65 billion and $130 billion each year. Less stress, illness, and death would enhance workplace productivity. The study concluded that it is "both mistaken and dangerous" to assume that a large uninsured population in the United States harms only those who are uninsured.[26]

In the global marketplace, American corporations would fair better if our country had national health insurance because the risk pool would be enlarged. This also applies to small businesses as well.

Arguably, we are not apt to achieve universal health insurance any time soon because health insurers, a strong lobby, make huge profits— $100 billion in 2005, for example. But pressures are growing. States are experimenting. Massachusetts has established an effort to make universal health care mandatory. All of its citizens must buy medical insurance, subsidies are available for those who cannot afford it, and businesses are required to offer insurance. Maine, Vermont, and California are developing plans to cover all their citizens.

Municipal and state governments (taxpayers) face some $1 trillion in unfunded present and future retiree medical costs (as of 2006). They have set aside practically nothing to meet these legal obligations. This will increase the pressure for the federal government to provide national universal health insurance. Medicare itself has a $33.4 trillion unfunded liability (as of 2006).

We could phase in lower age requirements *incrementally* (to age fifty-five to remove disincentives for employers to hire and retain older workers), cover children,[27] and then move to the middle years over time. It would be costly. On the other hand, competition, managed care, and marketing disciplines have not controlled costs. Centralized buying power could. Further, were the high costs of private insurance eliminated, universal health care could be provided without new funding beyond innovative technology and medicines, the introduction of which could be subject to evaluation.

Perhaps a growing demand for health care and greater longevity for all Americans will speed progress toward comprehensive, universal care. There hardly seems to be any other choice.

CHAPTER 13

LIVE LONGER, WORK LONGER: PRODUCTIVE ENGAGEMENT

Is it realistic for people to spend about 25 percent of their adult life-time in retirement? To spend half as much time in retirement as they spend at work? Can society afford it? Is it good for men and women? Can they afford it? Does it serve health, longevity, and quality of life for a person to be idle? Should millions of baby boomer retirees[1] have no work to do while collecting Social Security and using Medicare?

These questions are especially pertinent in the industrialized world. Early retirement becomes wasted productive capacity. Yet until very recently, people retired at an increasingly early age, despite declining birthrates and longer lives. Most people over sixty-five are not working full time; neither are 50 percent of those between fifty-five and sixty-four. For example, between 1950 and 1990 the median age at which people retired in the United States declined steeply from age sixty-seven to sixty-three. By 1996 the average retirement age was 61.5 years and people averaged about twenty years in retirement.

In the United States in 1900 a man spent about 3 percent of his life in retirement. By 1980 he spent 20 percent, or fourteen years. In the early 1900s, fully 75 percent of men over sixty-five were in the labor force; by the mid-1980s it was only 18 percent. Since the turn of the twenty-first century there has been a slight increase in the age people retire, influenced by improved life expectancy, economic downturns, inadequate savings, stock market losses, reduced health insurance coverage, and an improved outlook for older workers (e.g., looming worker shortage, age discrimination laws). The number of employed workers seventy-five and older grew from 669,000 in 1994 to just under one million in 2004, according to the Labor Department. Many of these people are working out of necessity. In 2007, the Bureau of Labor Statistics reported, surprisingly, that older people are receiving the biggest raises.

In Europe since the 1960s there has been a marked decline in labor force participation, especially by men ages sixty to sixty-four. It is below 20 percent in Belgium; 35 percent in Italy, France and the Netherlands; 50 percent in Germany; and 40 percent in Spain. There have also been declines in employment between ages forty-five and fifty-nine! The rates of labor force participation between fifty-five and sixty-four are less than 50 percent for both men and women in Finland, France, and the Netherlands.

But in Europe there has not been an upturn in retirement age comparable to that in the United States.

Given these trends in the developed world, how will people without considerable wealth finance their longevity in long retirements? Can we keep older persons healthy, reeducated to prevent job obsolescence, productive, and on the job? Can we ever hope to achieve a society in which everyone in good health who needs to work will be able to get a job?

Restructuring, downsizing and redundancy threaten workers of all ages, especially the youngest, the oldest, and the unskilled. Yet in truth, there is never really a shortage of work to be done. There are so many needs to be met. Rather, the private and public sectors have failed to establish mechanisms to link work with jobs and skills both on a paid and voluntary basis and to create full-employment societies.

RAISING THE AGE OF RETIREMENT

There are moves to elevate the age of retirement and pension eligibility in Germany, France, Italy, the United Kingdom, and Japan as well as recommendations to do so at the ministerial level of the Organization for Economic Cooperation and Development (OECD).[2] The thirty richest nations in the world are members of the OECD. They are committed to democratic government and the market economy.

Since we have had an increase in life expectancy, why not an increase in work expectancy as well? When Social Security was passed in 1935, the average life expectancy of a male was less than sixty and of a female about sixty-three. Since then the figures have changed considerably—it's now about seventy-five for men and eighty for women. Why not tie the age of full eligibility to Social Security to sixty-eight, or beyond, to changes in life expectancy?

In 1982, Alan Greenspan, chair of President Reagan's National Commission on Social Security Reform (known as the Greenspan Commission), proposed such a reform.[3] As director of the National Institute on Aging, my opinion was solicited. I said that Greenspan's proposal was reasonable under certain conditions—the availability of jobs in a full employment economy (otherwise many older people would be sentenced to poverty); full implementation of the Age Discrimination in Employment Act (ADEA); and an improved means of assessing disability so that individuals who qualified could receive Social Security disability earlier.[4] After all, retirement has been one of society's great achievements, especially for those whose work lives have been onerous and dangerous, such as miners. Congress did elevate the eligibility age to sixty-seven in 1983 (slowly phased in by 2027) and is considering further changes, perhaps up to seventy. Any such advancement in the age of eligibility should be phased in over a very long period of time (up to fifty years) to assist in individual planning.

LONGEVITY AND THE CHANGING WORLD OF WORK

Increasing longevity has come at a time when remarkable changes, both positive and negative, are taking place in the workplace, with growing efficiency, automation, and structural changes in the agricultural, manufacturing, and service industries. What happens to our workforce, including its older members, when we increase productivity that results from breakthroughs in science and technology? For example, what about automated teller machines (ATMs) that eliminate employees? What about other cost- and labor-saving devices, such as computers? Do we simply lose jobs? Will we see modern Luddites smashing technology?

Some policymakers and economists, like the Nobelist Wassily Leontieff, believe there should be a divorce between income and employment to ensure needed income in the consumer-based economy, whether one has a job or not. Leontieff suggested that the government should provide help if workers are displaced because of technological obsolescence or structural changes in the labor market. Others promote the idea of a basic citizen wage, which is higher than the federal minimum wage (some cities and states have established such a citizen wage). After all, prosperity depends upon consumer spending as well as productivity.[5]

The demands of an increasingly complex society and economy point to the need for lifetime learning, with educational programs that prepare young people who are entering the workforce while at the same time training older workers to avoid job obsolescence. Norway is experimenting with sabbaticals that enable workers to learn new skills throughout their years in the job market. Paid educational leave (PEL) programs are also in place in Austria, Denmark, Finland, France, Germany, Italy, and Sweden.

Sabbaticals should be available for everybody, and not be the exclusive privilege of academics or members of the corporate hierarchy. Instead of a specific number of years earmarked for retirement, I have suggested creating periodic retirements; that is, spreading retirement over the entire life span instead of stacking it up at the end, when it may be too late to enjoy.[6] These retirements, qua sabbaticals, could help individuals acquire new knowledge and perhaps new careers. In the United States, these periodic retirements could be financed through neutral reallocation of the Social Security trust funds and other pension funds *ahead of time* and be similar to a viatical[7] made available by the life insurance industry, for instance, to help AIDS victims.

Would we not be better served by replacing the three tight compartments of education, work, and retirement that evolved in modern times with an interweaving of all three throughout life to avoid programmed obsolescence? That would ensure an up-to-date, knowledgeable, and efficient workforce and reduce the fear that an aging workforce must inevitably result in economic stagnation. These changes in education, work and retirement would help women who, because of childbirth and family care, necessarily move in and out of the workforce, and help older persons remain productive. They would also provide the leisure time necessary for family life and civic engagement.

In Australia, the long service leave for employees has been tried for well over thirty years. Most Australians work under a "wages award" system and a basic safety net after completing fifteen years with one employer. Workers are entitled to three months of long service leave at full pay at the base rate. Workers in the health industry are entitled to six months of long service leave after fifteen years of continuous service. From the workers' point of view, the sabbatical seems to work quite well, although it is rather costly for employers, and its long-term success in retraining employees has yet to be determined.

One in ten major companies in the United States (e.g., McDonald's and IBM) offer extended job leaves or sabbaticals for some employees. Although it is often used to attract or keep workers, to help them deal with stress and burnout on the job, or to broaden their professional skills, the sabbatical has the potential to become an automatic part of the concept of shared work and phased retirement. Phased retirement makes it possible to work a reduced schedule before full retirement while simultaneously collecting pension benefits.

THE WORKWEEK

The twentieth century began with employees working almost sixty hours a week, and by 1920 the workweek was just under fifty hours. During the desperate years of the Great Depression and the widespread share-the-work philosophy to help alleviate unemployment, weekly hours fell below thirty-five in some businesses, but generally a forty-four-hour week was the norm (eight hours daily and four on Saturday mornings). As a result of the New Deal's Fair Labor Standards Act of 1938, a new law regulating hours and pay, after World War II America moved to the five-day, forty-hour week.

Today the shorter workweek and part-time work are potential means of dealing with unemployment and other issues. Some European nations enjoy longer vacations, often six weeks. For example, in 1985 the (West) German Metal Workers Union secured a 38½-hour week, and in 2000 France introduced a thirty-five-hour week. However, the consequences of these changes are controversial and still being evaluated.

PRODUCTIVE AGING AND PRODUCTIVE ENGAGEMENT

Aging often means a process or processes by which individuals deteriorate and lose capacity to be productive. The term *productivity* usually refers to work for pay and is measured in monetary transactions. *Productive aging*[8] is a term that draws attention to the fact that aging and productivity are not necessarily contradictory and that the older population is being harmed by stereotypes, prejudices, and discrimination. Productive older adults who contribute to the economy are likely to help shatter these pervasive stereotypes.

Productive aging or productive engagement refer to the capacity of an individual to serve in the paid workforce, partake in volunteer activities, assist in the family, and maintain oneself as independently as possible throughout the life cycle. To quantify productive aging in a population it is useful to measure output in the paid workforce. But that is not enough. If a *complete* statement of the monetary value of an older population is to be computed, values must be assigned to elements that usually are not monetized: volunteer service, service in the family (as caregiver, mentor, advisor, and in child-rearing),[9] and independent living that relieves family and society of caregiving and other support, entirely or in part.

Our society will be slowly drawn into productive engagement by cultural transformation and especially by economic and demographic pressures. Workers will be needed due to the declining birthrates. In addition, they will have to help support themselves in their old age. The rising eligibility age for Social Security benefits will add to the pressures.

We are confronted at present with major obstacles. They include:

- The attitudes and prejudices that affect employers toward older employees and even older employees toward one another. For example, a younger worker may think and say the older person should get out of the way.

- Employment policies may fail to define productivity on the basis of ability and function.

- Job tasks and the work place may require alterations to accommodate older workers.

- The design of transportation, housing, and the overall environment can impede their productivity.

- The Age Discrimination in Employment Act (ADEA) is not effectively enforced (see Chapter 4).

- There are costly employer disincentives to retrain, retain, and hire older workers.

- There are employee disincentives to work. For example, unemployment benefits, were they to be high and without a time limit, could be a disincentive to work. Moreover, there are tax disadvantages.[10]

- Requiring Medicare to serve as a secondary payer is a significant barrier to hiring or retaining older workers over sixty-five. In fact, it would be useful to lower the age of Medicare eligibility to sixty, or even fifty-five, to encourage the hiring and retention of older workers and eliminate the need for employers to provide health insurance. (Of course, as I argue elsewhere in this book, it is time to relieve business of the burden of health care costs that impede global competitiveness.)

The Americans with Disabilities Act (ADA), which was passed in the United States in 1990, has the potential to overcome the third and fourth obstacles. Moreover, ergonomics can help alter the workplace and jobs to accommodate disabilities as well as the prolonged reaction time associated with aging. We must construct more effective methods of evaluation for disability so that certain people unable to work are not penalized and so that those disabled but capable of work do not lose out by doing so.[11] Finally, much more money must be invested in biomedical and behavioral research to forestall the debilitating dysfunctions that conspire against productive aging and engagement.

In 2002, the Supreme Court ruled that to qualify as disabled, and therefore to be protected by the Americans with Disabilities Act, a person must have substantial limitations of abilities that are "central to daily life," and not only to life in the workplace. The unanimous ruling narrowed the broad terms of the 1990 law, which obligates employers to make reasonable accommodations for disabled workers.[12]

DISABILITY AND PRODUCTIVITY[13]

The Social Security Disability Insurance program aids people of all incomes who have worked for at least ten years before becoming disabled, helping 4.7 million beneficiaries and 1.6 million members of their families (see Chapter 14). As of 2005, people with disabilities may earn up to $830 per month before losing their federal benefits. Separately, the Senate passed legislation giving people with disabilities other incentives to work, including the ability to keep government health insurance.

Questions that require resolution: How to better define a disability—is it the inability to do any job *or* the job the person is trained and qualified for? What are the specific medical criteria? Other criteria? What are the possibilities and consequences of rehabilitation?

Some impaired individuals have voluntarily sought new skills. Severely visually impaired persons have taken up massage therapy in order to have work when and if they are no longer able to carry out their present jobs. Technology offers aids for those determined to work. For example, voice-activated technology is now available to help people with disabled hands.

OLD AT FORTY!

In 1929, people over forty were consigned to the human scrap heap. Even today the Department of Labor defines the older worker as someone forty and over. (In 2004, there were 72.8 million workers over forty in the United States, according to the Bureau of Labor Statistics.[14])

The stereotype equates aging workers with nonproductive drains on society, but, ironically, older workers who remain productively employed are more likely to remain healthy and able to contribute to society than those who retire. Indeed, there is a triangular relationship between aging, health, and productivity. Health promotes productivity, and productivity, in turn, promotes health throughout life. (Of course, these interrelationships are very complex.)

Studies of older workers in the United States conclude that although reaction time may be slower and more time may be required to complete certain tasks, older workers have less turnover and less absenteeism, higher job satisfaction, and are more dependable and experienced. And we know that in the workplace, experience matters. Studies also refute the aphorism "You cannot teach an old dog new tricks" because older workers have been shown to demonstrate flexibility and innovation. Learning ability, intelligence, memory, and motivation do not decline with age in the absence of disease, dementia, and depression and if one keeps active. Healthy older workers have fewer accidents than younger workers and often a stronger work ethic. Admittedly, perhaps one of the reasons studies of older workers are so positive is that only the most committed remain in the workforce.

STAYING PRODUCTIVELY EMPLOYED

In 1990, the Commonwealth Fund conducted a program known as "Americans Over 55 at Work." It focused on three companies: Days Inn, Travelers, and B&Q, a British company. Days Inn, the third-largest hotel chain in the United States, had a policy of employing persons age fifty-five and older as reservation clerks because they tended to be patient and courteous when dealing with customers. Travelers, an insurance company, created job banks for their retirees to provide part-time work and give the company a flexible workforce to handle insurance claims. The program was so successful that the company expanded its job banks to include retirees of other companies. B&Q, a do-it-yourself chain of stores, also reported a successful experiment in employing persons over fifty.

Wells Fargo has bused more than one hundred retirees from Sun City, Arizona, to its operations center in Tempe, where they helped process monthly statements several days a month. In addition, McDonald's created a successful McMasters program for hiring older workers because some franchises find that they can't attract enough younger ones.

Older persons have made extraordinary contributions in late life. Benjamin Dugger, a professor of biology at the University of Wisconsin and an expert on soil bacteria, was arbitrarily forced to retire. Lederly Laboratories of the American Cyanamid Corporation hired him at age seventy-two and gave him another chance to carry out his scientific interests. The company was well rewarded when he discovered Aureomycin, a profitable and valuable antibiotic.

Dr. Michael Heidelberger, one of the fathers of immune chemistry, discovered that antibodies that respond to infection are composed of proteins. At age sixty-five he was retired from Columbia University's College of Physicians and Surgeons and subsequently hired by the New York University School of Medicine, where he enjoyed a second productive scientific career from 1964 until his death at 103.

Some small biotechnology companies employ outstanding retired scientists on a part-time and short-time basis in keeping with changing needs and limited personnel. Other American companies could employ retirees and benefit financially.

CONSEQUENCES OF PRODUCTIVE ENGAGEMENT

Can productivity measurably enhance life expectancy and the quality of life? In studies at the National Institutes of Health in the 1950s and 1960s, my colleagues and I found, to our surprise, that people who had clearly specified goals and organization in their daily lives lived longer than those who did not. Similar findings were reported from a longitudinal study conducted in Tecumseh, Michigan. Whether work extends the length of a person's life requires further investigation, but its enhancement of the quality of life seems certain.

Studies show that cognitive health can be maintained by intellectual stimulation and social engagement as well as by physical activity (see Chapter 11).[15] For example, Professor Marian Diamond of the University of California, Berkeley, conducted studies in which she enriched the environment of rats.[16] Subsequent autopsy revealed positive changes in the structure of their central nervous systems.

As we hopefully move toward a more productive aging society, we need to learn more than we presently know about human performance. As noted earlier, we need human population laboratories that identify and measure human capabilities, including intellectual functions such as problem solving, reaction time, and coordination, and ergonomic studies. The industrial and postindustrial revolutions have reduced human drudgery, with a shift in emphasis from physical labor to productivity that relies on intellectual labor.

AGE DISCRIMINATION

With passage of the Age Discrimination in Employment Act (ADEA) in 1967 and its subsequent amendments, compulsory retirement has virtually been eliminated in the United States, theoretically replaced by functional criteria that are used to determine if an older worker is capable of doing the job. A landmark amendment in 1988 was fostered by Claude Pepper, the oldest member of the House of Representatives at the time, and signed by Ronald Reagan, the oldest president to date. At the heart of ADEA is the belief that employment in old age is a civil right, that people have the right to earn a living throughout their lives, and that hiring, promotion, and retention should be based on competence and function, not on age. A corollary is the understanding that society loses talented, experienced, and able individuals when it allows age discrimination.[17]

After passage of the last amendment to the ADEA in 1988, the Government Accounting Office optimistically estimated that millions of dollars would be saved in Social Security and Medicare payouts because older persons would remain more active and less dependent, and continue paying into the two systems. They also estimated that approximately two hundred thousand persons over seventy would stay in the work force as a result of the legislation. In fact, 1.6 million older Americans over the age of seventy are still in paid employment, out of a total of 23.9 million people past seventy, 448,000 in managerial and professional jobs.[18]

Age discrimination cases have shaped an active field of law and a number of cases have been won, among them by television reporters who have won back pay and job reinstatement after they were fired when image consultants reported that viewers wanted to watch younger persons.

A lawsuit filed by the Equal Employment Opportunity Commission in 1999 charged the Woolworth Corporation with firing dozens of employees in stores across the nation because they were over forty. The lawsuit sought monetary damages and an injunction to keep the company from discriminating in the future. The suit was settled in the workers' favor, and Woolworth agreed to pay $3.5 million in damages.

Although the United States is the only nation with its comprehensive Age Discrimination and Employment Act, European nations are beginning to make advances in this area. Gradual or phased retirement is practiced in Sweden, and the European Union mandated that age discrimination laws be in place by 2006.

Those young people who may complain of promotions being made later rather than earlier in life should be reminded that this is compensated for by a longer life!

THE VOLUNTEER SECTOR

The United States possesses an extraordinary informal network in the voluntary sector, and volunteerism is both a national and cultural norm. In the 1830s, the great French observer of the American scene Alexis de Tocqueville made note of our associative character.

The value of volunteer work in the United States is estimated to be $272 billion, or 2.5 percent of the nation's GDP, and older Americans provide millions of dollars of equivalent productivity in volunteer activities.[19] Jay Winsten, associate dean of the Harvard School of Public Health and director of the school's center for health communication, said he anticipated that when the baby boom generation begins to retire in 2011 they would expand the volunteer sector in the nation's $10 trillion-plus economy beyond almost eighty-four billion hours of volunteer work each year.[20] Florida has passed legislation to support programs that encourage volunteers to contribute services, for which they receive "time currency" or "time dollars" that can be cashed in when the volunteers themselves need services.

Individuals who have retired from the worlds of science, business, and academia, among others, could teach, mentor, and sponsor young people. A registry of retired scientists who want to continue working or to volunteer their services should be kept by the National Academy of Sciences and the Department of Education. They could lead after-school enrichment programs for children, in this nation of science illiteracy. Moreover, we should not lose the great gifts of craftsmanship and the arts carried out by older persons.

Major volunteering opportunities for older persons include:

- The Service Corps of Retired Executives of the Small Business Administration (SCORE) utilizes retired businessmen and women to provide assistance to self-employed persons with emphasis on minorities, widows, and young persons.

- The Peace Corps now encourages older recruits[21] because often they can adapt to projects that need low-tech solutions. For example, older civil engineers may do a better job building a bridge in an underdeveloped country than a young engineer who has been trained to work with advanced technology.

- Volunteers in Medicine is a medical clinic staffed almost entirely by unpaid physician retirees to provide free health care for the uninsured working poor and their families at Hilton Head, South Carolina.

- Experience Corps, now in more than two dozen cities, provides schools and youth organizations with older adults who help improve academic performance and foster the development of youngsters.[22]

- Mark Freedman, president of Civic Ventures, endeavors to foster the continuing voluntary contribution of older persons.

- ReServe was developed successfully in New York City by The New York Times Company Foundation. It encourages retired older persons to serve again.

- The AARP has a large volunteer program that uses a database to match skills with needs.

- The Corporation for National Community Service is the umbrella agency that supports a number of volunteer programs, including the Foster Grandparents Program, which pays minimum wage to low-income, older men and women who work with physically and mentally disabled children for twenty hours a week.

- National Senior Service Corps,[23] the Retired Senior Volunteer Program (RSVP) with over six hundred thousand volunteers, and the Senior Companion Program help frail older persons to live independently.

- Doctors Without Borders welcomes retiree volunteers.

A national youth service could usefully delay the entry of young people into the job market and encourage older persons to remain in it longer. Before entering the workforce young persons could be encouraged or required to contribute more extensively than they do today through voluntary activities to America. This should be accomplished through AmeriCorps, which pays tuition benefits in exchange for service. More young people are turning away from politics and concentrating instead on community service. In 1906, William James raised the idea of national service for young people. Idleness was listed among his five giant evils, along with want, disease, ignorance, and squalor. We need to expand the spirit of volunteerism with the creation of a national intergenerational voluntary service corps that could assist the frail aged, provide aid in family emergencies, and, through family service and community centers, contribute to quality of life.

DO WE NEED EARLY RETIREMENT?

Retirement is a great human victory when work is backbreaking, toxic, and boring. But now we face problems of retiring too early and living too long. Early retirement decisions are a function of health, disability, desire to retire, contract agreements, financial status, and the nature of the job. Pension plans can be an incentive to retire.

In the early 1940s and prior to World War II, just as Social Security began to kick in, older men began to withdraw from the world of work. They returned in force during the war when there were temporary labor shortages. Then in the 1980s we saw downsizing of corporations, often with "sweeteners" to encourage better-financed early retirement. But as James Burke, one of America's most imaginative CEOs and former CEO and president of Johnson & Johnson, noted, this may prove only to be a short-term savings measure.[24] It could be a long-term disaster. What are the long-term consequences of early retirement?

An American Management Association study in 1995 found that fewer than half the companies that downsized since 1990 saw improved profits. Kodak, in Rochester, New York, cut its workforce by half and its payroll by $1 billion over an eight-year period. However, this did not improve Kodak's fortunes. Downsizing is not a substitute for innovation.

If people remain in the workforce longer, will they take jobs away from young people? In recessionary times this question becomes even more important. However, the United Nations International Labor Office (ILO) in Geneva, Switzerland, has found that the continuing participation of older persons in the paid workforce does not measurably reduce opportunities for younger workers. Age Concern of Great Britain reports similar findings. It is a mistake to assume that a sixty-four-year-old will be instantly replaceable by a twenty-year-old in jobs and industries. As noted earlier, there is no validity to the idea of a "fixed lump of labor."[25]

Tenure

For decades, high-ranking academicians have enjoyed tenure, which is considered critical to security in the face of political pressures as well as providing intellectual freedom to pursue scholarly work without financial constraints. However, I believe that academics should be required to undergo periodic evaluation (e.g., every ten to fifteen years), rather than be given a job for life. Legal avenues to preserve intellectual freedom exist, and, unfortunately, a small but significant number of academics do not function creatively over extended periods. Seniority and academic tenure coupled with pay schedules should also be revisited. Just as evaluation precedes hiring, promotion, and firing, it should also be used to decide an academic's retirement.

In 1996, the University of Minnesota tried to make it easier to dismiss tenured professors, but it eventually caved in to pressure. Finally, the new requirements were only applied at the law school. Financial incentives to retire are offered at some colleges.

One institution that should retain tenure for life is the federal judiciary because of the demonstrated value of the separation of powers in our democracy to ensure restraint of the executive and legislative branches of government.

Worker Shortages

Rising populations could potentially add to unemployment. On the other hand, declining birthrates, as in Europe and Japan, will result in fewer people entering the job market. A vast transformation of work patterns is necessary that will affect people of all ages. *There will necessarily be an extension of life spent in the workforce.*

Most companies are not going to offer lifelong jobs,[26] and, therefore, employees must reinvent themselves regularly to stay employed. People have to take greater responsibility for their continuing education as they must also do for their health. The information revolution and electronic media will prove helpful. (Google may be compared to Gutenberg's invention of the printing press.) Companies also need to help in various ways. Benefits must include opportunities for continuing training.

LEISURE

Let us look at the promise and logic of the industrial/scientific revolution—freedom from drudgery, greater leisure, and an improved standard of living. Machinery brought services as well as relief to the lives of the toilers in the fields, at sea, and in mines, factories, and foundries. But has humanity in general actually gained the dividend of leisure promised, which would contribute to an improved family life and greater opportunities to exercise one's role as a citizen, pursue happiness, and enjoy the products of culture?[27] To effectively exercise citizenship, the primary office in a democracy, some leisure to participate in the civil society is necessary. Better parents also need more leisure time. Sweden and Norway have instituted shorter workweeks to enable workers to deal with family and civic responsibilities.

In 1908 Elie Metchnikoff wrote a treatise[28] that summed up what I consider to be the credo of optimistic gerontology: "It must be understood . . . that the prolongation of life would be associated with the *preservation of intelligence and the power to work. . . .* [my emphasis]. When we have reduced or abolished such causes of precocious senility as intemperance and disease, it will no longer be necessary to give pensions at the age of sixty or seventy years. The costs of supporting the old, instead of increasing, will diminish progressively. . . . We must use all our endeavors to allow men to complete their normal course of life, and to make it possible for old men to play their parts as advisers and judges, endowed with their long experience of life."

By the year 2000, U.S. unemployment (4.3 percent) was remarkably low, a situation only supreme optimists would consider permanent. Nonetheless, the tight labor market has provided opportunities for the disabled and older persons to get jobs.[29] Just as World War II helped bring women into the workforce, the contemporary labor market may do so for older persons. In 2000, Congress unanimously ended the retirement penalty under Social Security under which workers lost some of their Social Security payment if they worked. One reason for that was the evolving labor shortage. Boomers, who are up to a third of the nation's workforce, will be retiring soon. The average registered nurse is forty-five, and she will be retiring just as the medical needs of the aging population soar. The federal Bureau of Labor Statistics has identified other fields that will be particularly affected by boomer retirements, such as airline pilots and industrial engineers (Table 13.1).

It is time to overcome the factors that encourage early retirement, introduce shared work and more flextime and phased retirement, advance the ages of eligibility for Social Security, and secure more effectively the employment and rights of older persons and the disabled.

Productive aging and engagement will help quell the three great fears of longevity—that there will be an unprecedented number of economically dependent older persons, that old people will drag down economic productivity, and that there will be intergenerational conflict. Further, productive engagement helps to sustain Social Security, which will be discussed in the following chapter.

CHAPTER 14

SOCIAL SECURITY:
SELF-RESPONSIBILITY AND THE STATE

Today's average sixty-five-year-old man and woman can expect to live to be eighty-one and eighty-five, respectively. Over 17 percent of sixty-five-year-old men and 31 percent of sixty-five-year-old women are expected to live to ninety or beyond. Will they outlive their resources and be forced to lower their living standard?[1]

One consequence of longer life expectancy has been more years of retirement, which has resulted in mounting pension and health costs. Governments and employers, in turn, have reacted by reevaluating and reducing their commitments. Can we afford old age, especially in this new era of expanded longevity? How much can government and business be expected to provide for retirement? Will the growing numbers of retirees bankrupt nations? How much can individuals save and invest? How long should a person work?

At the same time that experts suggest a replacement level of 70 percent to 80 percent of a worker's last earnings[2] (a questionable and arbitrary idea in any case) in order to maintain a standard of living at retirement, there is growing pressure for people to predict their own needs and to bear some or all of the financial risks.

Some factors, such as anticipated life expectancy (especially living "too long"), tolerance of investment risk, the state of the economy (such as the rate of inflation or the occurrence of recession), health status, employable skills, living-cost requirements in retirement, education of children, and other responsibilities can only be encapsulated into an imperfect algorithm. Ultimately, much depends on the questionable integrity of the securities market and on the state, which varies in its loyalty to its citizens.

The next two chapters are devoted to a discussion of these complex issues. This chapter looks at various aspects of Social Security, both a venerated and maligned program; the next addresses the role of the private sector in financing longevity.

Having survived twelve presidents, at least as many political agendas, and the myriad fluctuations inherent in a free-market economy, Social Security has earned its place at the table as an enduring, well designed, and adaptable, albeit poorly understood, social program. (See Table 14.1.) Through war, recession, and even governmental shutdown, Social Security has always come through. Backed by Treasury bonds, the safest asset in anyone's portfolio,[3] and protected against inflation by having its benefits reflect the Consumer Price Index, Social Security is an extraordinarily efficient government program. Its trust funds are held separately. Administrative costs are less than 1 percent of benefits paid out, in contrast to private pension administrative costs, which range between 12 and 14 percent.[4] It is also fully portable; vesting is not required.

The Social Security system was expected to remain solvent through 2055 if the budget surplus left by the Clinton administration remained. The surplus, however, faded away due to Bush administration tax cuts, the terrorist attacks of 9/11, the costs of the Afghanistan and Iraq wars, Katrina, and the slowing economy.[5]

WHAT IS SOCIAL SECURITY?

Social Security is a lifetime annuity that is paid for as long as a person lives, assuring a minimal level of financial security even if all other resources are depleted.[6] Unlike a private annuity or pension plan, Social Security is not prefunded. The retirement benefits are paid mostly by the Social Security contributions of today's workers, an arrangement that has functioned successfully for more than sixty-five years. It is a universal compulsory savings program that is earnings-related with regard to both contributions and benefits, with 95 percent of American workers paying Social Security taxes. (Some four million municipal and state employees are not in the Social Security system.)

The program requires a regressive tax for low-income workers; however, earned-income tax credits lower taxes for low-paid workers or they are paid directly if they are too poor to pay taxes. If the combination of a couple's income and Social Security does not exceed $32,000, they do not pay Social Security taxes after retirement.[7]

Through its Old Age, Survivors and Disability Insurance (OASDI), Social Security protects children, survivors, and disabled persons, as well as retirees. It is a life and disability insurance program that provides a total of $14.5 trillion in life insurance—greater than the total of the private disability and life insurance markets.[8] For disability, the value of the protection is about $203,000; and for survivors life insurance, it is $295,000 for a twenty-seven-year-old worker who dies with two children and average earnings.

In 2001, it provided survivor benefits to young families who lost breadwinners in the terrorist attacks of September 11, 2001.[9] Survivor benefits are paid to dependent children of deceased workers, which includes unmarried children under age eighteen, full-time high school students under age nineteen, and adult children disabled before age twenty-two; a widowed spouse age sixty or older (if disabled, eligibility begins at age fifty); a widowed spouse of any age who is caring for eligible children under age sixteen or a disabled adult child; and the dependent parents of the deceased worker.

The initial Social Security benefits of a new retiree are determined in part by the overall level of wages in the economy, as well as past earnings. This is called wage indexing, and it assures new retirees that their living standard will keep up with that of the working population.[10]

Who Funds Social Security?

Wage earners pay 6.2 percent of their wages into Social Security, and employers are required to match that amount.[11] The self-employed pay the entire 12.4 percent. In 2003, $534 billion (84 percent) went to Social Security from payroll taxes and about $85 billion (13 percent) came from interest earned on the system's surplus. Thirteen billion (2 percent) came from taxes that working retirees paid on their benefits. The taxable wage base in 2007 was $97,000 for OASDI.

The overall program is progressive in that higher wage earners receive higher benefits, although that is a smaller percentage of their average earnings than low-income workers receive. For example, the average wage earner who retired in 1997 received about 42 percent of his/her average earnings, while low wage earners received about 60 percent and those with high incomes received about 28 percent.

For the majority of workers employed in low and moderate-income jobs, Social Security provides equal benefits for equal earnings.

Who Receives Social Security?

In the Social Security Act of 1935, the word *entitlement* was used simply to mean that the people were legally entitled to benefits if they were eligible and filed a claim. Today, unfortunately, it has come to be identified as welfare and is considered a pejorative reference. Who are these entitled individuals?

In 2002, payments went to more than forty-six million people each month, averaging $9,744 in yearly benefits. Couples received an average of about $15,000. Of the more than forty-six million recipients, more than thirty-two million were retirees, nearly five million were spouses and children of retirees and disabled workers, nearly seven million were survivors (of whom 1.9 million were children), and 5.5 million were disabled. *Altogether, 3,915,520 children were covered by Social Security,* most protected from impoverishment when wage-earning parents died or were disabled. Of the retirees, 26 percent relied on Social Security for at least 90 percent of their income. Forty percent of all income for people over sixty-five derives from Social Security (see Table 14.2). For 20 percent, Social Security was their only source of income, and 50 percent would have fallen below the poverty line were it not for Social Security.[12] This is especially significant for older women, who live longer than men and who receive 52 percent of benefits.

It must be noted that the official poverty line is calculated differently for older than for younger people, for reasons I do not understand. In order to be counted as poor, an older person has to receive 8 to 10 percent less money than a person up to sixty-five years of age.[13] In 2004 the poverty level for an individual sixty-five and above was $9,310 per year,[14] for a couple $12,490 per year.

How Does Social Security Influence the Intergenerational Contract?

The intergenerational contract has its origins in the earliest recorded models of organized society. It implicitly defines how the generations come together for their mutual benefit and for the overall benefit of society. Social Security represents one such intergenerational contract in that it is a direct income transfer from younger to older workers, their families, and the disabled. In effect, members of the younger generation, in a pay-as-you-go (PAYG) system, pay benefits for their parents. At the same time, more private money is transferred from old to young than from young to old; indeed, some beneficiaries return part or all of their Social Security to younger family members.

Some argue that Social Security promotes intergenerational conflict. For example, the Americans for Generational Equity (AGE) warned inaccurately in 1987 that payroll taxes could rise to between 23 and 42 percent to pay for Social Security and Medicare, compared with 15.3 percent at the time. Third Millennium in Manhattan, a lobbying group that works on issues affecting people under thirty-five, expressed similar concerns.[15] On the other hand, Harris Interactive,[16] Peter D. Hart Research Associates, and other poll takers indicate that old and young alike support Social Security and Medicare. It is noteworthy that Social

Security is itself an economic stimulus and stabilizing influence in the economy (see Table 14.3).

Should Social Security Require Means Testing?

Social Security does not require means testing; instead, funds are redistributed to low-income workers upon retirement. Peter G. Peterson, President Nixon's secretary of commerce and author of *Gray Dawn: How the Coming Age Wave Will Transform America—and the World*,[17] has effectively and usefully raised concerns about the sustainability of Social Security and Medicare. He has advocated a means test for retirees with incomes above $40,000. Only 8 percent of Americans over sixty-five have incomes in excess of $50,000 a year, and they are often still employed. According to the Census Bureau, 18 percent of men and 10 percent of women sixty-five years and older are currently in the civilian labor force.[18] On the other hand, Robert J. Myers, who was chief actuary of the Social Security Administration between 1947 and 1970 and executive director of the Greenspan Commission, has noted that if the redistributive nature of Social Security were eliminated, either the old and poor would be abandoned entirely or an extensive and expensive public assistance program with high administrative costs would be necessary.

Means-tested programs have negative repercussions. Australia, which instituted a means test in the 1940s, has reported a "perverse incentive" for people to deliberately choose to earn less or to spend down as retirement nears to minimize their income and assets in order to be eligible for retirement benefits.[19]

In view of the inevitable events that are beyond our individual control, I believe Social Security needs to remain universal and independent of one's employment, health, income, or assets.

Should Social Security Be Privatized?

President Franklin D. Roosevelt said of Social Security payroll taxes in 1936: "We put those payroll contributions there so as to give the contributors a legal, moral, and political right to collect their pensions and their unemployment benefits. With those taxes in there, no damn politician can ever scrap my social security program."[20] And during his term in office, Republican President Ronald Reagan emphasized that Social Security was not responsible for the federal deficit. Indeed, it helped conceal it. Today, two contradictory beliefs are held about Social Security:

- Older Americans do not get their money's worth and would do better in the private market, and

- Older Americans receive far more than they contributed in payroll taxes during their working years plus accumulated interest.

Which is true? Do people get their money's worth? Or do they get too much? Let's look at the numbers. Average workers who retired in 1996 regained what they paid in their Social Security tax contributions with interest after 6.2 years in retirement. Minimum wage workers regained theirs after 4.4 years and the maximum wage earner after 8.2 years. (Of course, the more well-to-do live longer and therefore collect benefits for more years.[21])

When we think about privatizing Social Security, we must also seriously consider the stock market's vulnerabilities. All investments carry the prospect of risk as well as profit; on the other hand, *average* active managers do far worse than index funds. (An index fund is a portfolio of investments that follows the stock market index and mirrors performance, rising and falling in response to fluctuations.) Only about 20 percent of managers do better (and not always the same 20 percent). Yet, in fact, the Standard & Poor's 500 stock index fell somewhat lower after the October 1987 precipitous stock drop than did actively managed stocks. Index funds are *not* safer than actively managed funds and gain only in a bull market.

In addition, more than half of Americans do not understand the difference between a stock and a bond.[22] Only 4 percent know the difference between an index and a managed mutual fund. Only 16 percent have a clear understanding of an individual retirement account and only 8 percent completely understood the expenses that their mutual fund charges.

Let's face it: Investing in the stock market requires knowledge, judgment, discipline, and a strong stomach—plus luck and a thriving economy. And not everyone who invests is successful. Most Americans cannot afford top-flight asset management and investor services. In general, since the 1920s, corporate bonds have yielded 7 percent, and stocks an average of 10 percent. But there are volatility and long down periods.

In 1996 Alan Greenspan asked, *"How do we know when irrational exuberance has unduly escalated asset values which then become subject to unexpected and prolonged contractions?"* (My emphasis.) In 1999 it was reasoned that a significant drop beyond a possible 20 percent correction that some expected in the stock market would dampen enthusiasm for privatization. In fact, two years later the stock market severely contracted as the dot-com speculative bubble burst[23] and scandals, such as at Enron, emerged.

Nonetheless, ideological, political, and business interests have seized upon any dire projections to promote the privatization of Social Security, and some organizations have misunderstood and/or exaggerated issues related to its solvency. Support for privatization has come from a number of sources and follows several lines of thinking. It has been promoted in part by those who oppose all public welfare programs. One example is the Cato Institute, a libertarian research organization, which conducted a $2 million, three-year campaign to promote replacing Social Security with privately managed individual accounts. Think tanks like the Heritage Foundation also encourage this crisis mentality in support of an agenda to discredit what some call the nanny state. Wall Street is divided on the issue, with one camp predicting a financial bonanza if part or all of Social Security is privatized. On the other hand, David Walker, controller general of the United States and an appointee of President Ronald Reagan, believes Social Security can be kept solvent without privatization.

In 2000, President George W. Bush established the Commission to Strengthen Social Security. It presented three options for workers to establish individual investment accounts by using a percentage of their Social Security payroll taxes. It is still unclear how these options would be financed in an era of large projected budget deficits outside Social Security. One alternative is to initiate large reductions in traditional Social Security benefits, but this option fails to restore long-term solvency to Social Security, despite previous pledges that any commission plan would do so.

Although the Social Security system is not in long-term actuarial balance, this is due neither to poor management nor to poor planning. According to one estimate, Social Security will be solvent until 2042, at which time the youngest surviving baby boomer will be seventy-eight and the oldest ninety-six. The Congressional Budget Office estimates that Social Security is solvent until 2052, when the youngest boomer will be eighty-eight and the oldest 106.[24] Then if current trends continue and if no changes are made, Social Security would be required to reduce benefits, add revenue, or effect some combination of the two. At worst there would be a 30 percent cut in benefits, with beneficiaries receiving 70 percent of those presently scheduled. However, this estimate, which is really pessimistic and which projects a declining economy, must also apply to any plan that involves privatization!

I believe that individuals share a responsibility to finance their own longevity. However, even with the best of intentions, today's family often requires two incomes simply to survive, to say nothing of helping to support children through college and pay the costs of unexpected illness. It is often very difficult to save for the future, especially in the face of unemployment or stagnant wages. A survey conducted in 1998 by the Employee Benefit Research Institute (EBRI) found that 40 percent of workers said they could not afford to save even twenty dollars each week for retirement. And, unfortunately, there will always be individuals who are shortsighted or imprudent with regard to their future.

These realities argue for Social Security continuing to ensure *a basic,* if modest, safety net.

ACHIEVING SOLVENCY

The Old-Age, Survivors and Disability Insurance (OASDI) trustees routinely report on the system's actuarial status for the next seventy-five years.[25] By law, they must provide three alternative projections: pessimistic, intermediate,[26] and optimistic. Ironically, longer life and population control are called pessimistic assumptions. Presumably, further medical breakthroughs that would extend life would also be bad news. *Expanding national productivity, happily, is an optimistic assumption.*

According to the Social Security Administration, the seventy-five-year funding gap is $3.7 trillion or about 0.7 percent of gross domestic product (GDP). Estimates made in 1997 calculated that over the next seventy-five years, *if nothing were done,* payments from the Social Security trust funds would average 1.92 percent[27] more than receipts from payroll taxes. In the 2001 report,[28] the pessimistic low total fertility rate over the long range was 1.7 per woman; the optimistic high was 2.2, and the intermediate level was 1.95. It is reasonable for society to encourage an intermediate level (which approaches zero population growth), or about 2.1.

The solution to Social Security least commented upon publicly is increasing productivity, because with an increase of 1 percent in the GDP the actuarial problem disappears. The history of the growth of the GDP suggests optimism. For example, in 1997, the trustees predicted that the trust funds would run out in 2029. However, without any changes to the program, the nation's improved economic performance added thirteen years to the estimate of when the system would face a genuine crisis.[29] But other steps can and should be taken.

Payroll Taxes. The simplest way to achieve solvency today is to raise payroll taxes[30] by approximately 1.92 percent to cover the shortfall over the seventy-five years, moving up payroll taxes for OASDI to 7 percent from 6.2 percent for both the employer and the employee, or from a 12.4 percent to a 14 percent total. In addition, the employer and employee would each pay 1.45 percent of total payroll[31] (2.9 percent combined) for Medicare hospital insurance. Thus, the employee and employer would have a tax rate of 8.45 percent of payroll, and if self-employed, 16.9 percent.[32]

The downside to a raise in payroll tax is that it is regressive and hits low-income workers hardest, although it is redistributive at the time of payment of benefits and, to a degree, as mentioned earlier, is offset by the earned income tax credit available to those workers.

Longevity and Work. Since people are living longer, it makes sense that they should work longer, increasing the amount of money going into the Social Security trust funds while simultaneously reducing the outflow.[33]

Increasing age eligibility will mean that people will get Social Security for fewer years of their life. Slowly phasing in the change will compensate but may stimulate additional claims for disability insurance and present the need for reforms in assessing disability.

The Dependency Ratio[34]

Like a stable economy, a stable population is desirable but difficult to achieve. The dependency ratio, as previously discussed, represents the ratio between the number of persons who are employed to those who are economically dependent. The decline in the birthrate and the reduction in the number of children helps offset the advancing number of people over sixty-five.[35] Moreover, it must be noted that productivity per capita has increased over the years, and fewer workers are needed to produce comparable goods and services. Clearly, the dependency ratio is a very uncertain and controversial notion. For instance, many older people have saved and invested, and approximately 15 percent still work; some children earn money before age twenty, and some people between twenty and sixty-five do not.

SOCIAL SECURITY REFORM: THREE SCENARIOS

The Advisory Council on Social Security, appointed by Donna E. Shalala, secretary of Health and Human Services in 1994, could not agree on recommendations to achieve Social Security solvency. So in 1996 it presented three competing scenarios.

Although more than a decade old, they are cited here because they accurately reflect the possibilities.

Maintenance of Benefits Plan

Robert M. Ball, former commissioner of Social Security (1962–1973), led the largest number of committee members in support of a plan that recommended only modest benefit cuts and up to a 40 percent investment in indexed stock funds. The Ball plan maintains Social Security in its present form. It proposed "a Federal Reserve-type board with two functions: one, to select the index to be used, probably 3,000 to 5,000 stocks; and two, on a bid-basis to select day-to-day fund managers who would buy and sell stocks to maintain the folio in accordance with the index. These managers would be experienced portfolio managers from the private sector, not civil servants."[36] The total investment of Social Security funds in equities could not exceed 5 percent of the total valued stocks on the open market unless authorized by Congress. Retirees would be insulated from stock market risks.

Commentary. Would there be investments in corporations that create environmental problems or in companies possibly subject to antitrust suits? Would the blue chips gain at the expense of start-up innovative companies? And what if the stock market experiences deep turmoil? Wouldn't the government come under strong political pressure to support the market, as Hong Kong's government did during the 1998 Asian financial crisis?

Individual Accounts Plan—Partial Privatization

Edward M. Gramlich, economist from the University of Michigan, presented a plan that would scale back Social Security benefits but keep the 12.4 percent payroll tax. This proposal would add an additional tax of 1.6 percent earmarked for government investment accounts in an unmanaged passive index fund. The Gramlich plan would keep the same payroll tax, but would add a mandatory supplemental privatization. This plan was supported by only one committee member, Marc Twinney, former director of pensions for the Ford Motor Company.

Commentary. Proponents argue that this is not a tax increase. A state-mandated contribution would go into a private account owned by the individual. However, it has the essential characteristics of a tax, the compulsory confiscation of money. If such confiscation of a private contribution is legitimate, how does it differ from adding to the Social Security payroll tax itself or turning to general revenues?

Personal Security Accounts Plan

Sylvester J. Schieber, director of the Research and Information Center of Watson Wyatt Worldwide, a benefits consulting company, and his group of five committee members proposed a two-tiered system that would provide a flat benefit set at 70 percent of the poverty line, with about 47 percent of the benefit paid to an average beneficiary. This plan, which is the most radical of the three proposed, would deduct 5 percent of the existing payroll tax and place it in a personal security account.[37] It would take about $1 trillion in transition costs and require the equivalent of a 1.52 percent payroll tax for seventy-two years.

Commentary. This controversial proposal would mark a major transformation of the system created in 1935 in that *Social Security would become more of an investment program* conducted on an individual basis and not an insurance system. It would presumably give a lift to the stock market. There would be massive administrative costs—given the millions of individual plans[38] and the cost of private retirement accounts—and management fees and commissions would be enormous, reducing the presumed additional income for beneficiaries beyond present-day government return. (Trading securities and commodities costs investors more than $100 billion in transaction costs taken from returns each year.)

From the perspective of the world of business and government, the advantage of privatization is clear—it transfers risk to the individual.[39] Moreover, privatization of Social Security might be a source of economic expansion and general prosperity. Depending on which of the three types of plans, if any, was adopted, Social Security could hold 5 to 15 percent of the total market value of stocks. I do not know what it would mean to introduce billions of *new* dollars per year into the market. Presumably, it would be a real advantage to Wall Street, but might it have some unexpected and undesirable effects? Would it be inflationary, for example?

In the present system, low-income workers benefit, but such a subsidy would not exist or have the same effect under privatization. Typically, low- and middle-income investors do not choose to invest in the stock market but favor lower-risk investments, knowing they cannot afford risk. When they can save, they tend to choose government-insured bank deposits, savings and loans, and credit unions.

If the Social Security surplus were not available, the government would have to turn to the private market to borrow, at possibly greater costs. If the stock market did fall cataclysmically, especially over a long period of time, who would absorb the losses—would it fall on pensioners? On taxpayers? This would mark a shift to risk, not a shift to security, for the individual. With the baby boomers constituting 20 percent of the population and 30 percent of the voters between 2020 and 2030, would the government be likely to let older persons, predominantly older women, starve? This would be the true Social Security crisis, and a reason for panic!

In a report published in 1997, Dean Baker[40] of the Economic Policy Institute warns that the present trend in stocks cannot be sustained. He argues that the suggested 7 percent real return to investors over the last seventy years is incompatible with a long-term mainstream prediction of less than 2 percent growth in gross national product. This might sharply curtail one's enthusiasm for privatization of Social Security and for individual investor savings. Put bluntly, the stock market's price-earnings ratio would have to rise to 34 by 2015 and to 485 by 2070 to sustain a 7 percent average return.

A number of other solutions have been proposed. For example, Lawrence Kotlikoff, Boston University economist and inventor of generational accounting, suggests the use of a federal retail sales tax to cover large amounts of unfunded liabilities of the Social Security system. And, as already noted, Peter Peterson proposes a means-tested system.

Least destabilizing is the proposal by the 1939 Social Security Advisory Council that the income tax be the third party, if necessary, along with the employer and the employee.

A FOURTH SCENARIO

Let's envision a fourth scenario, which is comprehensive (involving health as well as economic considerations) and goes beyond the traditional conceptual approach to Social Security solvency. It includes no benefit cuts but instead raises the taxable wage base and, as necessary, payroll taxes in measured steps and only to the degree required. This scenario is based upon actuarial advice and evaluation of annual productivity,[41] taking the least painful steps first.

Here are the main points in this scenario:

Increase the wage earnings limit on annual payroll taxation for Social Security from $97,000 (in 2007) to $150,000.[42] The intent here is to cover 90 percent of all wages. This would solve about 60 percent of the shortfall and affect only 6 percent of the population. It is compatible with Adam Smith's vision of progressive taxation and would add security as well as progressivity. Expanding the earnings base is consistent with other taxes—there is no ceiling on sales, income, property, inheritance, and capital gains taxes.

Study the effects of the rise in eligibility age to sixty-seven to determine if employers make adjustments and retain workers longer. If justified, slowly raise the eligibility age to sixty-eight, as discussed in the previous chapter. Eventually, retirement age should be indexed to changing life expectancy under ideal labor-market conditions; that is, freedom from age discrimination and available jobs. This requires that government increase employer incentives to employ older persons by reducing their costs[43] in a program that is phased in over an extended period of time and that takes into account racial and gender disparities in life expectancy.

I believe some changes to Social Security should be introduced immediately, in a system of "partial advanced funding,"[44] saving today for the needs of tomorrow. This is preferable to an exclusive pay-as-you-go system that requires increased future taxes to maintain benefit levels.

Individuals who retire at sixty-two have a 20 percent reduced benefit under Social Security,[45] but there is an incentive to work longer under Social Security since benefits grow by 7.5 percent per year as long as one chooses not to draw from the fund. A one-year increase in the retirement age is equal to a 7 percent cut in Social Security, says Ron Gebhardtsbauer, senior pension fellow at the American Academy of Actuaries. Such an increase would eliminate about one-third of Social Security's projected $3.7 trillion shortfall over the next 75 years, not counting the extra tax revenue it would generate from people working longer.

Support productive engagement. The majority of workers in this country retire involuntarily, for example, for health reasons or job loss. Steps should be taken to encourage continuing work as discussed in the previous chapter. In addition, delay the age of early Social Security eligibility from sixty-two to sixty-three, and gradually to sixty-five (with the assurance that disability insurance is available).[46]

- Make older employees more attractive and less expensive to business by lowering Medicare eligibility to age fifty-five, while simultaneously making Medicare the primary payer. (Today, for the "working aged," Medicare is the secondary payer after the employer health plan has paid.) This plan would put more money into Social Security as people work longer, and less money would be paid out of the Social Security trust funds. Parenthetically, and compatible with data that suggest that productivity and health are related, there would be increased national productivity and lower health costs.

- Offer other tax incentives to employers who retain older workers, and tax incentives for retraining throughout the working years.

- Increase health through improved lifestyles and medical research.

Increase the time period for computation of benefits under Social Security from thirty-five to thirty-eight years. Since this would have untoward effects on women who are in and out of the workforce because of maternity leave, child care, and other unpaid family work, establish Social Security family-care credits.[47] France's system of providing family-care credits could be a model for study.[48]

Authorize the Bureau of Labor Statistics of the Department of Labor to revise the Consumer Price Index (CPI).[49] The CPI measures inflation and consumer prices by noting monthly price changes in a "market basket" of ninety-five thousand goods and services that match typical consumer purchases. To recompute the CPI, it is necessary to study substitution by consumers when prices rise. Throughout its history, the CPI has been constructed as if no substitution occurred. Both Alan

Greenspan and the late Senator Daniel Patrick Moynihan recommended revision downward. In any case, it should *not* be greater than a 1.1 percent decrease in the cost-of-living adjustment (COLA), which on average would cost $96 per year of Social Security benefit. By just reducing the CPI by 1.1 percentage points, the Social Security system would move further toward actuarial balance. As Senator Moynihan pointed out, such revision would reduce the payout of the trust funds by $1 trillion in twelve years. Reduction of the CPI alone would lower the long-term deficit by two-thirds. If the CPI were reduced, it would slow the rise of the federal tax bracket and deduction amounts, effectively raising taxes and cutting the deficit. Part of the tax code and a third of the federal budget outlays have been indexed by the CPI. The CPI is also used to determine wage increases in many union contracts. Because of the CPI, Social Security is the only source of income protected against inflation, in keeping with increasing longevity.[50]

Tax the benefits of retirees after they have received Social Security income beyond their initial contributions. (People in the upper income brackets already pay back one-third of their Social Security benefits in taxes.) This includes taxing employers' contributions, so that about 85 percent or more of the Social Security benefit would be subject to tax.

Since the payroll tax is regressive and affects low-income people the most, increase the payroll tax only if absolutely necessary and only for as long as necessary to make up for any residual shortfall after other suggested steps have been taken. The alternative and socially wiser choice might be to use general tax revenues.[51] I advocate a tax increase for the future, if needed.

Introduce reforms in Social Security that eliminate inequities in payment between men and women. In 1935, when Social Security was inaugurated, the conventional American family consisted of the breadwinning husband and the homemaking wife. But today, in the majority of cases, the economic unit is the two-paycheck family, and it should be treated as such in any Social Security reform. Left undone in Social Security is development of a modified earning-sharing system, which combines the earnings in a marriage by economic unit and helps reflect the new family in which both partners work. This reform should be carried out without cutting the benefits of widows who have not had a work history. Divorcees who were married fewer than ten years do not receive the benefit of their husband's Social Security but should receive a prorated cutoff. Currently, these individuals receive no Social Security benefits accruing from their years in the marriage.

Undertake a major reconstruction of the Social Security disability evaluation and awards to speed up the process and avoid errors in either direction. Eligibility for disability benefits begins five months after the disabling illness or event, while eligibility for Medicare begins only after twenty-four months if cash benefits have been received! The disability insurance trust fund (under OASDI) is projected, under the intermediate estimates, to pay full benefits until 2015 and requires a fiscal solution.

Encourage employers to provide both defined-benefit pensions and/or defined-contribution pensions, 401(k)s. Social Security became necessary in the first place because the free-enterprise system—including small employers as well as large corporations—did not, and perhaps could not, always provide pensions. (See details in next chapter.)

Mandate a supplementary private pension plan. I favor an *additional* mandatory payroll charge of approximately 1 or 2 percent of covered earnings to be dedicated to individual retirement accounts. They would be best managed by competitive Wall Street companies under broad general governmental supervision. This might enhance both national savings and investments[52] and augment pension adequacy, as well as send an important psychological and symbolic message to people to further provide for themselves. It must be acknowledged as possible, however, that any mandated savings could simply result in a shift in savings. This would give a lift to the stock market.

Those with low incomes (under twenty thousand per year) should be given the choice of being excluded from this requirement or subsidized. A sliding scale should be put in place on incomes between twenty thousand and a hundred thousand dollars. I envision a flexible plan so that the mandated savings could be used in genuine emergencies without penalty.

Encourage private savings and investments spurred by tax-favored IRAs. For example, in 1997 a new federal law allowed nonworking spouses to set aside two thousand dollars each year in tax-sheltered individual retirement accounts (IRAs). A separate law created the Roth IRA tax-free retirement distributions, especially useful for younger people with lower incomes who anticipate having higher incomes later in life. However, the program would need to be carefully monitored to see if this leads to additional savings or simply shifts them.

Educate the public about Social Security. "Social Security is the most successful program that the government has run," the *New York Times*[53] quoted Marc E. Lackritz, president of the Securities Association, a Wall Street trade group, as saying.

Conduct a federally supported health initiative to reduce early retirement. This would encompass public health efforts to improve lifestyles, support new investments in medical research, and expand the power of the Occupational Safety and Health Administration (OSHA) to reduce frailty, dementia, and other diseases. Those who enjoy general physical and cognitive health can work longer. Declining disability rates reduce disability insurance costs and facilitate working longer. Special attention should focus on racial disparities in life expectancy that adversely affect African and Hispanic Americans.[54]

.   .   .

The fourth scenario adds up to a revision of the very limited U.S.

welfare state, based upon an understanding and evaluation of the twentieth-century European, Japanese, Oceanic, and North American experience. We need to understand the thresholds of occurrence of what the insurance industry calls a moral hazard. This refers to the likelihood of tax and work evasion, and a culture of dependency.

There is good reason to be optimistic about Social Security, which has proven adaptable. Historically, Social Security has changed and evolved since 1935. Four years after its inception, the revolutionary family protection policy was instituted, which expanded coverage to dependents. Since then, any problems of solvency have been dealt with. After deliberations by President Reagan's National Commission on Social Security Reform and followed by congressional action in 1983, Congress imposed income taxes on the benefits of higher income retirees, revised the tax rate schedule, and phased in an elevation in the eligibility age for full benefits to sixty-seven by the year 2027 (for those born after 1959).[55]

An important point to remember is that in a system as large as Social Security, small adjustments result in big changes. As a pension and insurance policy, Social Security should draw upon what little we know about the human psychology of risk and providence, taking into account the entire context of income maintenance in late life. Policy should not be entirely driven by political possibilities, but by ideals that could invent a different future.

Most people find it difficult to be prudent; at the same time, government must protect itself from unexpected costs. This is the psychological and realistic advantage of a compulsory system that provides only a basic safety net. However, it is obvious that neither the government nor business provides ideal worker protection. Certainly, the individual can't do it alone. We each must do our part. Both defined-benefit pension plans and Social Security may incentivize earlier retirement.[56] We need a balanced partnership of responsibilities involving the individual, business, the civil society, and government. We need a commitment to the intergenerational contract between young and old fundamentally embedded within the framework of the Judeo-Christian tradition in America and within the thinking of those who dramatically influenced the development of our country—the theorists of democracy and of the social contract, especially Thomas Hobbes, John Locke, and Jean-Jacques Rousseau.

Society as a whole would benefit from an explicit move toward such a multibased social protection system. This would constitute a kind of neo-welfare state—a new covenant—that promotes individual responsibility in alliance with the voluntary sector, the market, *and* government. But a secure, tax-based safety net must still back up such a reform.

CHAPTER 15

THE PRIVATE SECTOR: PROVIDENCE AND RISK—THE ROLE OF THE MARKETPLACE

From time to time in the life of societies, extreme circumstances elicit community responses designed to improve the lives of its citizens. Passage of Social Security at the time of the Great Depression of the 1930s and the GI Bill of Rights after World War II are examples of this phenomenon in the United States. They reflect the ideals of community and typically will remain intact until affected by economic factors or ideological shifts. Over the last two decades in this country, however, the ethos of shared community responsibility has given way to one in which, increasingly, citizens are expected to assume nearly total personal responsibility for whatever befalls them.

Defined contribution or do-it-yourself plans, such as 401(k)s, and the proposed privatization of Social Security mark the newest changes in the dynamic balance of American values between the sense of community we associate with the Pilgrims and FDR's New Deal and the ideals of rugged individualism and self-reliance. If carried out, these changes would represent a significant movement away from society's acceptance of specified intergenerational income transfers, toward the expectation that each generation fund its own retirement. They represent a sharp departure from the social contract epitomized by the generation that now receives Social Security—the generation that helped create the physical and social capital that benefits present and succeeding generations.

But what is wrong with rugged individualism? Isn't self-reliance the best economic incentive? Shouldn't people be pushed to greater self-responsibility? To even greater risk?

Perhaps. But how far?

The five main pillars of retirement in the United States are life annuities (i.e., Social Security and a traditional occupational pension), personal savings and investments, home ownership,[1] and continuing to work. As discussed, only a modest addition to the length of time people spend working would significantly improve the financial status of Social Security; still, recommending continued paid employment after the retirement age has become the new third rail of politics, and it was advocated by neither candidate in the 2004 presidential campaign.[2]

Although there is no such thing as a risk-free society, enormous disparities exist in late-life financial outcomes among Americans with similar incomes.[3] People are beginning to realize they could live longer than they had anticipated—and could outlive their resources! So, is there a middle ground? Is there some basic floor of universal security, coupled with an expectation of greater personal and social responsibility to prepare for life's contingencies and old age?

LIFE ANNUITIES

The two main sources of life annuities for retirees in the United States are the Social Security system and employer-provided defined-benefit (DB) pensions. A third, less commonly utilized option is a privately purchased annuity from an insurance company. Finally, and least used, is the charitable remainder trust.[4] Defined-contribution (DC) plans, mostly 401(k)s, generally provide lump-sum payments at retirement, and can be converted into annuities. They are regulated by state insurance law, which allows annuity payments to be distributed differently for men and women. (A woman who annuitizes her pension receives a lower benefit than a man, due to her longer life expectancy.) More than 70 percent of participants in defined contribution plans are now offered a life annuity as a payout option.

American employers are not required to provide pensions. In 2004, only 17 percent did so, compared to 44 percent in 1960.[5] Moreover, in 2004 only 33 percent of retired men and a meager 13 percent of retired women received private defined benefit (DB) pensions.[6]

The availability of 401(k)s[7] and individual retirement accounts has become a huge national experiment, which, so far, has failed the test of providing people with a secure old age in the face of a declining economy and dishonest dealings both in the corporate world and on Wall Street. Social Security has always paid off; private pensions have not.

Private, individual accounts that substitute or supplement Social Security would drain money from Social Security—meaning less money to pay guaranteed benefits. Private control ("it's your money") and the the ownership society sound appealing, but individual private retirement accounts would exchange the inflation-protected lifetime-guaranteed benefits provided by Social Security for market risk and inflation. A relatively modest annual inflation rate of 3 percent will cut the real purchasing power of a fixed nominal income stream by 45 percent in twenty years!

If responsibility and control for one's health care and retirement were actually turned over to citizens, it would be appropriate to speak of the "ownership society." In fact, up to a point, *Social Security is owned by or belongs to the participant,* for it is a legal entitlement. Of course, under Social Security there are no individual investments. However, it covers survivors and is a social contract across the generations. To that degree, wealth is passed on.

Privatization does not provide survivor or disability benefits. In addition, privatizing Social Security would cost at least $1 trillion to $2 trillion in transition costs over ten years (up to an estimated $4.5 trillion over two decades). This is because the taxes that workers pay today to provide Social Security benefits to current retirees would be reduced by letting young workers invest a portion of their Social Security taxes in individual private retirement accounts.[8]

Traditional Private Pensions

Traditional defined benefit (DB) company pension plans pay a set monthly income for life based on wages and years of service. Since the mid-1980s, DBs have begun to disappear, with the number falling by more than 72 percent. (Most of these plans weren't even financially troubled.) In the past, companies that provided pensions were required to set aside a designated sum to cover pensions, but between 1986 and 2002 they have cut their retirement spending by 22 percent.[9] In 1985, about 115,000 American companies had traditional pension plans, compared to about 31,000 in 2003. In 1998, DBs covered 40 million employees, more than half still working. Still, in 2003, three-fourths of America's five hundred largest corporations and thousands of small businesses still offered defined-benefit pensions. Thirty-four million workers were covered with $12 trillion in assets (as of 2004). (See Figure 15.1.)

Figure 15.1—With and Without Private-Sector Workers' Pension Plans

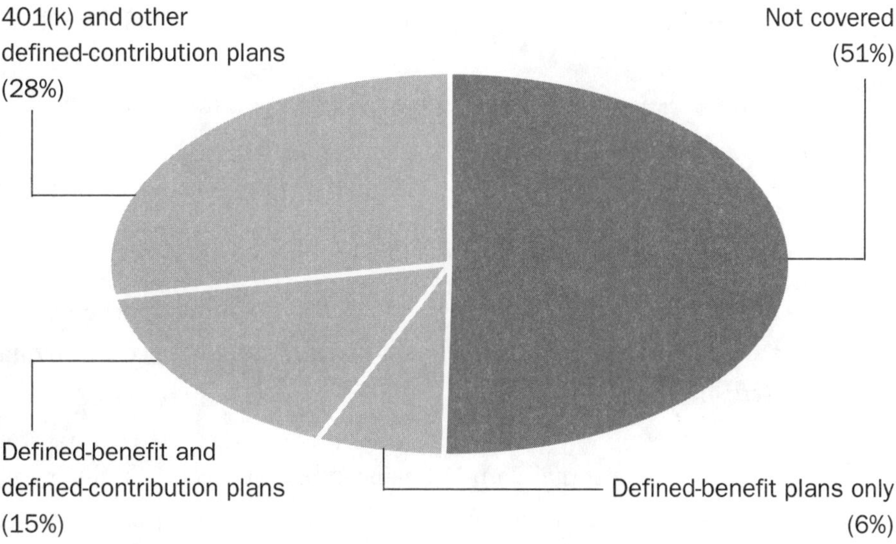

401(k) and other defined-contribution plans (28%)

Not covered (51%)

Defined-benefit and defined-contribution plans (15%)

Defined-benefit plans only (6%)

Labor Department: Vanguard Group.

Private pension benefits for men average about $7,500 per year but just under $4,000 a year for women. Ten to 20 percent of our workforce is now part-time and contingency workers, who are usually not covered by pensions (nor provided life and health insurance). Employees of small companies, nonunion members, those with low-paying jobs, women, and minorities are often uncovered. So, too, are workers in service industries, such as fast food and home care.

Despite their popularity in the past, DB pensions have *not* always paid off, either because the plan is underfunded, the company is losing money, or because of fraud.

So-called legacy costs of pensions and health benefits of corporations reflect the failure of the latter to save and invest appropriately to meet their obligations to their employees, who in old age cannot make up their losses. Moreover, companies and unions should not have made promises they could not keep. This would protect taxpayers as well as workers.

The Employee Retirement Income Security Act (ERISA), which was passed in 1974, provides limited coverage for failed DB pensions.[10] When companies fail and ERISA has exhausted its resources, taxpayers ultimately bear the cost, as they did at the time of the savings and loan crisis.

Defined benefit pension plans have other disadvantages:

- A person who continues to work after sixty-five can no longer contribute.

- Benefits do not protect against inflation, are not fully portable, and are based on wages earned when both salaries and the cost of living were lower.

- Companies must maintain DB plans in bad times and liabilities can increase. The downturn in the stock market during 2001 had a major negative impact on the financial performance of pension plans.

- Companies can freeze or terminate pension plans, as major companies such as Verizon did in 2006.

- While federal laws forbid reducing benefits already earned, companies can reduce benefits to be earned in the future and eliminate them completely. They can change formulas and average income over ten years or over the entire period of employment instead of the usual highest three years of pay. Companies also find other ways to cut pensions that are subtle and difficult to detect.

- Companies may subsidize early retirement in their pension plans but avoid paying this benefit by selling divisions and treating the workers as if they had resigned.

Some unionized workers as well as municipal police and firefighters have been successful in their efforts to retain DB pensions, but many have not, and companies have little incentive to pursue this course. On the other hand, the Economic Growth and Tax Relief Reconciliation Act of 2001 made defined benefits attractive for self-employed, high-income earners who own companies with no more than five employees.

The Pension Benefit Guaranty Corporation

In the 1980s the savings and loan industry collapsed, costing taxpayers between $150 billion and $200 billion. In the future, the collapse of the defined benefit pension system could cost them the same or even more. The Pension Benefit Guaranty Corporation (PBGC), created under ERISA, is designed to protect taxpayers from such a debacle. In its first thirty years the PBGC assisted employees of 3,200 failed pension plans, and today it guarantees defined benefit pensions to forty-four million people. The PBGC estimates that pension plans are presently underfunded by some $450 billion (as of 2004).

The PBGC is not funded by taxpayers. It collects insurance premiums from employers that sponsor insurance pension plans. It earns money from its investments and receives funds from the pension plans it takes over. However, the PBGC has a more serious immediate problem than that faced by the Social Security trust fund. Douglas J. Elliott, president of the Center on Federal Financial Institutes, has warned that if current conditions continue, PBGC will be broke by 2020.

While other industries, such as steel and automobiles, are challenged by pension shortfalls, the airline industry has been the most affected so far. If every airline with a traditional pension plan were to default, it would cost PBGC some $31 billion. Moreover, airline pilots typically retire with pensions of a hundred thousand dollars or more, based on their final salaries and years of service. With limits set by Congress, the PBGC usually matches the payments only for workers who retire at age sixty-five and earn pensions of up to about $49,000 a year (as of 2007), it pays less for younger retirees, and it does not cover the full pension of retired higher income workers whose pensions fail. United Airlines alone carries some $13 billion of pension debt on its books. *Pension law says the dollars owed workers cannot be taken away, but bankruptcy law says workers are unsecured creditors with respect to the PBGC billion-dollar shortfall. Unsecured creditors usually lose in a bankruptcy case.*

Experiences with Privatization

Portions of the $2 trillion in state and local pension funds are steered into high-risk investments by pension consultants and others who often have business dealings with the very money managers they recommend. The structure of public pension funds leaves them particularly vulnerable. Fund boards are responsible for investing hundreds of millions of dollars, but only the largest can afford professional investment staffs. Public trustees often have little financial training and are frequently drawn from the ranks of firefighters, teachers, and other public employees whose retirements they are protecting.

As of 2002, more than half of public pension plans for teachers, firefighters, and other state and municipal employees were underfunded and suffered stock market losses. Funds were often diverted to build highways and repair schools. For example, former Republican Governor Christie Todd Whitman offered a plan in 1997 to borrow $2.75 billion through a state bond sale to cover the New Jersey state pension system's unfunded liability. The bond sales produced an immediate budget addition of $575 million. She lowered the state's regular contributions to the system and did not have to raise taxes or cut services. Over time this failed, and by 2002, the state had to pay the 7.64 percent annual interest on the state bonds and recoup stock market losses.

Whitman is not the only governor to manipulate the pension system. Democratic Governor Mario Cuomo of New York also did, luckily without disastrous effects.

The 127 largest state and local plans are $200 billion short of funds needed to pay beneficiaries, according to the National Association of State Retirement Administrators. In California, the California Public Employees Retirement System (CalPERS) accounts for $2 billion of the state's $8 billion budget deficit. This led Governor Arnold Schwarzenegger to push for workers hired by the state of California after June 2007 to be covered by 401(k)-like private accounts.

Defined Contribution Plans

Defined contribution (DC) plans, such as 401(k)s, are tax-deferred[11] voluntary payments, made by employees and sometimes employers, to individual employees' accounts. Employees agree to have a set amount deducted from their paychecks each pay period and placed into accounts that are managed by investment brokers. While these accounts are called self-directed accounts, they are *not* totally under an individual's control. In 2004, the maximum legal annual investment in a 401(k) was limited to $13,000 for workers up to age fifty and $16,000 for people fifty and older.

The average employee has ten 401(k) investment options but, inexperienced and fearful, chooses fewer than three.

About 55 percent of all the money that goes into 401(k) plans is company-matching contributions, most in company stock. If the company stock falls (e.g., following a merger or acquisition), or if the company fails, the employee has lost retirement savings. In addition, the company risks legal action on the basis of conflict of interest. The Enron disaster[12] (2001–2002) and the loss in value of Lucent stock are illustrative.

In the 1990s, DC pension plans such as 401(k)s promoted the bull market, and in turn the bull market popularized 401(k)s. The stock market rose in part because DB pensions were being phased out as employer-sponsored 401(k) plans took their place.[13] Money from these defined contribution accounts initially fueled a bullish market. When this conversion itself began to stabilize, it could have contributed to the recession that followed.

In 2004, fifty-five million Americans[14] were voluntarily enrolled in DC plans. While total assets were $1.9 trillion, almost half in mutual funds, the median account of each investor was only about $13,000. The average worker age fifty-five to sixty-four had a balance of only about $42,000 in 2001.

Cash Balance Plans

This is a new type of pension plan, a conversion of the traditional defined benefit plan. Cash balance plans are somewhat like 401(k)s: employees receive accounts with annual contributions and are vested after five years of employment in one company.[15] The accounts earn interest and employees can take the money with them when they leave. Younger workers like cash balance plans because they can build up benefits faster and the plans are portable.

Unlike the traditional pension based on the highest pay level achieved, cash balance, defined benefit plans provide for a low, steady accrual throughout a work career. Corporations save millions but the benefits of older workers are cut by as much as 50 percent, since the plan is not based on the last years of highest pay. The plan indefinitely relieves companies from taking on long-term obligations to a retiree. While these plans may be legal under the tax code, they were declared illegal under the Age Discrimination in Employment Act (ADEA) in *Cooper v. IBM Personal Pension Plan and IBM Corporation* (2004). In 2006, however, a federal court ruled that they did not violate ADEA.

Unlike defined benefit pensions, there is no governmental protection under ERISA for defined contributions, and typically 401(k)s have no provisions for disabled employees.

Stock market plunges and a declining economy pose inherent dangers to defined contribution plans. Inadequate, biased information and poor financial training add to the risk. The recession of 2000 to 2002 that markedly reduced 401(k) value is a sad reminder that any decline in the stock market puts 401(k)s *and* individuals at risk.

Workers, especially lower-income employees with greater debts and fewer assets, often jeopardize retirement security by borrowing from their DC plans. According to the General Accounting Office, by 1997 one-third of all workers had dipped into their retirement savings. Because this borrowed money is not being invested and earning interest until it is returned to the account, these workers typically end up with as much as 30 percent less retirement income. And despite tax and penalty payments intended to discourage people from withdrawing all their 401(k) savings when they change jobs, workers often cash out.

Lower-paid and young workers do not participate in 401(k)s to the same degree as older and higher paid workers. In 1994, the *New York Times* reported that, on average, only 35 percent of a company's workers put money in the company's 401(k) plan, compared to 70 percent of management employees. These disquieting numbers have not changed.[16]

Following the Enron pension scandal, and the crimes and bankruptcies of other companies which robbed hundreds of employees of their retirement income, reforms[17] were proposed, but not all were implemented. A cap on company stock in employee 401(k) plans was one of these. However, some progress has been made. Under the Sarbanes-Oxley Act of 2002, accounting firms are restricted from performing consulting services for their clients, and the Financial Accounting Standard Board calls for public companies to record options as an expense at the time of award to an employee. This began after June 15, 2005.[18]

Mutual Funds

The mutual funds industry has been among the fastest-growing seg-ment of the securities business, and today it controls about 40 percent of the nation's retirement assets. (It constitutes one of the great sources of capital formation.) Mutual funds are used both as an individual in-vestment and as a component of a 401(k). In 1980 only 6 percent of American households invested in mutual funds, compared to nearly one-third in 1995. In 1986, there were about eighteen hundred funds with $716 billion in investments; in 1996, there were seven thousand funds representing over $3 trillion in investments. In 2004, mutual funds were valued at $7.6 trillion with ninety-five million individual investors and about 45 percent of 401(k) plan assets. (However, the median household income of mutual fund buyers is $50,000 and lower paid workers were not usually participating in mutual funds.)

Early investors were attracted to mutual funds because they offered inexpensive access to the stock market through no-load mutual funds. But many new mutual funds have sales charges and these loads have a big impact on investment returns, particularly index funds. In addition, many funds now carry an initial charge independent of any load. Unfor-tunately, the cost of mutual fund fees to investors is not fully transparent. Fund companies often charge individual investors a quarter percentage point more than institutional investors. Overall costs at many mutual funds are high. Some five hundred mutual funds still charge more than 2 percent to cover their expenses, according to Lipper, the fund research firm, whereas the average expense ratio is 1.37 percent.

Much of the problem results from the failure of the mutual funds business to regulate itself, and the SEC has failed to effectively oversee the industry. The Sarbanes-Oxley bill put in place federal accounting measures, but it was weakened by the mutual fund trade organization the Investment Company Institute, which succeeded in getting impor-tant exceptions for the industry.[19]

Individual Retirement Accounts (IRAs)

These accounts were originally established solely as a tax-deferred mechanism for people who had no other pension coverage, but in 1981 eligibility was expanded to include all workers, regardless of pension status. The contribution limit is $2,000 per year, principally representing rollovers from both defined benefit and defined contribution plans.

A variant, the Roth IRA, which is limited to a contribution of $3,500, is useful to those with low incomes who expect higher incomes later. They pay taxes now and not upon withdrawal, while the interest earnings accrue tax-free.

By 2002, roughly forty-two million households (40 percent of the total) owned an IRA. In 2001, IRAs had $2.5 trillion in assets, more than either defined benefits or defined contribution pension plans, mostly due to rollovers from company plans.

PERSONAL SAVINGS AND INVESTMENTS[20]

Depending upon specific age and resources, it is generally suggested that people save 10 percent of their income, but average annual savings per capita in 1996 was only $2,388.

In *Nickel and Dimed: On (Not) Getting By in America,* Barbara Ehrenreich[21] writes of her experiences as a worker trying simply to survive, let alone save, on wages of $6 to $7 an hour. This is corroborated by the Economic Policy Institute, which reports that fourteen million people work for these bare-bones wages. Given these realities, how can a poor person save?

What makes people save? Is it primarily fear of personal disaster and of old age? Is it the desire to buy a home and start a family or to pursue pleasure through vacations, entertainment, and recreation? Is it the need to educate children or to carry out gift-giving during life and at death (the bequest motive)? Probably all of the above?

But what can we do to get people to save? We could set defaults for automatic enrollment in 401(k) plans, for example. In 2001, only 8.4 percent of people with 401(k) accounts made the maximum contributions, according to Alicia Munnell, director of the Center for Retirement Research at Boston College. Some fifty-three million households representing 91.2 million people—47 percent of all Americans—owned stocks or mutual funds, however modest, as of July 2003, according to the Investment Company Institute. Yet by 2003, the net savings rate had fallen to just 1.2 percent. Moreover, household debt has risen by 39 percent since the 2000 presidential election, and in 2004 it stood at $9.7 trillion.

The median financial assets of the fifty-plus population are only $38,598 (as of 2004). Edward Wolff, a New York University economist, analyzed eighteen years of household financial data collected by the Federal Reserve. He found that the average net worth of an older household grew 44 percent (adjusted for inflation) from 1983 to 2001, to $673,000. But much of that growth was in the accounts of the richest households, which pushed the averages up. When Wolff looked at the net worth of the median older household, the picture changed. That figure declined during the period to $199,900.

A good way to get started on a savings program if you are some years away from retiring is to use a retirement calculator. The Retirement Probability Analyzer, for example, doesn't seek to tell people how much money they might need to retire; instead, it aims to help retirees and near-retirees estimate—based on the amount they have now—how long their funds might last.[22]

People who are unsophisticated or too busy to handle 401(k) plans themselves could benefit from automatic enrollment in lifecycle funds (or target-date funds), which invest in stock and fixed-income securities and adjust the asset allocation in line with the investor's planned retirement date. These can be expensive, but there are low-cost options.

Investments should become, if possible, mainstays and reflect *prudent diversification* essential in one's portfolio. These include:

- home ownership and other real estate assets (land and buildings for farming),

- mutual funds,

- fixed—*not* variable—annuities,[23]

- stocks, especially blue chips, those that pay dividends, and noncyclical stocks,

- corporate and tax-free municipal bonds,

- money market funds,[24]

- certificates of deposit (CDs),

- the new federal ten-year, inflation-indexed bonds, Treasury Inflation Protected Securities (TIPS),[25]

- other government securities,

- cash,

- gold or other precious metals, and

- art and antiques.

One excellent example of the virtue of diversification: When the Federal Reserve cuts interest rates, people earn less on bonds and CDs. Ideally, one wants one's investments to mature in a layered fashion. Obviously, there are benefits to paying for professional financial planning services if one can afford it, but make sure a prospective financial planner is certified.

HOME OWNERSHIP

Americans still have most of their wealth tied up in the very old-fashioned investment of home ownership, and it remains the cornerstone of household wealth. Because housing wealth has greatly appreciated, and many retirees did not save very much and/or lost money in the market, many retirees are using their homes to finance their retirement.[26] One such arrangement is to secure a reverse annuity mortgage. A homeowner sells the property to a bank. The former homeowner remains a tenant, and the bank provides an income for the homeowner for a designated number of years. However, when the reverse mortgage comes due, the former owner must vacate the property and it is no longer in one's estate. In 2005, forty-three thousand older homeowners took out reverse annuity mortgages insured by the Federal Housing Authority. This marks a six-fold increase since 2000.

Conventional mortgages are preferable,[27] although the best option, of course, is to pay off a home mortgage: optimally, one should face neither mortgage payments nor the possibility of losing one's home in old age. In 2004, however, 32 percent of persons sixty-five to seventy-four were carrying home mortgage debt, and people frequently borrow against their home.

RETIREMENT IN OTHER NATIONS

Other countries face the same issue of income security in old age as we do, and we can profit from their experience. Compared to the European welfare state, the United States never developed a comprehensive system of social protections; it offers only modest protections of Social Security, Medicare, Medicaid for the medically indigent, and workers compensation and unemployment insurance. While the European system is under some revision, it is not abandoning the core welfare state, and the social contract remains intact.

Europeans enjoy a more secure retirement than Americans. About 25 percent of European Union pension fund assets are invested outside the home countries, according to the European Federation for Retirement Provision (EFRP)—14 percent in other countries of the union and 11 percent outside the union. The funds operate under various controls, which reflect a risk averse attitude.[28] People can take their state pensions and Social Security benefits as well as private occupational and company pensions and use them in the original fifteen European Commission countries. Moreover, in Europe, in general there is higher replacement of earnings in retirement, ensuring greater pension adequacy.

But the pension time bombs in Europe are set to go off, due to its aging populations. In the twenty-first century, Europe (and Japan) will be the oldest region of the world, with 25 percent of its population over sixty-five. How will these nations, whose peoples have enjoyed generous retirements, cope? Does this truly constitute a crisis, or are there simple measures that can be taken to adapt to these challenges? Is increasing productivity the answer? Or working longer?

United Kingdom

By 2005, Britain's twenty-five-year experiment with pension reform, which included substituting private investment accounts for a portion of government pension benefits, was regarded as a failure. The Confederation of British Industry (CBI), the functional equivalent of the U.S. Chamber of Commerce, called for a higher state retirement benefit to be paid for by raising both taxes and the retirement age, from sixty-five to seventy. Employers are burdened and want the government to play a greater role. Said the chief executive of the National Association of Pension Funds, an employers' group, "It's actually cheaper for the state to carry the risk."

Sweden

Sweden changed its pension law in 1999. Similar to Social Security, Sweden's old system paid out a defined benefit based on salary and years of employment using contributions from current workers to support retirees.

The new plan added mandatory individual accounts. Of the 18.5 percent payroll tax that workers set aside for retirement, 16 percent goes to a defined benefit program. The other 2.5 percent must be put into individual investment accounts. Workers could choose from some 650 funds, but some 90 percent accepted the government-managed default fund.

Developing Nations

If income maintenance in old age has become an issue in the wealthy industrialized world, in much of the world, government financing of old age does *not even exist.* As Julia Alvarez, former ambassador to the UN from the Dominican Republic, said, "Social security in our developing nations is not a problem, it is a fantasy." According to a report of the International Labor Organization in 2000, 90 percent of the world's working-age population had inadequate pension savings. In many poor countries, people bear children so they have workers in the field and caregivers in old age.

Some developing nations have established a compulsory defined contribution program known as a provident fund. Regular contributions are withheld from wages and invested on behalf of the worker. Workers generally receive the money upon retirement, often in a lump sum, but they can also make early withdrawals to meet special needs. Employers may match or exceed the employees' contribution. Malaysia was the first nation to establish such a provident fund in 1951, followed by Singapore in 1955. By the mid-1990s, some twenty nations had provident funds; however, these funds are vulnerable to economic fluctuations and have not always performed well.

Singapore's Central Provident Fund (CPF)[29] is the largest in the world. Singaporeans pay 20 percent of their wages into the government-managed funds, but they rely on the fund for investment, schooling costs, and mortgages, as well as for retirement income. At 52 percent of the GDP, some economists believe Singapore has an excessive savings rate.

Under Gen. Alfredo Pinochet's dictatorial regime, Chile introduced the world's first mandatory, fully funded, privately managed pension scheme in 1982.[30] It was able to privatize reasonably easily because at the time it enjoyed a budget surplus and could provide substantial prior service credits and minimum pension guarantees.[31] Through tax revenues, the government contributed to the transition from the governmental social security system to privatization. Pension privatization may have modestly boosted Chile's economic growth by improving both capital and labor markets, but many workers remain outside the pension system altogether, and today just over half of the labor force is covered.

Currently, employee Social Security contributions amount to about 20.5 percent of taxable income: 10 percent of payroll for retirement; 3.5 percent for disability, survivor benefits, and administrative costs; and 7 percent for health benefits. But by 1997 about 20 percent of workers who should have been contributing were not doing so. Many others were not contributing on their full wages.

Aside from a mandatory, one-time pay raise employers were required to give when the program was started, employers in Chile contribute nothing toward the old-age pension. The Chilean system does not ensure against poor investment decisions, nor can the system guarantee a particular rate of retirement income, except for a minimum pension payment.

Although the Chilean experiment had annual returns averaging some 12 percent during the 1980s, it has done substantially worse in recent years, and there are serious problems, such as high administrative costs. In 1994 more than half of the funds incurred losses. Returns averaged 2.5 percent in 1995 and 1.8 percent between 1996 and 2001.

Would Americans accept such a large tax burden and risk placed on workers and taxpayers?[32]

By 2006, the failures in the Chilean system became a major campaign issue, and the new government under President Michelle Bachelet led reforms in which the government would play a greater role.

PRIVATE INVESTMENT: PERILS OF THE MARKETPLACE

According to Jeremy J. Siegel of the Wharton School at the University of Pennsylvania, since the early nineteenth century, stocks have provided returns after inflation of about 7 percent per year, *if held over very long periods.*[33] *When the market falls, however, it can take long periods to recover.* In comparison, long-term government bonds have yielded only 2 percent to 3 percent in interest, but they are more secure.

The Social Security system has two extraordinarily important advantages over private pensions. Social Security is backed by the taxing, borrowing, and spending authority of the federal government. Moreover, it can spread its costs over a much larger population of contributors and beneficiaries including those in several generations. This constitutes the universal risk pool.

In contrast, private investing concerns the tradeoff between risk and reward. This gets especially scary later in life.

The Business Cycle

The U.S. stock market peaked in March 2000; funds that tracked the Standard & Poor's 500-stock index lost investors 40 percent by October 2002, compared with 37 percent for an average managed and diversified stock fund.[34]

As Robert J. Schiller, Yale economist, said in 2000, "Stocks have not always outperformed other investments over decades-long intervals, and there is ultimately no reason to think they must in the future.[35] Schiller expects something comparable to the twenty-year doldrums that followed each of the three most prominent peaks in equity valuations of the twentieth century—the 0.2 percent each year in performance during the two decades after January 1901.

A 20 percent downward correction in the stock market defines a bear market. A recession is defined as a significant decline in activity across the economy.

Scandals[36]

Lester C. Thurow, MIT economist, has written that "the Enrons, WorldComs and Tycos are not abnormalities in a 'basically sound system.' Scandals are endemic in capitalism. Government can't make the market fair."[37]

Wall Street and corporate America are not pure, and the scandals of 2002 proved shocking. Analysts rarely advised investors to sell. "Research was biased, often corrupt. The value of companies was often made higher than justified by accounting devices. Extraordinary executive compensation, conflicts of interest, insider trading, incomplete disclosure, and other practices eroded confidence in corporate America, accounting firms and Wall Street. Some financial service companies allowed favored investors to undertake inappropriate trading. The New York Stock Exchange, a quasi-public organization, failed to police the stock market.

- Lawsuits emerged. In 2002, CalPERS, America's largest public pension fund,[38] sued WorldCom and various banks over 2001 bond issues, alleging fraud. Individual investors also sued, claiming their money managers lied about risk, bought and sold stock without permission, and engaged in making excessive trades to gain commissions, known as "churning."

- Scandals involved conflicts of interest of consultants[39] who advise pension funds on asset allocation, selection of money managers, and other investment matters.

- The Securities and Exchange Commission (SEC) reported "widespread evidence" that brokerage firms directed investors to certain mutual funds because of payments they received from fund companies or their investment advisers as part of sales agreements. Federal securities regulators, scrutinizing the mutual fund industry, are examining whether some funds are paying retirement plans to be included in their available funds.

- Labor Department investigators reported 1,269 instances of missing 401(k) money in the fiscal year 2004.[40]

- Moreover, 401(k)s often have high administrative costs and hidden fees.

In addition, other troubling issues include:

- About one-third of big companies do not pay dividends.

- All index funds, basically run on autopilot, are not created equal nor are they fully representative of stock market performance.

- Price/earnings ratios are built upon reports presumably based on a careful evaluation of a company. Ultimately, the value of stocks and bonds depends on a company's infrastructure, goals, and leadership.

- People are living longer and should invest more aggressively, which constitutes a peril because they may still not have time to recoup their losses.

- The largest investors in hedge funds are pension and welfare funds for unions and other groups. In 2007 there were considerable losses.

WHERE DO WE GO NOW?

Advances in genomics and regenerative medicine will give us advances in longevity. *But longevity will mean little if we cannot finance it.* We must:

Reestablish Public Confidence

We must establish new attitudes and policies that reduce public concern about financial security as people grow older. Many boomers (and generations X and Y especially) fear the government will break the Social Security contract, forgetting their own political safety in numbers.

The 1999 governmental surplus reduced anxiety, and the U.S. economy expanded in 1996, 1997, and 1998 by nearly a 4 percent pace. But with the recession, baby boomers and generations X and Y became aware that they needed to take more responsibility and work longer. At best, any governmental pension should serve as a fail-safe protection against penury.

Reform

- The SEC should be strengthened and given more power to act as a watchdog agency, independent of politics and resources.[41]

- The Department of Justice should be proactive in investigation and legal actions.

- At the state level, attorneys general must weigh in through investigation and legal action.

- An independent association should represent stockholders to overcome cynicism and understand legal liability issues following losses or crashes in the stock market.

- The Internal Revenue Service (IRS) should vigorously crack down on abusive tax shelters (through so-called shell corporations) set up by accounting firms.

- Financial Accounting Standards Board (FASB) rules should be enforced. The FASB sets U.S. accounting rules requiring that companies provide a breakdown of assets held by their pension plans, such as equities, real estate, and debt. State and local governments should be forced to provide realistic estimates of their pension liabilities.

- ERISA and PBGC should be greatly strengthened; e.g., the premium rates for PBGC should be raised in accordance with risk, the size of the company, and liabilities.[42]

- Consultants to pension funds should have no connections with brokerage firms.

- It should be made widely known that socially responsible investment decisions, such as avoiding tobacco stock, have generally paid off and have not proved to be a loser.

- One should check on a broker's record through the National Association of Securities Dealers' Internet site at www.nasdr.com as well as ask for references.

- Philanthropic funding for the Pension Rights Center (located in Washington, DC), which works for pension rights and broader pension coverage for all workers, should be expanded.

- Investment in innovations that depend upon the raising of capital should be fostered. Longevity derives, in part, from the applications of medical discovery, which are, in turn, dependent upon investment. Indeed, any crisis in business in America and beyond subverts prospects for medical successes and advancing longevity. By 2002, the biotechnology industry was struggling for support. Capital markets are essential and to work effectively must be policed and regulated.

- Human irrationality and corruption[43] are so universal that regulation is always necessary. Self-regulation does not work. The trick is to fine-tune regulations while reducing unnecessary paperwork and bureaucracy. This will be difficult.

Establish National Pensions

In 2002, General Motors (GM), America's biggest automaker, had a pension liability of $61 billion that was not fully recorded on its balance sheet. More than 450,000 GM retirees depend on their pensions, more than twice the number of present GM employees. GM is also confronted with gigantic obligations for health care for its workers and retirees.[44] Essentially, GM runs its own Social Security program.[45] In comparison, Japanese manufacturers have little to no such liabilities since the Japanese government provides pensions and health insurance. The effect of this disparity is that a GM car[46] must be priced at $1,000–$1,400 more than a foreign car from nations that provide national pensions and health insurance. In 2007, the United Automobile Workers agreed to a proposal by GM to establish a health trust, which relieved the company of health costs for its employees and retirees.

Given the global economy and the vagaries of the market, American business may decide that it is in its best interests to support national pension benefit programs that utilize the expanded universal risk pool in which all citizens participate.[47] Similarly, corporate and small-business America may come to favor a single payer or some kind of national health insurance for the same reason.

Create a Children's Demogrant

Consideration should be given to the provision of a modest demogrant (perhaps $5,000) given at birth to all children to help level the playing field in this land of opportunity. One proposal would finance this in part from a net-worth tax on wealth over $1 million and on large estates. The government would add a few hundred dollars annually to low-income children. The fund could be used for college, a first home, and old age. This idea has proponents all along the political spectrum.[48]

Expand Social Security

Social Security should be expanded to benefit businesses as well as individuals and their families. As discussed, the payroll taxes (on employers and employees) should be raised by 1 or 2 percent to include mandatory supplemental personal saving accounts (not subtracted from the current payroll tax nor voluntary as in the Bush plan) to be invested in the market under strict conditions. The need to expand Social Security is all the greater given the decline in private pensions.

The market economy has many virtues—the support of entrepreneurialism, competition, and innovation among them—that have proved inspirational, drawing upon the best of human intelligence, energy, and imagination.

But it also has its limits and drawbacks. Not everyone is equal to the challenges of the marketplace because of disadvantages of birth or nature. There are market failures.

James Heckman, conservative economist at the University of Chicago, has said the first market failure is that one cannot select one's parents (with variations in genetic endowment and wealth). Market forces do not aspire to solve all social problems, rare diseases, and the unpredictable vicissitudes of individual existence, natural disasters, and adverse socioeconomic conditions.

These domains are the responsibility of the state—to ensure the best for the least of us, to inspire human fellowship, to protect us in times of danger and economic duress. Of course, there are ways the state and the market can work together. Even our national defense is built upon the state and the private sector working together. This must be monitored, however; indeed both the state and the market require regulation. Of greatest concern is when the state and business are in too close a league, for underlying the marketplace and the state is human character and behavior, given to corruption and selfishness but also the best in us. Transparency of relationships and regulation are necessary.

But so far in the history of man, the marketplace has not been able to ensure decent care in the face of disability, disease, and frail old age.

Nor has civil society or philanthropy been able to do so. In the more glorious moments of the European welfare state, the government has done best. At its finest, it is the state that has the taxing and spending power to look after the welfare of both the nation and its citizens. Of course, governments can (in the United States through an act of Congress) reduce Social Security benefits. Ultimately, all social protections are political.

The best solutions to financing longevity are to continue to foster intergenerational solidarity and personal responsibility, provide financial education, and increase productivity, as well as enhance Social Security and carefully regulate private pension plans, whatever the type.

It all requires an informed, politically active citizenry, which I will discuss in the next chapter.

CHAPTER 16

THE POLITICS OF AGING AND LONGEVITY

The politics of aging is not just about the voting booth and voting blocs. It is about the politics of health, Social Security, ageism, and increasingly about what I call the politics of longevity—which include end-of-life issues (e.g., physician-assisted suicide and the provision of food and hydration), the science of longevity (e.g., embryonic stem cell technology), and the quality of life in our culture. These are obviously matters of common interest to all generations, but younger persons don't always appreciate how much their ultimate welfare is affected by what passes as the politics of aging and longevity.

In earlier chapters I wrote about the human interest in longevity and the health politics of anguish. I emphasized the critical importance of biomedical research, especially conducted and supported by NIH, the role of the FDA, and the need to redesign Medicare and protect Social Security. Here I intend to be more concrete and lay out an agenda that speaks to an equitable distribution of society's resources to all the generations to help assure an equal chance for longevity and freedom from poverty in old age. Is it possible for politicians to campaign with a genuine agenda for aging and longevity? Would it have broad appeal? The sanctity of life (or culture of life) issues[1]—abortion, stem cells, euthanasia—deserve debate and evaluation, but they have overwhelmed discussion of significant health and economic disparities which, in turn, affect or are affected by the sanctity of life issues, embraced in the politics of longevity as I define it.

One aspect of the politics of aging and longevity is safety in numbers. Whatever fears younger generations may have about their future, such as concern over the solvency of Social Security and of Medicare, they should realize that so long as we enjoy social democracy,[2] their benefits will be reasonably secure following modest fiscal reforms in Social Security and major structural reforms in Medicare and the health care system as a whole.

On the other hand, as the numbers of older persons grow, their political prospects will improve. Their growing number and proportion as well as their economic power constitutes the real world politics of aging and longevity, but only so long as they are well informed and exercise their power. Older persons, in many cases with time on their hands, could inspire a cultural maturity, directing attention to some of the great problems of humankind, such as nonviolent resolution of conflict, the conservation of nature, and the care of children. Further, older persons could participate more actively every day in community affairs, public schools, and politics.

The mainstream political parties are dominated by money. We live in another Gilded Age,[3] even more profound than the first, which evolved during and following the presidency of Ulysses S. Grant in the last quarter of the nineteenth century and ended in 1915–1916, when gross materialism and political corruption overwhelmed this once-egalitarian agrarian country. There was minimal regulation and no income taxes.

This nation, now barely two hundred years old, imported and refined European ideas, principally those of ancient Greece, England, and France, and still feeds upon them. *Neither of the major parties confronts the new Gilded Age, for they are part and parcel of it.* Other than the so-called cultural divide, there is less and less to differentiate the two major parties. There are differences regarding the degree of regulation of the workplace, environment, and marketplace. Both parties subscribe to a smaller role for government.

So far, the Democratic Party is prepared to protect Social Security and Medicare and advance the minimum wage and liberal policies. These distinctions are not unimportant but, at bottom, government today is too frequently one of special economic privilege. It has failed to produce pragmatic solutions to many growing national issues, such as the deteriorating infrastructure and public education. The poor are nearly off the radar screen. The environment and energy independence are only now receiving some attention. Health care is emerging as a critical issue. There are few differences between the political parties[4] in foreign policy, but both supported the war in Iraq. The lobbyists of K Street, many of whom are former members of Congress, advance special interests. Local and state governments are not much different. Lobbyists are also the fund raisers for the lawmakers.[5] Both political parties and the states make it very difficult for new parties and independent candidates to be on the ballet of all fifty states.

Congress as a body today neither supports nor reflects majority opinion, which declares itself more independent than loyal to a party. Most Americans think of themselves as moderate and nonjudgmental. Most say they would pay higher taxes to support much-needed universal health insurance and high-quality child care. Most support the present Social Security system. Most believe workers should be paid a living wage and have the right to join unions. Most support stronger environmental protections. But Congress is not completely in line with these wishes of the electorate.[6]

Old people could become more powerful were they to form a bloc. The ten states that have 50 percent of all older persons have nearly half of all electoral votes, just shy of the 270 needed to win a presidential election (see Table 16.1). This population, particularly those between sixty-five and seventy-four, is the most likely to register and to vote. In fact, 72 percent of them voted in the 2000 election.

AARP

This organization seeks to advance the social good and contribute to the issues of global aging, as well as advocating for and serving the needs of its constituency, persons fifty and over. Today, AARP's membership—thirty-six million people, most of whom are middle class—includes about 40 percent of the 61.8 million Americans over the age of fifty. Its annual budget is nearly half a billion dollars and it has nearly two thousand employees and four thousand local chapters. It has extensive contacts with the executive and legislative branches of government and has necessarily directed almost exclusive attention to preserving Social Security and Medicare, obviously essential to longevity. It is exploring support of long-term care. AARP has not vigorously lobbied to gain support for the development of geriatrics, the National Institute on Aging, or Alzheimer's disease research, while it has vigorously supported Medicare and its expansion into drug coverage.

AARP sells health insurance, insurance annuities, and mutual funds and runs a mail-order pharmacy. In order to separate its business from its social mission, AARP settled with the IRS for $135 million and set up a so-called firewall between itself and a corporate subsidiary. Profits go to serve its membership.

AARP is an important and unique force on Capitol Hill, but its power has been exaggerated in the media and elsewhere.[7] When one looks at banking, insurance, agriculture, energy, and manufacturing industries, it is obvious that the AARP is not the most potent lobby in Washington. In money terms, AARP pales before the major lobbies (see Table 16.2). It spent $9 million in 2004.

It has never been a grassroots organization, and when it has acted on critical but controversial issues, such as the 2003 prescription drug legislation, there have been membership protests and resignations. In 1983, the Medicare Catastrophic Act, adopted by the AARP, was repealed under membership pressure. On the other hand, on central issues, such as the preservation of Social Security, there is unity within the huge membership that carries great weight in Congress.

If mobilized politically, the boomers could foster a powerful political group within the AARP or elsewhere. This could certainly happen if they face severe cutbacks in benefits, especially if they become sophisticated enough to recognize the dangers as well as their own political power. The AARP is nonpartisan and does not have a political action committee or support specific candidates or political parties.

In 2007, AARP campaigned on the theme of Divided We Fail, in unity with the Business Roundtable and the Service Employees' International Union. This is unique in the history of AARP, mobilized now under Bill Novelli.

LEADERSHIP: DOES AGE MATTER?

Although there have been remarkable leaders of every age, in general, effective leadership requires accumulated knowledge, judgment, and experience, and, indeed, most political power in the world today is wielded by people over fifty. In 1996, ageism reached new heights when the age of Bob Dole, the seventy-three-year-old Republican presidential nominee, was used against him in both crude and subtle ways. Although Senator Dole had 10.7 years of average life expectancy left, it was hinted that he was too old and frail to assume the mantle of the presidency.

At present there are no members of Congress who have chosen to represent the older generation in the tradition of the late Claude Pepper. Moreover, the House Committee on Aging, which he led, has been eliminated. (The Senate still has a Special Committee on Aging.)

ARE WE MOVING TOWARD A GERONTOCRACY?

With population aging, political leadership is likely to grow older, and some pundits have expressed an unreasonable fear that this will lead to a gerontocracy. This is unlikely, because even with fresh new additions to longevity and further declines in birthrates, the old would still remain a distinct minority.

Although the fear of a gerontocracy is ill founded, there are clear advantages for older people to increase their numbers in political offices. In fact, I believe that older people should be better represented through election of older persons, which is not to say that politicians should grow old on the job (like Senator Strom Thurmond). It should also be noted that older people are not conservative per se. Their vote is tied to generational and historical circumstances (e.g., New Deal liberal Democrats grew older). Politicians may grow more radical with age, as in the case of William Gladstone, the nineteenth-century British prime minister.

THE POLITICS OF TAXATION

An important element in the politics of aging and longevity is the politics of taxation in America. Despite the notion that taxes and government spending (high or low) harm the economy, no one has actually been able to prove it.[8] After reviewing many studies, economist William Easterly of New York University and Sergio Rebelo of Northwestern have concluded that "the evidence that tax rates matter for growth is disturbingly fragile."[9] Tax specialist Joel B. Slemrod of the University of Michigan notes that in the twentieth century, a rising tax burden in the United States and other developed countries occurred simultaneously with rising prosperity.[10] Periods of strong productivity growth have occurred when the top tax rates were the highest. High-tax countries (as in Europe) are the most affluent countries. Tax-financed enterprises like education, research, health, and infrastructure contribute to growth.

The United States has one of the lowest tax rates (see Figure 16.1). As a result, among the thirty rich OECD nations, only four spend less on social programs as a share of the economy.[11] Of course, the U.S. finances some social programs with tax breaks, but adding these in, still only six nations spend less on social programs than we do.

Plutocracy

The richest 1 percent of Americans makes up 40 percent of individual federal campaign donations over $2,000. The George W. Bush administration's ten-year tax cut, to be completed in 2011, benefits mostly the richest 1 percent of taxpayers. As of 2002, by the administration's own estimates, 40 percent of the $4 trillion surplus decline in the ten-year outlook is due to tax cuts.[12]

Figure 16.1—Increasing Cost of Government: The Share of Gross Domestic Product Going to Taxes Has Increased Since 1975 in Most Countries

		Tax revenue as a percentage of G.D.P.		Percentage-Point Change
		1975	2006 Preliminary	
1.	Sweden	41.6%	50.1%	+ 8.5
2.	Denmark	38.4	49.0	+ 10.6
3.	France	35.4	44.5	+ 9.1
4.	Norway	39.2	43.6	+ 4.4
5.	Finland	36.5	43.5	+ 7.0
6.	Italy	25.4	42.7	+ 17.3
7.	Austria	36.7	41.9	+ 5.2
8.	Netherlands	41.2	39.5	− 1.7
9.	Britain	35.3	37.4	+ 2.1
10.	Spain	18.4	36.7	+ 18.3
11.	Germany	34.3	35.7	+ 1.4
12.	Portugal	19.7	35.4	+ 15.7
13.	Canada	32.0	33.4	+ 1.4
14.	Turkey	16.0	32.5	+ 16.5
15.	Ireland	28.7	31.7	+ 3.0
16.	Switzerland	24.5	30.1	+ 5.6
17.	United States	25.6	28.2	+ 2.6
18.	Greece	16.9	27.4	+ 10.5
19.	Japan	20.9	27.4 (2005)	+ 6.5
20.	South Korea	15.1	26.8	+ 11.7

Organization for Economic Cooperation and Development. *New York Times,* 2006.

The gap between rich and poor in America has widened since 1970, which is a concern of many of the rich, who recognize the need for a more equitable spread of resources and the importance of the consumer to the economy. It is no longer the land of opportunity and cannot be contrasted with class-bound Europe. Economists and sociologists report that the typical child starting out in poverty in Europe (or in Canada) has a better chance at prosperity than in America.[13] The United States and Britain appear to stand out as the least mobile societies among the rich countries studied. France, Spain, and Germany are somewhat more mobile than the United States; Canada and the Nordic countries are much more so.

Proposition 13: Transforming the Political Landscape

Let us not doubt the ability of a few citizens to effect change. By 1977, local property taxes across California were high, which particularly hurt older persons on fixed incomes. Howard Jarvis, who ran an aircraft parts company, and Paul Gann, a bankrupt real estate salesman, were in their seventies when they campaigned successfully, mobilizing older persons to pass Proposition 13, which held property tax rates at 1 percent of assessed valuation. Proposition 13 had many unintended effects. It made the state increasingly reliant on less stable sources of revenue, like the income tax, and since property taxes are a principal source of public school funding, it cut funding for education. Proposition 13 influenced the entire nation. It stands as a potent example of the politics of aging.

THE POLITICS OF SOCIAL SECURITY (AND MEDICARE)

Anti-tax advocates, free-market proponents, and Wall Street interests pushed for President George W. Bush's private accounts plan for Social Security. They were represented by two coalitions made up almost entirely of trade associations: the Alliance for Worker Retirement Security (AWRS) and the Coalition for the Modernization and Protection of America's Social Security (Compass), which represents nearly one hundred state and national trade associations, including banking. In addition, the lobbying group USA Next claims 1.5 million members. It seeks to attract one million members away from AARP, by presenting itself as a conservative, free-market alternative. Formerly known as the United Seniors Association, USA Next was founded in 1991 by Richard Viguerie, a Republican master of direct mailings. He raises millions from older Americans by sending scary messages about Social Security.

Most Wall Street firms haven't taken a position on private accounts. The AWRS had just three financial firms as contributors—Charles Schwab, Wachovia, and Waddell & Reed. The latter pulled out of the alliance after pressure from unions. The Financial Services Forum, an association of major financial institutions, withdrew in 2005 because members weren't ready to support a specific solution. The two main lobbying groups for Wall Street, the Securities Industry Association, which represents Wall Street firms, and the Investment Company Institute, the trade group for mutual funds, organized no big lobbying efforts on behalf of privatized Social Security accounts.

On the opposite side of the Social Security battle are organizations like the National Committee to Preserve Social Security and Medicare and the Alliance for Retired Americans, a three-year-old political organization that claims three million members. The Campaign for America's Future; MoveOn.org; and AARP spent millions to run advertising opposing personal accounts. Three nonpartisan women's groups—the Older Women's League, the American Association of University Women, and the League of Women Voters—also oppose private Social Security accounts. Women especially depend upon Social Security.

Unions help oversee $400 billion of pension assets and serve on the boards of public pension funds, but the AFL-CIO opposes private accounts. Its president, John Sweeney, sent letters to financial firms, stating that the push to privatize Social Security is a conflict of interest between the financial services industry and the public. The AFL-CIO has organized protests. Backed by labor, Americans United to Protect Social Security planned forums around the country. However, it must be noted that labor runs lucrative financial services businesses, including life insurance and credit cards. Sixteen million union households make up this market. Thus, unions also must avoid conflict of interest.

The Longevity Lobby

When I was director of the National Institute on Aging, I became aware of one very interesting political pressure group I labeled the longevity lobby. It is not surprising that such a lobby would settle so nicely within America. Following Benjamin Franklin, Andrew Carnegie, and Horatio Alger, we Americans espouse the gospel of success and renewal. We hear the promises of salvation, a long life, and wealth through faith, hard work, frugality, scientific miracles, and good works. Little wonder, in this cultural context, that the term *successful aging* became popular, and the hope for a longer life is strong.

Contrast the lives of billions of the world's peoples living in squalor, hunger, and illness with life expectancies of less than fifty years with the relatively small group of wealthy and influential people living in the industrialized world who seek further longevity. There are those who, having it all, yearn for the elusive prospect of a still longer life, if not immortality. By "life span extension" is meant discovery of knowledge of longevity genes, senescence genes, hormones, embryonic stem cells, drugs, or other means of breaking the longevity barrier. But most of the members of the longevity lobby, often gifted and successful men and women, do not expect the *full* miracle. The wealthy and certain politicians seek to know scientists who are working on life span extension and who, in turn, may themselves be seekers after longevity. Or in other instances, scientists may simply exploit the longevity seekers to gain funds for their laboratories.

However, some of the longevity seekers have formed the informal lobby that besieges scientists and Congress to support life extension research. Many, happily for them, have also altered their habits to maintain their health; travel to health spas; hire cooks to follow the Pritikin, Ornish, or other diets; run daily; stop smoking; and so forth. Some also go to expensive Swiss clinics for cell treatment and seek plastic surgery and cosmetology. On the other hand, of course, many who go to the surgeon and the cosmetologist have no awareness of research on aging, longevity, and life extension. A few longevity seekers have made arrangements for cryonics—to have their bodies or severed heads preserved by freezing, to be awakened in the future when the diseases that afflicted or killed them can be cured.[14] Longevity seekers and their lobby are not unique to the United States. In my role as director of the National Institute on Aging, at the height of the cold war, I received enthusiastic communications from the USSR concerning life extension.

The longevity lobby group has its leaders who honestly believe great scientists doing life extension research have been unfairly denied governmental support for their research. They do not recognize the inferior quality of much of the work and ideas of many of these researchers. Some of these scientists, in turn, have deluded themselves concerning the quality of their work. When challenged to do so, they would not submit proposals for funding to the National Institutes of Health, where they would be reviewed by scientific peers who work in the private academic sector just as some of them did. It is not the government that reviews proposals and makes funding decisions but the peers of scientists in the academic community.

Happily, there are some fine scientists who are interested in life extension research. And there is probably some small advantage to the longevity lobby. It helps push the field of aging research along and emphasizes that gerontology should be dedicated to understanding longevity as well as aging. Some longevity lobbyists hold high positions in academia, industry, and government, including some members of Congress, such as the late Senator Alan Cranston, who was convinced that "biomedical research will soon increase not only life expectancy but life span."

However, the longevity lobby has posed problems. Its cadre of scientists, who are outside the mainstream, are perceived by some decision makers and by other scientists as weird. Some members of Congress, for example, worrying about the costs of increasing numbers of older persons, fear and oppose the possible success of life extension research. It is important to differentiate for these key decision makers that the goal of gerontology is to extend the prime years of life, not length of life per se. On balance, the more intense members of the longevity lobby have unwittingly held back the development of gerontology.

Gerontology is not a lunatic field. And longevity is an appropriate goal of scientific inquiry, especially with the tools of molecular genetics now available and the beginnings of regenerative medicine. But the subject matter is only beginning to be developed. And, like all scientific undertakings, there should be social and ethical perspectives that help guide science in what it does, just as society must do for other fields of human activity.

WHITE HOUSE CONFERENCE ON AGING

The White House as well as Congress has shown an interest in aging. The first White House Conference on Aging (WHCOA) was conducted in 1961, drew renewed attention to the issue, and helped set the stage for the passage of Medicare. The 1971 conference supported the creation of the National Institute on Aging. These approximately decennial events promote awareness of both issues and opportunities related to the aging of the population.

The White House conferences on aging in 1961 and 1971 were the most effective. While those in 1981 and 1995 helped sustain interest in older persons, they did not result in new policies or new funding for the field.

The 2005 WHCOA was the first time a president did not appear, although he was only an hour or so away speaking to an upscale audience of older persons about the Medicare Prescription Act. It was ironic, indeed absurd, that the administration killed funding for the training of teachers in geriatrics at the same time that the WHCOA was in process.

The Fourth Branch of Government

White House conferences usually result in the presentation of up-to-date information. Modern scholarship, scientific knowledge, and databases constitute as much a concentration of power as do the legislative, judicial, and executive branches of government. In a sense, we are seeing the beginnings of the rise of a fourth branch of government, which should be formalized. Knowledge serves all the people. Knowledge also serves business, and corporate investment in science in general and gerontology in particular has increased. Certainly, the issues of an aging workforce compel greater corporate attention.

It is essential to ensure the accurate collection of information concerning demography, health, vital statistics, commerce, and socioeconomic status such as poverty, and make a more nuanced, systematic, and discrete analysis of those sixty-five years of age and older and a myriad other categories. To accomplish these goals, we need a separate, non-political branch of government with secure funding.

Parties in power may try to change the way the statistics are compiled. For example, since about 1930 the start and end dates of recessions have been set by the National Bureau of Economic Research (NBER), a private nonpartisan research group. The George W. Bush administration decided not to use the NBER definition.

The statistical agencies take many precautions to ensure confidence in the government data. Almost all the agencies' work is done by career employees protected by the civil service system; the Bureau of Labor Statistics has only one political appointee; the Census Bureau, four; and the Bureau of Economic Analysis, none. Despite the nonpartisanship and professionalism of the employees of federal statistical agencies, public skepticism toward government statistics runs deep. The word *statistics* originally meant "state arithmetic." Objective, credible statistics are essential for government, business, and personal decisions—and democracy itself. Census data determine the allocation of general aid and Congressional representation by state. The census and other agencies should become components of a new fourth branch of government, shielded from politics.

CONTEMPORARY POLITICAL STATUS

There is a disturbing crisis in American democracy, evident in the longevity of incumbents. Only thirty-five races of 435 House seats were competitive in 2004. There were only seventeen competitive races among the fifty states. About 98.2 percent of congressmen were re-elected to the House of Representatives in 2004! Voter turnout has declined steadily in the four decades since 65.4 percent of those eligible cast ballots in John F. Kennedy's 1960 victory over Richard M. Nixon. In the Bush-Gore contest of 2000, only 53.8 percent of Americans voted.[15]

The congressional races in 2006 were somewhat reassuring. The unpopular Iraq War made the races more competitive, there was a greater voter turnout, and the Democratic Party won both houses, the Senate narrowly. But the loss was within the norm of midterm elections in a president's party.

Governments and politicians rarely succeed in dramatically improving the human condition. Rather it is science, technology, public health, and the arts, various forms of creativity, that give us health, prosperity, quality of life, and, yes, longevity too. Of course, governments and leaders can foster favorable conditions.

In general, genus homo sapiens has proven unable to manage his aggression and sexuality, handle conflicts and avoid war, discipline his lifestyle, be tolerant and plan ahead—all behaviors that the word *sapiens* would suggest. Rules and laws, constitutions and covenants, societal norms and expectations have only helped modestly. The Hobbesian jungle remains very much with us. Some argue for a different kind of democracy—a deliberative democracy, negotiating competing special interest groups through debates concerning policy choices and moral dilemmas.[16]

Policy sets the stage for good governance through careful analysis and provides the basis for effective politics. But it is politics, its understanding and manipulation, that ultimately achieves policy changes. Politics reflects the multiple interests in society, achieves compromises, and articulates policies. The growing population of older persons and their many special interests will necessarily advance, but hopefully not at the expense of other age groups.

Older persons and all who would be old—the boomers and generations X and Y, for example—should be civically and politically engaged and should study and pursue the politics of aging and longevity. For example, wealth and political power possessed by some older women have yet to be effectively exercised.

A political activist agenda for aging and longevity would:

- Support equitable distribution of resources across the generations.

- Work to reduce the widening gap of income and wealth and the inequality of longevity.

- Support universal health care.

- Sustain social security.

- Combat ageism through strong legislation and enforcement as well as education.

- Enforce nursing home standards and regulate assisted living.

- Establish geriatric departments or equivalents in all medical schools.

- Dramatically increase the support of the national and local health agencies—a strengthened Office of the U.S. Surgeon General, the Centers for Disease Control and Prevention, the Health Resources and Services Administration, and the National Institutes of Health (which for fiscal year 2007 received only a 2 percent budget increase).

- Register everyone to vote, including older persons. Secure valid voting procedures, including voting machines.

- Promote political grassroots activism of older persons. Encourage older persons to run for office at all levels.

- Advocate for term limits and systematic rotation of the seniority system. Campaign finance reform and term limits are necessary to control stagnant incumbency and corruption. Create an independent nonpartisan commission to avoid gerrymandering.

- Encourage public service through a mandated youth corps and a voluntary intergenerational service corps.

- Strengthen public education. Include economics, civics, the arts, music and humanities, and physical fitness from the life span perspective.

- Create women's and men's longevity groups to raise consciousness of the issues of aging and longevity.[17]

- Support harmonization of social benefits in the process of globalization.

- Protect the human rights of older persons and the disabled (and all persons) against abuse and other forms of discrimination.

- Maintain progressive taxation and estate taxes to avoid dynasties and promote a level playing field for the newborn.

- Support freedom of science (e.g., regenerative medicine) while providing an ethical framework.

- Introduce sabbaticals for all to maintain skills and knowledge and promote career change.

- Strengthen unions, since labor is the true basis of society and prosperity and to counterbalance the excesses of capital.[18]

- Establish an agency and counselor in the White House to coordinate policies concerning old age and longevity across all governmental departments.

- Recruit congressional advocates on behalf of the older population with the intellect, drive, and charisma of Claude Pepper.

CHAPTER 17

POPULATION SOLUTIONS TO LONGEVITY: TRANSITION TO STABLE POPULATIONS

At the birth of Jesus Christ there were about two hundred million people in the world.[1]

At the settlement of America in the 1600s, there were five hundred million people.

In 1850—One billion

In 1950—Two billion

In 1990—5.3 billion

In 2000—6.1 billion

In 2025—8.5 billion (projected by the United Nations Population Division)

In 2050—9.3 billion (projected by the United Nations in 2001)— reduced to 8.9 billion in 2003 as a result of declining birthrates and deaths resulting from AIDS.

Zero population growth has not been achieved. The world's couples average 3.4 children at present.

Historically, population growth was a result of reproduction—when more babies were born, the population boomed. Today, the overall population is growing because people are living longer and at the same time birthrates are decreasing. The population of the world is projected to reach 8.9 billion by 2050, with Africa and Asia experiencing the greatest increases. We are witnessing a historic demographic shift. How nations respond will have a huge impact upon many generations to come.

We do not know the active "carrying capacity" of Earth, for its ultimate ability to sustain life depends upon the future and on as yet unknown technological, political, and economic factors. However, we do know that grasslands are shrinking, that seafood, the world's principal source of animal protein, has declined through overfishing and pollution, that water is in short supply, and that quality of life is a critical issue. The carrying capacity of the earth is profoundly affected by asymmetry between resources and peoples that is the result of geographic realities as well as of manmade events. Illustrative is the greater use of energy and other resources by the developed world compared to the developing world.

An intricate relationship exists between humanity, the Earth's natural environment, and the human-built environment. In the 1980s the field of sustainability science arose, which "seeks to identify the fundamental character of interactions between nature and society."[2] Finding the means to reduce asymmetry would help resolve the problems brought about by the overall growth of populations.

POPULATION CONTROL

Let us begin with a look at the numbers. Table 17.1 details the most populated nations in the world. India, China, and the United States are the top three. Table 17.2 shows the effectiveness of population control in Europe and Japan. Table 17.3 breaks down the success and the failure of controlling infant mortality. These figures make it clear that population growth must be considered in the context of longevity.

Malthusian Theories

The Reverend Thomas Robert Malthus was an English economist, a pioneer student of population studies, a mathematician, and an Angli-can clergyman who lived in the midst of England's Industrial Revolu-tion, when there were fewer than one billion people in the world. In 1798 he wrote his famous essay, *The Principle of Population as It Affects the Future Improvement of Society.*[3] Writing that the drain of popula-tions on available resources can be catastrophic, Malthus contended that poverty was unavoidable because populations increase geometri-cally, outdistancing the means of subsistence, which progress only arithmetically. Personally opposed to birth control on moral grounds, ironically, Malthus lived at a time when expanded (albeit primitive) birth control and lower mortality rates came into being and the indus-trialized West achieved the "demographic transition"[4] to fewer children we know today. Malthus believed three events kept populations in check—famine, pestilence, and war.[5] His theories have profoundly af-fected the way people think about population and economy to this day.

As we see, population control has been quite successful, but because there are still millions of people who are fertile, "population momen-tum" still exists.

Voluntary Depopulation

One obvious population solution is to voluntarily limit the number of children people have to less than two; that is, below the replacement level. The goal would be to attain zero population growth, followed by stabilization *at a level compatible with resources and quality of life,* indi-vidualized to geographic locations. Once population decreases, popu-lation stabilization or simple replacement would be the goal, meaning two children per family.

Negative population growth, or depopulation, is naturally occurring in most industrial societies.[6] With the exception of China, whose strin-gent laws seek to prohibit the birth of more than one child per family, contemporary depopulation in most industrialized societies is not a national or coercive policy but the collective result of decisions by cou-ples. Rosella Palomba at the CNR National Institute for Population

Research in Italy says that "Italian couples feel strong pressure to become parents, but one child is enough to fulfill this social duty." In the United States, an estimated 15 percent of women have intentionally decided not to have children.[7]

Population stabilization at replacement levels (that is, a couple having 2.1 children) would provide predictive power to better manage the economy as well as preserve renewable and, most important, nonrenewable resources, thereby reducing poverty and the degradation of the environment. Successful population planning[8] would reduce struggles over resources that contribute to domestic turmoil, migrations, and war. It would help build stable economies that in turn would promote population stability.

Full-time life equivalents. In theory, were the world to attain a level replacement ratio or *lower overall population growth*, societies could concentrate on enriching the lives of children and upon population aging. Some population control has been successful in Europe, and in the developed world in general, this is happening. An equivalent number of *total human life-years* (TLY) is conceivable if fewer people live longer lives. Since newborns now live to about the age of seventy-five in the developed world, compared to the average life of less than age fifty in 1900, any three older persons, so defined at seventy-five, would today be the full-time-life equivalent of one newborn, who on average would also live to be seventy-five. Predictable, established, longer life-years could become the standard.[9] And as long as health is maintained, the tradeoff would be longer lives for fewer persons without significant new demands upon resources. We would learn much more about the characteristics, needs, and contributions of older populations and begin to identify the optimal age profile of a population and the ideal balance among the generations. Comprehensive multigenerational family planning might contribute to the cultural maturation of societies, with less intergenerational tension.

Population control and wealth. We can make ourselves richer by making ourselves fewer. Increased population is not essential for the growth of an economy.[10] Productivity per capita has increased, and fewer workers are needed to produce goods and services. In addition, with increasing wages the capacity to buy those goods and services increases. The old idea of increasing population to increase productivity is false, as noted by economist Charlotte Muller. This is important because, otherwise, an infinite growth of populations would be required to sustain productivity and prosperity. In fact, three of the nations of the world that have the highest productivity have fewer than five million in population: Norway, Switzerland, and Singapore. The only one that is larger is the United States. Consider also the fact that France and Italy have the fourth and sixth largest economies, yet both have low birthrates.

Furthermore, with less investment needed in extensive infrastructure, from school buildings to housing, more attention could focus on the individual. And there would be fewer mouths to feed using a declining pool of available land. Historically, small "birth cohorts," such as the Great Depression generation, are known to have enjoyed a more favorable life financially than large ones such as the boomers.

Global population control. The 1990 world population was estimated to have been 5.3 billion, and it was projected to reach 6.25 billion by the year 2000. This is equivalent to adding a second People's Republic of China in one decade! The Thai annual population growth rate dropped from 3.2 percent in 1972 to 1.2 percent in 2005, one of the world's sharpest reductions. Most of the world shows a decline in population, except for Latin America and Africa, which account for 9 percent of the world's population; by the year 2100 they could account for 27 percent. AIDS, a tragic form of birth control, could slightly lower that percentage.

By 1988, sixty-seven nations, which account for 85 percent of the developing world's population, considered their rates of population growth too high. The United Nations reports that half the couples in the third world are now practicing some form of modern contraception, compared to only 9 percent in the early 1960s. The fertility rate has fallen from 6.0 children to 3.4 children per fertile woman. Western couples attempting to control fertility are estimated to have increased from about 12 percent in the 1880s to 96 percent in the 1990s.

Population stabilization or reduction provides us with the opportunity to more effectively address other issues in addition to health and longevity—among them are the world economy, the physical and social infrastructure of nations, and the development of equitable relationships among them. While efforts to attain population control are gaining momentum, the question remains how rapidly planning can be put into effect in order to achieve stabilization.

Contraception. Contraception counseling and education are obviously sensitive issues. In addition to the Roman Catholic Church, some developing countries also oppose population birth control, viewing it as a reflection of either racism or the Western desire for dominance or both. They see a growing population as increasing their power. But these views are changing, The 1994 United Nations Cairo[11] assembly on population sought to deal with the soaring world population. Reproductive health services and reproductive rights were central in the final declaration of the assembly, which embraced a new concept of population policy based on advancing the legal rights and economic status of women. It also served to allay the fears of underdeveloped nations that industrialized countries were trying to weaken poor nations by promoting limitations on their populations. Moving away from traditional family planning, the declaration was endorsed by 179 nations.

In the nineteenth century, John Stuart Mill observed that families with fewer children enjoyed a higher standard of living. For that reason, young couples around the world today are opting to limit family size, and, in fact, voluntary depopulation strategy is typical in most of the industrialized world. Couples have fewer children because both spouses must or want to work, because of housing restrictions that prevent them from providing an adequate home for many offspring, or because they do not have the financial means to provide many children with adequate educational and other benefits.[12]

In the 1970s and 1980s, many baby boomers in the United States and Europe lived in a highly sophisticated, well-to-do urban society where maternity and the hard work of child rearing and homemaking was less attractive than a higher quality of life. They learned that a smaller family size was advantageous. Moreover, to live the good life it was necessary for the wife to work. This marked the advent of the two-paycheck marriage. Women in the workplace fed the rise of feminism and vice versa.

OPPOSITION TO POPULATION CONTROL

In many parts of the world a large family remains a status symbol and often is connected to longevity. In the absence or inadequacy of public programs, children become a personal form of social security. In many societies in the developing world, parents purposely give birth to a child who is dedicated to their care in old age. Indeed, there could be an inverse relationship between the availability of social security and the birthrate.

Religious leaders who are opposed to birth control vary in their approaches. American Christian evangelical groups tend to consider the population crisis a myth and family planning a code for forced abortions. On the other hand, the late Pope John Paul II acknowledged that overpopulation is a problem, and although opposed to artificial contraception, he stressed improvements in education and living standards as indirect means to regulate population size.

Pronatalism

Western nations had 22 percent of the world's population in 1950, 15 percent in 1987, and are projected to have 9 percent in 2025. These numbers contribute to fears concerning depopulation that fuel an exaggerated concern over the supposed limitations of an aging workforce and the dependency ratio.

In 2001, *The National Post Saturday Night* (Canada) published an article by Peter Shawn Taylor entitled "Be fruitful, or else. How having more babies can solve all our problems. Population aging has re-ignited pronatalist thinking." And in a letter to the editor of *U.S. News & World Report,*[13] Steven W. Mosher writes, "Humanity's problem is not too many children, but too few children . . . to fill the schools and universities, too few couples buying homes and second cars. In short, too few customers and producers to drive the economy forward and to provide support—through their tax dollars—for the ballooning popu-lation of elderly. It is time to end population-control programs. Why spend hundreds of millions of dollars a year trying to further reduce the populations of dying countries?"

A number of countries have advanced pronatalist policies. For ex-ample, Chad and Kuwait outlaw contraceptives and abortions and pro-vide incentives to have babies. During the reign of Nicolas Ceauşescu, Romania had one of the most repressive pronatalist policies, severely limiting access to family planning services and contraceptives. In Hol-land, the Dutch Calvinist Church is officially pronatalist, whereas the government is not. But both appear to hope for a pronatalist, rather than an immigration, solution to that country's declining birthrate. Singapore[14] cites population aging as a rationale for pronatalism, and Malaysia wants to increase its domestic market and maintain its major-ity ethnic Malay population. France and Finland also have a pronatalist stance.

French demographer Alfred Sauvy (1898–1990) was strongly pes-simistic (to say nothing of ageist) about population aging and pre-dicted that Europe would eventually become a continent of "old people in old houses with old ideas."[15] Such sentiments were already current in France at the end of the nineteenth century, when they grappled with the question, Can we afford older people? Could France support two million people who were over seventy?

In *The Prolongation of Life* immunologist Elie Metchnikoff[16] ad-dressed this pressing issue. It was Metchnikoff's view that society could work to prolong a healthy life and a modified old age. Although written nearly a hundred years ago, his words presage today's national alarms over rising social and health care costs of older persons. He aptly described the widely held misgivings regarding the aging of the population while considering the very real possibility for further increases in human longevity.

The inhabitants of Europe's twenty most populated nations will decrease in number from 449 million in 1993 to 342 million by about 2050, according to the European Population Conference in Geneva—more than a hundred million fewer Europeans in just sixty years. The birthrates of all Western European nations except Ireland are below replacement level. Soon, 20 percent of Europe's population will be over sixty-five.

This has led to lamentations similar to those heard at the turn of the twentieth century in the United States, when voices cried "racial suicide."[17] This not-too-subtle ethnic and religious prejudice reflected the concern of Americans of northwest European extraction that America would be overtaken by persons from southern and Eastern Europe. Similarly, in the 1980s and 1990s the great immigrant waves from Latin America and Asia also prompted hostile reactions. In the United States, most calls for pronatalism are made by "native-born" Americans—a shifting adjective. In 1992, even the majority of Hispanic Americans, the more recent "natives," reported that they did not favor more immigration, and in the 1994 elections, anti-immigrant feelings were strongly mobilized by candidates promoting nativist agendas.

Pronatalism and Women

Can women be persuaded to have more children? It is unclear how effective pronatalist incentives actually are. There were modest increases in sixteen European countries (including Iceland and Italy) from 2004 to 2006. But birthrates have fallen in the Netherlands, Norway, and Austria despite family allowances,[18] tax exemptions,[19] outside help for families, day-care centers, maternity leaves, specific housing for young people, and other family policies. The experience of Western

Europe suggests that these family supports do not increase birthrates markedly or on a sustained basis. (Of course, birthrates could fall further without these policies, and some are excellent in their own right.) For example, despite pronatalist efforts, the native ethnic German population will actually diminish by the middle of the next century if its birthrate remains at its present rate. It has been projected to fall from 82 million to some sixty million by 2050. Parenthetically, there has been an extraordinary drop in the birthrate in former East Germany following reunification.[20]

Pronatalism in Perspective

There are obvious disadvantages to pronatalism. As economist and Nobelist Robert Solow has pointed out,[21] "The initial effect [of an increase in the birthrate] would be to increase the dependency ratio. Second, if the birthrate eventually reverts to the maintenance level, the end result will have been a second baby boom; the new boomers will, for a while, lower the dependency rate when they are all working; they will then worsen it again when they retire and eventually disappear. Faster population growth brings with it an increased pollution, loss of open land, and pressure on food and water supplies. It seems an odd way to solve a different problem."

Incidentally, pronatalist forces seem more interested in producing babies than in what happens to them after birth. Are they fed, clothed, educated?[22] What kinds of lives do they, can they, live?

IMMIGRATION AS A SOLUTION TO POPULATION AGING

If pronatalism does not constitute a prudent, effective, long-term means of dealing with population aging, what about immigration? It is no better. One UN report suggested that massive global migration could stem population aging and save the economies of the rich nations.[23] This proposed solution to offset population aging relies on nations opening their borders and even recruiting young immigrants. It has also been suggested that pronatalism and immigration together could be beneficial. Immigration has become a political, economic, and cultural issue in Europe (as in the United States) with rising nativism and racism.[24] The European Union has cracked down on illegal immigration. The immigration issue is a more violent and politically explosive demographic issue than is aging. It approaches in significance the other great and related demographic issues of race and racism.

Is there a threshold, a level of immigration needed to avoid social tension? The former Socialist president of France, François Mitterrand, once spoke of a "threshold of tolerance,"[25] a phrase he later regretted. Yet, there is a great need for tolerance of differences. All of us have a responsibility to promote ethnic and religious harmony rather than stoke the fires of ancient tribalism. Powerful forces of ethnic and religious tension can overwhelm tolerance, education, law, even intermarriage.[26] Should such a threshold depend upon the time required for absorption in society—education, common language, employment, cultural identification? Can such assimilation happen if the immigrant group refuses cultural re-identification, as is said to be the case of the Muslim immigrant population in France?[27] Should immigration be planned more thoughtfully to avoid concentrations in the cities?

In Sweden, Germany, France, Italy, and the United States, among others, politicians play the anti-immigration card, raising fears regarding jobs, culture, and crime. Demagogues such as Jean-Marie Le Pen in France promote xenophobia. In Germany, the influx of foreigners prompted a rising tide of right-wing violence by neo-Nazi skinheads. In one episode Germans burned Turks. Already by 1994, 7 percent of the German population were immigrants and 10 percent of all marriages were between a German and a foreigner. But former Chancellor Helmut Kohl once said that "Germany is not an immigrant country," and resident foreigners are not considered immigrants. Ethnic Germans who have lived in Russia and Eastern Europe for generations are accepted in Germany. Yet, only about 2 percent of Turks, Serbs, Croats, Italians, and other non-German immigrants have become naturalized.

In truth, Europe cannot easily become an effective fortress against immigration, especially with immigration pressures from the former USSR, Eastern Europe, the Balkans, Turkey, and North Africa. Moreover, as noted, Germany and the other nations (except Ireland) have falling birthrates. Despite unemployment rates, employers seek cheap labor from abroad. Reflecting these realities, in 2000 the European Commission issued a policy paper stressing the need for more proactive immigration to ensure the continent's future economic growth and the viability of its social welfare system, and in 2001 the European Union's policymaking bodies prodded member states to take the lead in fighting racism and xenophobia.

Some see immigration as a costly burden. In fact, it has proven economically valuable *in the longer term*, although there may be significant short-term costs. Immigrants are willing to bring in the harvest, wash dishes, and make clothes. They pay Social Security, income, and other taxes. Whether legal or illegal they receive fewer social benefits. Immigrants are more likely than the rest of the population to start their own businesses, such as restaurants, and to hire workers, buy goods and services, and pay taxes. Citizens of host nations worry about immigrants' overwhelming needs for health services, education, and welfare, and some politicians exploit the image of immigrants exhausting tax dollars, utilizing it to attack all entitlements. Immigration works best when national birthrates are falling and there are worker shortages—and when there is dull, dirty, difficult, dangerous, and/or hard work to be done. Finally, however, immigration offers only a temporary solution to the challenges of population aging, since immigrants also grow old. (And since they are usually young, their departure adds to the median age of their country of origin.)

It is difficult to create population and immigration policies appropriate to individual nations independent of international considerations,[28] especially with massive migrations of peoples seeking economic opportunity, religious freedom, and political asylum, or fleeing the three Malthusian demons of war, famine, and pestilence. Recipient nations may become engulfed in civil unrest, always prompted by xenophobia near the surface, and emerge with economic downturns and social insecurity, as occurred in the 1990s in Europe, especially in France, Germany, Great Britain, and Italy, as well as in the United States.

The courage of individuals who uproot themselves and seek to build new lives in strange cultures is admirable; however, such free movements should not be the consequence of unbearable political and economic conditions. The industrialized world needs to factor in its decisions on aid to the developing world with the dangers to itself if massive tides of migrants to their shores continue unabated. A 1993 UN study reported that 2 percent of the world's people—over one hundred million—were in the process of migration. And the majority were not migrating to the industrialized world but to neighboring countries. But when the educated and ambitious migrate to the developed world, this becomes a brain drain, which only worsens conditions in the native country.[29]

Neither pronatalism per se nor immigration per se is a sound policy to solve the challenges of population aging (on a national or international basis) over the long term.

EUTHANASIA

Another theoretical solution to population growth would be age-based active euthanasia. In his farewell address to Johns Hopkins University Medical School in Maryland, the famous physician Sir William Osler (1849–1919), an icon in American medicine, suggested ironically that persons over sixty be retired, given a year of contemplation, and then be chloroformed. Based upon British author Anthony Trollope's novel *The Fixed Period,* this famous speech was undoubtedly intended to provoke thought rather than advocate policy. Today, however, ethicist Daniel Callahan has suggested age-based medical rationing that ignores functional status. In Kurt Vonnegut's novel *Welcome to the Monkey House* overpopulation leads the world government to encourage ethical suicide.

Global Considerations

Let's consider the impact of population aging on a number of different cultures.

Japan. The world's fastest aging society will have thirty-one million people over sixty-five in 2015, or 25 percent of its overall population. Since 1947, when the average total fertility rate (TFR)[30] of Japanese women was 4.5, the TFR has been declining. In 2000 it dropped to 1.32—one of the lowest in the world. Currently, it is on par with Spain, Germany, Italy, Poland, and Russia, and below that of the United States (2.1), France (1.9), Australia (1.7), and the United Kingdom (1.6). The Japanese government and some leading Japanese thinkers[31] regard declining fertility as a national crisis, and in response the government has adopted the Angel plan to encourage marriage and childbearing. Economic, cultural, and political, as well as personal motivations intertwine in national birth customs and policies. Factors that come into play include concerns over a labor shortage, immigration, and the greater independence of women.

In Japan, unlike in many industrialized nations, immigration is minimal. Since total fertility rates are falling, maintaining a stable population becomes an important issue. There are some Koreans and other foreigners in Japan, but only *nikkeijin,* or foreigners of Japanese descent,[32] may work there as unskilled laborers. The foreign-born account for 1.4 percent of the country's population. Japan still remains a nearly classless monoethnic society, but it is involved in its first serious experiment with immigration. The government is under pressure to bring in foreign guest workers to fill manufacturing jobs that Japanese do not want, as well as to address a shortage of service workers. Overcoming prejudices against foreigners and opening the doors to immigration could be a matter of national survival. In the meantime, the age of retirement has been extended, and many older Japanese continue to work.

The Former Soviet Union and Soviet Bloc. In the world at large, birthrates fall in climates of economic uncertainty among reasonably educated people who have access to birth control. Sudden reductions in birthrates follow severe economic disruption as well as war and famine (shades of Malthus), which is what occurred in the former Soviet Union and among the former Communist nations of Eastern Europe. These countries are experiencing political and economic instability and demoralization, and some social benefits designed to encourage births are disappearing. There is also a trend toward more deaths than births and a fall in marriage rates, especially in Russia, Belarus, and Ukraine.

The United States. The fastest growing nation in the industrialized world will double its population in 145 years, due to immigration as well as its birthrate.[33] It is projected to have a population of 570,954,000 by 2100.[34] The United States has the highest fertility rate among rich nations. Between July 1, 2002, and July 1, 2003, its population grew by 1 percent (2.8 million people). Using one set of assumptions, the Bureau of the Census projects that 21.8 percent of the population (65.6 million people) will be age sixty-five and over by the year 2030. When all the members of the baby boom generation have died, nearly 25 percent of the population will still be sixty-five and over. The United States faces a permanent increase in the median age of society. This assumes no dramatic increases in either fertility rates or immigration, and no wars, plagues, or other disasters.

Some Americans worry that the melting pot isn't melting at all but has become an ethnic mosaic of distinctive, insular groups. Some oppose the retention of immigrant identities by the use of hyphens. On the other hand, 1991 data showed that 80 percent of Italian Americans and Polish Americans and 45 percent to 50 percent of Jewish Americans were marrying outside of their groups. Among Hispanics and Asians, one-third of the present generation was intermarrying. Japanese Americans have a high rate of intermarriage and one of the lowest birthrates in the United States, and there are only a small number of Japanese immigrants each year.

China. Without the Draconian one-child-per-family policy it started in

1979,[35] by the year 2055 China would have acquired an estimated population of 5.2 billion—approaching today's world population! If it did not control its population, it would suffer with famine and disease, as in its painful past. The Chinese had an average of six children per family in 1965. In 1994 they had 1.7 children per family. China met its own goal of not exceeding a population of 1.2 billion by the year 2000. Researchers forecast that by 2040–2060 people sixty-five and over will have reached 22 percent in China and will still be increasing. It is estimated that the proportion will reach its peak in the 2060s.

Today, pressure to control the birthrate has eased, especially in the countryside. If China's one-child-per-family policy increased longevity, the family pyramid would be turned upside down—with one child caring directly and indirectly for up to possibly six adults (parents and grandparents). The government will need to create new forms of family support and substitutes to care for the fewer children and for frail older persons. China's children born after 1979 have been called the generation of little emperors.

We are seeing rising numbers of persons of all ages living alone, especially older women and single young people in China and throughout much of the world. According to China's fifth national census of 2000, 7 percent of its population is sixty-five or older, which meets the UN criteria for an aging population.

POLITICS OF NATALITY AND IMMIGRATION

Political manipulation of natality and immigration goes back to ancient times. The decline of Rome may have been due to uneducated Germans who slowly immigrated into the Roman Empire[36] and outbred the small families of educated Romans.

Ethnic groups within some nations want to sustain or increase their birthrates. We could call this political pronatalism, and one can find many examples in the world today. For example, pronatalist Québec presses French-speaking families to have more children. The Orthodox

Jews of Jerusalem, the Israeli Arabs, and the Palestinians of Gaza all have much to gain by population growth.

Population Stabilization: The Ultimate Balance

The promise of longevity depends upon a healthy, vigorous family supporting healthy and welcome pregnancies and deliveries. This implies the presence of excellent prenatal care and nutrition and the absence of disease, alcoholism, addictive drugs, and tobacco abuse.

Family planning policy is not just welfare or humanitarian aid, but of developmental importance for the individual and society. It is a vital instrument for a nation that aspires to achieve equal opportunity and a prosperous economy, and a necessary force to assure worldwide population stability. Nations with advanced family planning (contraceptive services) and empowerment and education of women enjoy both greater prosperity and greater longevity.

In the unlikely prospect that nations and religious groups would suddenly agree, voluntary reduction and stabilization of population could not be put into operation instantly because of *population momentum*. The present (even a smaller) number of fertile couples will keep the population growing in any case. Present immigration levels are influential in the United States, for example, and they now make up about 25 percent of the growth rate of the population. From the new global perspective, no nation can isolate its population from the larger world. Because of population momentum, voluntary depopulation by lowering the birthrate is all the more necessary.

Population considerations cannot be divorced from the so-called four Es: economics, energy, the environment, and esthetics. Economic development needs to be balanced against the depletion of energy and the environment. Development and growth consume resources. How many terawatts of total global energy can we afford without destroying ecosystems? These factors bear upon the interest the industrialized world should show in assisting the developing world in family planning. More and more developing nations perceive the importance of population stability and family planning to their own welfare.

The issues of sovereignty and international security within the global economy are relevant here. Multinational corporations transcend boundaries and put nations and national identity in question. Both capital and labor have become more mobile, producers and consumer markets even more intricately interdependent. The availability of social benefits and general social circumstances are powerful incentives to the mobility of both capital investment and labor.

If further breakthroughs lead to longer life, a new kind of population challenge will emerge—indeed, this is already happening. Those interested in a longer life for themselves and their progeny had best support family planning and population reduction and stabilization. They must support the extension of the worklife and promote employment. They must recognize and exploit the economic opportunities that the growing market of older persons offers. They must support research and methods to equalize the life expectancy of the sexes. They must explore the implications of *longer life-years*.

Conventionally, we think of the newborn as the future, a new start or renewal, and a critical social investment. The corollary view is that the older person is less or no longer valuable. In fact, the old represent years of investment, experience, and knowledge; the majority constitutes an extraordinary national resource. Children involve an expensive, new, and uncertain investment in education and health care. (Fortunately, because of child labor laws in many countries, they cannot be put to work as they were in the past.)

Futurists, among others, can argue that the ultimate method for dealing with increasing population, including the rising numbers of older persons, is the colonization of space. But we are a long way from this questionable and fanciful solution.

In truth, as mentioned, we do not need as many people to accomplish the same amount of work; for example, we can feed people with a minimum number of farms and farmers. Many industries have become more efficient due to automation and other technology. There are always transitional problems, but in the long run scientific and technological progress generally augurs the greater good for the greater number. Standards of living rise. Inventions may spur consumer demand for further productivity. The invention of the cotton gin, for example, did not ultimately result in the loss of jobs but in new products consumers desired.

Certainly, there is no shortage of work to be done to maintain and expand the quality of our lives—better housing, cleaner environments, and improved educational programs are examples.

Voluntary depopulation with stabilization brings predictability and a better balance between humanity and nature. It would also help overcome complex issues that follow from worker dislocation and the changing age profile, such as funding Social Security and occupational pensions. It would offset unemployment and social conflict, reduce disease and starvation and unhealthy urbanization. At the biologic level, geneticist James V. Neel believed there should be no more than two children per family to stabilize the gene pool.

One scientific strategy should be made an international priority for predicting the timing of ovulation. Physiological measurements such as of the luteinizing hormone surge, or better markers of ovulation, could be helpful to those whose religion opposes contraception other than the rhythm method.

Those countries that worry about declining populations, such as Italy, Germany, and Japan, should take heart that their dwindling numbers may further enrich their people, their culture, and quality of life. Neither unemployment nor worker shortages need be a reality any longer. Zero population growth would help balance population size with manmade and natural resources. Plato advised restricting population within the means of subsistence as one means of keeping peace. I would not favor either state-sanctioned efforts to measure the proper population balance or to prescribe birthrates. Voluntary and easily available family planning, coupled with national goals, is a different matter. In fact, most of the world's governments have policies pointed toward the reduction or stabilization of their populations.

Especially in the nineteenth century, Europe's overpopulation could go to the new world of the Americas to make new homes. Between 1870 and 1910, 40 percent of Italy's labor force and 45 percent of Ireland's immigrated there. With the marvels of modern transportation and communication, it is easier in some cases to move back and forth across lands and oceans and enjoy the special opportunities that varied cultures and economies provide citizens of the world. But many families are torn apart in the process.

At present, the world's population is doubling every generation. The wretched of the earth are experiencing widening disparity in economic status. With voluntary depopulation, the developing world could join the developed world in escaping the Malthusian demons—and also make way for the new age wave and enhance the quality and length of life.

The intersection of birthrates, death rates, and immigration constitutes the age structure of a society and attracts powerful cultural, religious, political, economic, and tribal concerns. Population aging and declining fertility rates already constitute a critical geopolitical issue, their global reach entering every bedroom and family hearth.

Of the various solutions presented here, the one most sensitive to theological and cultural diversity and personal feelings is voluntary family planning that results, in some instances, in depopulation and in reasonably predictable population stability. *The goal is a natural replacement of population.* This is happening in much of the world. Family planning that does not involve abortion but uses scientific means of identifying hormonal rhythms and birth spacing is the most desirable solution. Especially important to it are the rights of women to education, access to health care, jobs, and property rights so that they can control their reproductive capabilities.

The continuing growth of science and technology requires an educated, healthy, smaller workforce, which, in turn, furthers productivity and prosperity.

These ideas are essential for people to fully enjoy the triumphant prolongation of life.

PART V: CAUTIONS

CHAPTER 18

WORLDWIDE DEMOCRATIZATION OF LONGEVITY: OVERCOMING FAMINE, WAR, AND PESTILENCE

The revolution in longevity that began in the twentieth century has not been fully realized in the developing world, for not everyone enjoys *equality in longevity*. The developing world as a whole has a life expectancy of fifty to sixty-two years; in the developed world, the average is more than seventy-five years. Infant, childhood, and maternal mortality are largely responsible for this disturbing inequality, with millions of "years of life lost in total worldwide."[1] In this chapter I will discuss the inequality of longevity around the world; in the next, the threats to longevity that affect us all.

It is in the best interests of the developed nations of the world to take action against the massive social, economic, and health problems confronting them. Rich nations ignore the plight of poorer nations at their own risk. Environmental threats, such as global warming, can affect nature's food chain, and what damages the plankton at the bottom of the sea can damage us all. The disabling and fatal diseases of the developing world also destroy the overseas consumer markets of the industrialized nations—another powerful reason for the developed world to be attentive. (See Table 18.1.) The developing world could become both productive and a growing market for the developed world. To be valuable customers, the peoples of the poor world must prosper.

How might the family of nations join together to promote the health and longevity of all people? How might this be in the common interest? The developed world could, if it would, do a great deal to confront many preventable and treatable diseases.

Although, of course, great progress has been made in combating medical conditions, remnants of the neolithic can be found in many parts of the world: A thorn that becomes lodged in a person's foot can still kill. A baby can still die when the placenta precedes it at birth. In neolithic villages diseases that killed were seen as supernatural occurrences. There are places in the world where preventable illnesses are still perceived as mysteries.

Although the rapid trafficking of viruses and other pathogens by way of modern transportation and immigration has prompted the industrialized nations to offer some aid to poorer nations to prevent pandemics, new or reemerging infections, from AIDS to cholera to malaria to gastrointestinal and respiratory diseases to TB, still demand greater attention.

And while infections and parasitic disease remain leading killers in the developing world, the disorders of longevity, such as heart disease, stroke, and cancer, are becoming major killers. Chronic diseases such as diabetes account for more than twenty-four million deaths annually, almost half of the world's total, most of which occur in the developing world, with tobacco and other lifestyle factors, such as fat-rich diets and the sedentary life, playing a significant role.

And there are still other public health issues that contribute to the global inequities in longevity, including mental illness, violence, malnutrition, and famine.

Mental Illness

In 1996, the Harvard School of Public Health reported on a study coordinated with the World Health Organization, the World Bank, and health officials in a number of countries. Noncommunicable disease and accidents were found to be replacing infectious diseases and malnutrition as the leading causes of preventive death and disability. A ten-volume survey entitled *The Global Burden of Disease and Injury Series*[2] found that mental illnesses (depression, alcohol dependence, and schizophrenia) are hidden diseases and impose a major burden. In this study, disability was factored in, not just death. While psychiatric conditions are responsible for little more than 1 percent of deaths, they account for almost 11 percent of the disease burden worldwide. The concept and measure called a Disability-adjusted Life Year, or DALY, is used to measure the burden of disease and injury. (See Table 18.2.)

Violence

In 1996, the World Health Assembly declared violence "a major public health issue."[3] Violence killed 1.6 million people in 2000, matching tuberculosis, according to a World Health Organization report. Suicide accounted for half of the violent deaths, homicide for 30 percent, and war for 20 percent. Violent fatalities represent about 3 percent of all deaths in the world. The scope of violence against women, children, old people, young men, and communities in general is extraordinary.

According to one UN estimate, between 113 million and 200 million women are missing. This gender gap is the result of the abortion of female fetuses and infanticide in countries where boys are preferred; a lack of attention that goes instead to brothers, fathers, husbands, and sons; so-called honor killings; and domestic violence. Each year between 1.5 million and three million women and girls are lost to violence, especially those from fifteen to forty-four.

Despite the overpowering extent and impact of violence as an expression of pathological anger and rage, neither the International Classification of Disease nor the Diagnostic and Statistical Manual of Mental Disorders (of the American Psychiatric Association) recognize anger/rage, the basis of which is usually powerlessness, as a mental disorder. Of course, controlled anger in a legal form is justified and healthy when provoked by injustice.[4]

Malnutrition and Famine

Hunger is still a major killer. Some twenty-five thousand die every day because of hunger and poverty and over eight hundred million people do not have enough to eat. Fifty-four nations do not produce enough food to feed their own people. In 1993, the World Summit for Children sought to establish several realizable nutritional goals: the elimination of vitamin A deficiency[5] (this nutrient is essential to the immune system and for optimal vision. Each year 250,000 to 500,000 children go blind as a result of vitamin A deficiency); universal iodization of salt to prevent iodine deficiency; and encouragement of breast feeding by eliminating the practice of supplying infant formula for free or at low cost, where possible. While there has been progress, the goals have not been reached. The United Nations Children's Fund battles iodine deficiency, which can cause brain damage, mental retardation, and deafness. Some six hundred million people, many in China, suffer from this deficiency, and many of them with goiter.

THE UNITED NATIONS

It is impossible to imagine overcoming the Malthusian demons of pestilence, famine, and war without an effective global union of nations. At the United Nation's founding in 1945 there were fifty-one member nations; by 2005, there were 191. Most UN financial and personal resources assist the poor. Ninety percent of program funds in the United Nations Development Program were allocated to sixty-seven countries that are home to 90 percent of the world's impoverished people.

The UN has had major accomplishments since 1945, taking on global issues that include armed conflict, genocide, starvation, refugees, disease, the environment, human rights, management of decolonization, nuclear proliferation, population, and the protection of women and children. Another major achievement was its help in abolishing South African apartheid.

Excellent, high-performance UN agencies include the High Commission for Refugees, the UN Development Program, the World Food Program, the International Atomic Energy Agency, and the World Health Organization.

In 1948, the United Nations established the World Health Organization (WHO), stating that "the health of all peoples is fundamental to the attainment of peace and security and medicine is one of the pillars of peace." The present WHO budget, which covers the entire world, was only $1.65 billion in 2005, less than 5 percent of the budget for the National Institutes of Health. Unfortunately, the WHO has not recognized population aging as a major priority.

Each year the WHO makes recommendations concerning the composition of the vaccine for the forthcoming flu season. It does this work through a worldwide network of 118 national influenza centers in eighty-one countries. These centers isolate the virus from suspected influenza cases and determine what type it is in order to follow the spread and the gravity of the disease. There are three major WHO collaborating centers, one each in Australia, the United Kingdom, and the United States.

If the vaccine against strain A(H5N1) of the avian influenza virus announced in 2005 is as successful as believed, the critical factor will be time—the time necessary to produce the vaccine and distribute it if there is a pandemic. This becomes a political as well as an economic and logistical issue. A possible worldwide pandemic of this seriousness requires an organized worldwide response, which, in turn, necessitates political agreements and special concern for poor countries. The vaccine industry today cannot produce the needed amount of vaccine. Also needed is a massive supply of an antiviral drug, Tamiflu. Finally, the strain A(H5N1) could mutate, making, in the worse case, the present vaccine worthless, thereby forcing even more rapid responses.

New protocols with respect to the environment are also needed and could follow the reasonably successful model effort[6] against the chlorofluorocarbons (CFCs) that cause the hole in the ozone layer. The World Trade Organization should serve the continuing expansion of guided free markets and regulated capitalism. International efforts must continue to combat racism, for which there is no scientific basis.[7]

The Graying of Nations

Although there is an unacceptable inequality of longevity among nations, nearly all are graying (Table 18.3). The over-sixty group is the most rapidly growing in the world: its population was seven hundred million in 2007 and is estimated to increase to two billion by 2050.

Infant mortality is the main, but not the only, basis for this disparity in life expectancy around the world. The world's highest life expectancy is 81.9 years in Japan and the lowest is 34 in Sierra Leone, according to the UN population division. Currently, the six nations that possess more than half of the world's oldest-old (eighty years or more) are China (twelve million), the United States (nine million), India (six million), Japan (five million), Germany (three million), and the Russian Federation (three million).[8]

The continuing increase in both world population growth and in population aging is a geopolitical issue of immense importance. In 1970, UN representatives from Malta first encouraged the UN to consider the growing numbers of the aging in the world. This initiative led to the first UN World Assembly on Aging in 1982, a two-week conclave held in Vienna, Austria. Attended by representatives of 123 nations, its final report detailed an excellent international plan of action, which, with the exception of Japan, governments generally ignored. One of the few tangible, if modest, results of the assembly was the creation of the UN International Institute on Aging, which has, among its other activities, assisted in training personnel to provide services to the aged in the developing world.

Peace Making and Peace Keeping

However incorrect the Malthusian view that population is *inevitably* checked by famine, war, and pestilence, all are significant destroyers of human life. War, for example, is always with us. The historian Will Durant calculated that there have only been twenty-nine years in all of human history during which there was not a war in progress somewhere. War would appear to be normal, and peace aberrant. On average, nearly half a million soldiers have died each year in wars during the twentieth century, raising the war-related military mortality rate from 19 per million population in the seventeenth century to 183.2 per million population in the twentieth century. Eight million died in the First World War and twenty million in the second. Along with this mountainous rise in military deaths, war-related civilian deaths accounted for 90 percent of all deaths in warfare in the 1900s, up from 14 percent in World War I to 67 percent in World War II.[9] Fifty-five million people across the globe died in World War II.

The development of a world tax or a weighted investment by nations to support world peacekeeping and peacemaking forces and to provide minimal universal social protections to help overcome root causes of conflicts would require individual nations to give up some degree of sovereignty and resources. Currently, only fifteen nations provide 80 percent of the UN budget. Assessment should be proportionate to a nation's share of the world economy,[10] but in 2004, the United States gave only 0.16 percent of its gross national product to foreign assistance. France spent 0.42 percent, Germany 0.28 percent, and Japan 0.19 percent. The Nordic countries have made the greatest financial contribution per capita to the UN for all fifty years of its existence.

New forms of additional financing, such as sharing in the mineral riches in the world's common deep-sea beds or a levee on foreign-exchange transactions and international air tickets, should be adopted.

Such a proposal is compatible with the 1945 UN Charter's preamble, which states the primary UN mission as "to have succeeding generations, free from the scourge of war, which twice in our lifetime has brought untold sorrow to mankind."

Perhaps a world volunteer peace force, originally suggested by the first UN secretary general, Trygve Lie of Norway, or an expanded professional voluntary corps, such as the French Foreign Legion, might be appropriate. This might obviate the need for individual nations to assign any of their own military.[11]

The United Nations celebrated its golden anniversary in 1995 under a dark cloud of criticism about Somalia, Angola, Bosnia, and Rwanda, and about its expensive, inefficient bureaucracy. Some criticism of the United Nations is justified, and reforms are mandatory; however, it is simplistic and hypocritical to blame the UN for failed results when member nations have not given it the power, structure, funds, or the tools needed to carry out its extraordinary, original mission to maintain peace. Member nations are reluctant to give it sovereignty or, for that matter, to pay their dues on time.[12]

The UN bureaucracy and management definitely require streamlining. The discredited Human Rights Commission is being reconstructed as a council, hopefully denying membership to countries with egregious human rights records. A peace-building commission is also a necessary reform.

In fact, the UN has proven useful in both peacemaking and peacekeeping. Its blue helmets have been kept busy. There have been successes at peacemaking in El Salvador and Guatemala (though there have been failures as well, painfully illustrated by Rwanda). The United Nations' success ultimately depends on all of the world's nations.[13]

Ideally, the UN should belong to the people, to a world citizenry. It should work more closely and be more responsive to the nongovernmental agencies, NGOs, which should exert greater influence independent of governments. Mostly run by women, NGOs are the world's grass-roots energy for change. They should enjoy special tax status comparable to the 501(c)3 in the United States and be given broad opportunities for advocacy and political activities. They could work to improve the UN's public image, push reforms, and help the WHO and other agencies foster health and longevity. NGOs such as Doctors Without Borders and HelpAge International play extraordinarily constructive roles.

In addition, the UN must grow beyond its origins and reflect the rise of the developing world in the globalization of political life. There must be changes in the structure of the Security Council to reflect the growing power of the developing nations. The last Security Council expansion was in 1965. It is only a matter of time before the developing world, which represents 75 percent of the world's peoples, secures more power in international affairs.

The UN and National Sovereignty[14]

The original concept of sovereignty was as a self-governing, independent, supreme individual, such as a divine monarch. In modern times, sovereignty has come to refer to a state and not an individual. Now the concept may move toward an expression of collective humanity. Even as nations insist upon promoting their own unique sovereignty, they want to influence and even intrude upon each other. The United States, for example, wants to curb human rights abuses; the former Soviet government wanted to export communism. International law, by necessity, affects sovereignty through a variety of treaties, all of which are efforts to restrain the power of sovereign nations for the overall common good.[15]

People's fear *and* their experience leads them to distrust international government and its processes, just as history has taught them to be wary of their own governments. Switzerland and Norway worry about losing independence should they join the European Union. Intrusive human rights efforts offend China. World Trade Organization rules make evident the problems of sovereignty. Rich nations do not abide by the rules of free trade in their dealings with the developing world. If there is to be free trade, U.S. hormone-fed beef would presumably be admissible in Europe, which its peoples do not want. Of course, public health concerns must be respected. Continental Europe denied the import of possibly mad-cow-diseased beef from Great Britain in 1996.

Some multinational corporations are richer and more powerful than some countries and, even within major countries, they have great power and influence. One example: Citigroup has positions in over a hundred nations outside the United States, and 40 percent of its operating income in the first three-quarters of 2005 came from there.[16]

National sovereignty is not going to be given up easily and soon. But over the next century, imaginative new geopolitical arrangements and global institutions will have to be put into place to build a new international order that combines the best of a regulated market economy with social protections as well as protects distinctive cultures, peoples, and local governance. Longevity will benefit.

Human Rights

There is much concern, as well there should be, for human rights in many parts of the world, but too little attention has been given to the basic rights of all persons to health and longevity. Although the Universal Declaration of Human Rights in 1948 expresses the deepest of human aspirations, it lacks any mention of the rights of older persons. In 2002, at the second UN World Assembly on Aging in Madrid, I wrote and presented such a document.[17] (See Preamble to The Declaration of Human Rights of Older Persons at the end of this chapter.)[18]

The Convention on the Rights of Children was approved in 1989 but was never ratified by the United States and Somalia, the only holdouts. The universal declaration of human rights may not immediately be restructured, but further radical changes are in order. The growth of citizen rights with a new bill of human rights should evolve first: that one has a right *not* to lay down one's life for one's country (conscientious objection), to be a citizen of the world, to health care and to education. Such rights promote longevity.

POVERTY AND FAMINE

Nearly one out of five persons in the world lives in poverty—over one billion persons. Seventy percent are women. A wealthy person lives twice as long as a poor, sick one. Twenty percent of the world's population receives 83 percent of the world's income, and there are between 120 million and two hundred million people unemployed or underemployed globally. Every year in the developing world, millions of children die, mostly from preventable causes, many slowly and painfully from chronic starvation. Julia Alvarez, the former Dominican Republic ambassador to the UN, has said that "if you want to age, you have to eat." Fifty-two percent of Indians live on incomes of less than a dollar a day, and one-third of the world's poorest people live in India. Nearly 75 percent of all Indians live in the less affluent countryside.

In the megacities of the developing world, thousands of children live and work in the streets without family support. They may be afflicted by malnutrition and disease and subject to drug abuse, prostitution, and criminal exploitation. The global Dickensian portrait is disturbing. There is a huge underclass in the world, still living the lives that Thomas Hobbes described as brutish and short. The world's landless farmers and peasants only vaguely understand the idea of property. They are undereducated, ill, and disabled, often demoralized, and enraged, dying without leaving a trace, but also endangering the unsuspecting rich and powerful of the world. Diseases can all too easily spread. Social unrest can result in random terrorism. Out of self-interest, developed nations should lead a worldwide public health campaign to protect their own citizens and promote health and longevity for all the world's peoples.

One successful means of confronting the poverty of the developing world was introduced by the Nobelist in economics Muhammad Yunus. He invented the concept of microcredit for the poor in Bangladesh in 1976. It is based on the premise that poverty is not a personal problem caused by laziness, sin, or lack of intelligence but a structural one due, in part, to lack of capital. The Grameen Bank enabled enterprising individuals to start their own businesses, and this idea has been tried in fifty-one nations around the world.

POPULATION

Although population control has advanced somewhat through family planning and the empowerment of women and through education and economic opportunities, Africa, Asia, and Latin America still have exponential population growth. There is a growing water shortage in eighty nations with 40 percent of the world's population, and this results in food shortages and health crises as well as armed conflicts over what water and food is available.

Countries that worry about declining population should take heart. Depopulation is not the unhappy fate of Europe and Japan, but possibly is their salvation. Successes against diseases as well as the population control and economic productivity that are characteristic of the industrialized world have promoted the revolution in longevity *and* have reduced poverty. People do make themselves richer by making themselves and their families fewer.

Japan, Korea, and most of Europe have reached the so-called demographic transition,[19] where *both* the birth and death rates fall, probably due to increasing income and education, as captured by the phrase "rising standard of living." China is also approaching the demographic transition that marks the beginning of depopulation. The term *demographic transition* is technical and yet gentle. It really means that a particular society now has sufficient income, social protections, and secularization so that it does not need babies to grow up to work in the fields and provide security for their parents' old age.

Migrations

Along with religious fundamentalism, racism, and tribalism, movements of peoples have led to or followed conflicts and wars. Migration pressures from Eastern Europe, the former Soviet Union, and North Africa into Europe continue. The United States is the destination of throngs from Latin America and Asia. But what happens at the new address? What is the welcome? As the world becomes more and more global, all are endangered, even the rich and powerful, by rising expectations and mass migrations of peoples and diseases. Of the more than two hundred nations, perhaps no more than thirty have achieved some measure of successful economic development.

Another worldwide demographic issue is the emigration of those out of nations previously subjected to colonization, imperialism, and humiliation. They enter Europe and North America to seek their economic fortunes and, at times, political asylum. It has not been easy for the privileged, industrialized nations to assimilate the rich, diverse cultures and religions that immigrants bring with them. In turn, it is painful and courageous of people to leave their ancestral homes. People-smuggling across borders is now big business. Immigration has become a powerful political issue, exploited by demagogues who play on social insecurity, on the worries of working class people who have suffered stagnant and often paltry wages, and on the fear and reality of unemployment. Still another great demographic issue is the continued ancient tribalism, marked by bloody conflicts, that we have seen in the Balkans, the former USSR, Sri Lanka, the Near East, and Africa. Tribalism, religion, and racism continue to tear apart the human family. Since 9/11, fear of terrorism has become a realistic concern.

Finally there are family matters. As pointed out in "Families in Focus," a report issued by the Population Council, changes in family life are not unique to the United States, but are common, worldwide trends—the growing numbers of persons living alone, rising divorce rates, cohabitation rather than formal marriage, smaller households, and feminization of poverty. Suicide rates among young people appear to be rising in both developed and developing countries. America leads among many industrialized nations in poverty, divorce, and poor maternal health.

The 1995 UN Fourth World Conference on Women in Beijing adopted a platform of action. Its preamble is of enormous importance for women of all ages. Violence against women was a predominant issue. Although slavery was abolished some two centuries ago, women are still treated as chattel and almost as slaves in many societies. They have not fully achieved freedom from ecclesiastical authority, ancient tribalism, laws, and marital control.

Urbanization

More than half the world population, 3.3 billion people, lives in towns and cities, according to the UN Population Fund.[20] Nearly one billion live in slums, without adequate water, sanitation, and power. Megacities, defined as having populations of ten million or more, include Buenos Aires, Calcutta, Mexico City, Sao Paulo, and Seoul. Greater attention must be paid to cities to ensure quality of life.

The Brain Drain

According to one study, one hundred thousand nurses have left the Philippines to work overseas since 1994, half of them since 2000, principally to care for the sick, aging populations of the developed world. The rich countries should be paying nurses and nurse's aides more and providing career ladders, according to the theory of the marketplace. By 2005, the health care system in the Philippines had almost collapsed.[21]

There has also been an exodus of doctors from Africa and the Caribbean to four wealthy, English-speaking nations, the United States, Britain, Canada, and Australia. These countries depend on international medical graduates for a quarter of their physicians, because they do not train enough doctors. Ghana has only six doctors for each hundred thousand people, having lost three of every ten it has educated to these countries, each of which has more than 220 doctors per hundred thousand people. The United States has about seventeen thousand medical school graduates each year to fill twenty-two thousand first-year slots for residents.

Debt Cancellation

In his report entitled *North-South,* Willy Brandt,[22] the former German chancellor, urged overcoming the disparity between the rich and the poor nations of the world and laid out a plan of action. The wealthier nations, like it or not, will eventually be forced to share their riches. The World Bank and International Monetary Fund were invented to achieve greater economic equality in the world but have had only limited success. The demands of these institutions for austerity in indebted nations often create shocking conditions, especially affecting older persons *and* children, in societies most vulnerable, whose benefits are cut back in the name of budgetary efficiency and credit worthiness. An important goal is debt relief without severe austerity. Robert Fogel argues the United States and other OECD nations should transfer funds to the developing world.[23]

In 1998, the World Council of Churches called for canceling the foreign debts of impoverished nations. The council joined a growing number of religious leaders and some secular groups who wanted to tie the millennium to the ideal of a jubilee year. The Book of Leviticus has a visionary description of such a period, which would occur every forty-ninth year, in which slaves would be freed, debtors released from their obligations, and land restored to its owners.

WHAT DO WE SHARE?

The United Nations estimates that the total annual sales of the 350 biggest multinational corporations equals one-third of the combined gross national products of the industrialized world and exceeds by billions that of the developing world. Many multinationals often function outside public accountability. Robert Reich, former secretary of labor, has said that "the very idea of an American economy is becoming meaningless, as are the notions of an American corporation, American capital, American products and American technology."

Big nations are not without hypocrisy regarding free trade. While 67 percent of nations agreed to freer markets and communications in 1997, the United States was still imposing trade restrictions on the banana and the apparel industries in Latin America. In his book, *A New*

Name for Peace: International Environmentalism, Sustainable Develop-ment, and Democracy,[24] Philip Shabecoff said, "The market has been the principal device for which humanity has been divided into rich and poor." Humankind has not demonstrated it is responsible enough to have its activities totally unregulated, with the positive but limited exceptions of creativity and the expression of ideas. International commerce, airlines, nursing homes, and banks all need regulation, though it should not be excessive and stiflingly bureaucratic.

The world's consumption bill is $24 trillion a year. The richest fifth of the world's people consume 86 percent of all goods and services, while the poorest fifth consumes just 1.3 percent. The three richest people in the world have assets that exceed the combined gross domestic product of the forty-eight least-developed countries. The world's 225 richest individuals, of whom sixty are Americans, have a combined wealth of over $1 trillion, which is equal to the annual income of 47 percent of the world's population. Americans spend $8 billion a year on cosmetics—$2 billion more than the estimated annual amount needed to provide basic education for everyone in the world.

It is estimated that the cost of maintaining universal access to basic education and health care for all, reproductive health care for all women, and adequate food and clean water for all is roughly $40 billion a year, or less than 4 percent of the combined wealth of the 225 richest people in the world.

The vast inequalities in wealth, health, and longevity in the world decisively leave the developing nations behind economically. The advancement of the health of nations has not loomed large in foreign policy considerations. But individual nations are becoming aware of the importance of the health of their people and their essential contribution to a nation's wealth.

A truly massive worldwide initiative to deal with health, longevity, and population aging is necessary. Wealthy nations should be joined by multinational corporations, NGOs, and an ecumenical religious movement in support of the UN and WHO in general and to inaugurate specific UN initiatives. Priorities include completion of the international immunization program by UNICEF and the WHO and the work of the UN Food and Agricultural Organization (FAO) designed to erase malnutrition and hunger. The UN needs to be strengthened to deal with the roots of conflict and to build an effective peacemaking and peacekeeping force. The UN needs new tools and experiments for innovative, proactive diplomacy to resolve budding conflicts.

Paralleling the peacekeeping and peacemaking forces should be epidemiological surveillance teams stationed around the world to protect societies against new and reemerging infections. To deal with the risks of infections, an overseas science and medical corps to teach and do research throughout the world should be established with multiple support and under the auspices of the WHO and various national science and technology organizations and governments. Telemedicine around the world could bring the finest specialty medicine to the most distant bedside. Innovative and model clinics could be established, such as one-stop family service centers with urgency and emergency services. Vaccine production could be subsidized.

Wealthy nations and wealthy people can no longer enjoy greater life expectancy while insulating themselves from modern hazards—not only the pathogens trafficking around the world, but pollution from manufacturing and the impact of mass migrations and massive population growth. Moreover, the twentieth century of the common man and the birth of the one world has resulted in a revolution of rising expectations which is not likely to be denied.

None of this can happen without nations bartering some measure of sovereignty in exchange for the prospect of overcoming famine, war, and pestilence.

PREAMBLE TO THE DECLARATION OF HUMAN RIGHTS OF OLDER PERSONS

At the first United Nations World Assembly on Aging in 1982, some consideration was given to human rights issues and in 2000, Mary Robinson, United Nations Commissioner on Human Rights, emphasized the importance of protecting the human rights of older persons. However, no official United Nations document has ever identified and specified what these rights are and why they are important. In April 2002, the second United Nations World Assembly on Aging was held in Madrid, Spain. In conjunction with the International Longevity Center-USA I submitted the Declaration of the Rights of Older Persons with the hope that it would become the basis of action as well as discussion at the Assembly and beyond.

Coming at a time of misery and chaos for many older citizens who have lost children and grandchildren in armed conflicts, who are often homeless and destitute, who suffer from malnutrition and ill health, and who live in societies that cannot provide them with the basic necessities of life, the Declaration was put forth to advance the struggle for human rights of all societies. Its message continues to be that we must not simply bear witness:

We must compel change.

Declaration of the Rights of Older Persons

Whereas the recognition of the inherent dignity and of the equal and inalienable rights of all members of the human family is the foundation of freedom, justice and peace in the world,

Whereas human progress has increased longevity and enabled the human family to encompass several generations within one lifetime, and whereas the older generations have historically served as the creators, elders, guides and mentors of the generations that followed,

Whereas the older members of society are subject to exploitation that takes the form of physical, sexual, emotional and financial abuse, occurring in their homes as well as in institutions such as nursing homes and are often treated in cruel and inaccurate ways in language, images and actions,

Whereas, the older members of society are not provided the same rich opportunities for social, cultural and productive roles, and are subject to selective discrimination in the delivery of services otherwise available to other members of the society,

Whereas the older members of society are subject to selective discrimination in the attainment of credit and insurance available to other members of the society, and are subject to selective job discrimination in hiring, promotion and discharge,

Whereas older women live longer than men and experience more poverty, abuse, chronic diseases, institutionalization and isolation,

Whereas disregard for the basic human rights of any group results in prejudice, marginalization and abuse, recourse must be sought from all appropriate venues, including the civil, government and corporate world, as well as by advocacy of individuals, families and older persons,

Whereas older people were once young and the young will one day be old, and exist in the context of the unity and continuity of life,

Whereas the United Nations Universal Declaration of Human Rights and other UN documents attesting to the inalienable rights of all humankind do not identify and specify older persons as a protected group,

Therefore new laws must be created, and laws that are already in effect must be enforced to combat all forms of discrimination against older people,

Further, the cultural and economic roles of older persons must be expanded to utilize the experience and wisdom that come with age,

Further, to expand the cultural and economic roles of older persons, an official declaration of the rights of older persons must be established, in conjunction with the adoption by non-government organizations of a manifesto which advocates that the world's nations commit themselves to protecting the human rights and freedoms of older persons at home, in the workplace and in institutions, and offers affirmatively the rights to work, a decent retirement, protective services when vulnerable, and end-of-life care with dignity.

Robert N. Butler, M.D.
Declaration of the Rights of Older Persons,
The Gerontologist **(2002)** *42:* **152–153.**

CHAPTER 19

THREATS TO LONGEVITY:
COULD WE LOSE THE LONGEVITY REVOLUTION?

The effects of the scientific-industrial revolution threaten the Longevity Revolution it helped create. They include industrial pollution and environmental spoliation by depletion of the ozone layer and by the greenhouse effect that causes warming of the earth, as well as the dangers of nuclear, chemical, and biological warfare and terrorism. In some areas, population growth has overwhelmed available resources and poses a serious threat. Finally, infectious diseases, both old diseases like tuberculosis (TB) and new ones like AIDS, are still with us. People, animals, food, animal products, and disease vectors all travel by air to anywhere in the world within thirty-six hours. We know that tuberculosis and highly contagious meningococcal disease have been transmitted during air travel. A malarial mosquito can hitch a ride on a plane out of Africa and land anywhere in the world in a matter of hours. And birds can carry the possibility of an avian flu pandemic. Our interests in general do not stop at property lines and national boundaries. The world has contracted.

"SHORTGEVITY"

It is painfully evident that some nations have never gained modern longevity and that modern nations can lose it. To be purposely provocative, I have decided to label this phenomenon "shortgevity." The Russian Federation and other nations of the former Soviet Union as well as some Soviet Bloc nations lost unprecedented life expectancy in the 1990s. In Russia's chaotic passage toward a market economy, for example, life expectancy of men there declined from 65.5 to 59.9 years. Russia's death rate surpassed its birthrate, which also declined sharply, with its population shrinking from 146 million to 130 million in fifteen years. Both birthrates and death rates were affected by the disordered economy, as well as by heavy pollution, unemployment, tobacco use, alcoholism, drug abuse, and AIDS. Social instability, violence, crime, unfettered greed, and massive disillusion that followed the disintegration of the Soviet Union also contributed to this profound reversal of longevity, which no modern industrialized nation has ever experienced. Infectious disease reappeared with a vengeance due, in part, to a destabilized health care system.[1] Homicides, accidents, and AIDS accounted for much of the decline in younger persons; in older persons, heart disease, chronic obstructive pulmonary diseases, and pneumonia were the leading causes of death.

Thyroid cancers and other diseases are found there, especially among the children of Belarus, Russia, and Ukraine, probably the consequence of the Chernobyl disaster in Ukraine. For example, the death rates have risen by 20 percent for men and women. The woods surrounding the Chernobyl nuclear plants are still polluted by fallout, although the disaster occurred in 1986. The average life expectancy for men in the area has fallen from sixty-two to fifty-nine.

ENVIRONMENTAL DESTRUCTION

As early as the 1890s, Swedish physicist and chemist Svante Arrhenius, winner of the Nobel Prize, predicted that massive coal burning would result in climate warming. Today, as a result of sunlight and its interaction with chemicals from automobile exhaust and industrial plants, Europe experiences toxic summers. Beautiful Paris is bathed in pollution. Most worrisome is the warming of Earth and the damage to the ozone layer.

The Greenhouse Effect. Earth's surface has been gradually warming since the last ice age peaked some eighteen thousand years ago (by 5 to 9 degrees Fahrenheit), *but many fear that human activity is accelerating the process,* because of atmospheric chemical changes caused by the burning of wood, coal, oil, and natural gas. Their byproduct is carbon dioxide[2] (CO_2), which blocks heat from being radiated back into space and acts like the glass panes in a greenhouse.[3] Little is being done to lower the levels of heat-trapping smokestacks and tailpipe gases. By the time of the 2002 report by the Intergovernmental Panel on Climate Change,[4] very few were disputing the role of human activity and the gravity of the situation.

One disastrous consequence of continued global warming is the melting of polar ice, leading to expansions of the seas that threaten to flood small islands, coastline cities, and coastal nations, including New York and Singapore. Some areas of the world would benefit, of course; cold wastelands might become fertile farmland. But with global warming, some tropical diseases such as cholera, dengue, yellow fever, and malaria could advance into new areas.

Climate change could result in an irreversible cycle of self-destruction of ecosystems such as the Amazon rain forest. Carbon management is essential.

Fearing a negative economic impact, the United States did not sign the 1997 Kyoto Protocol, which aimed to cut combined emissions of greenhouse gases to about 5 percent below their 1990 levels by 2012. In 2001, approximately 178 nations meeting in Bonn, Germany, decided to implement the Kyoto Protocol anyway, with some new provisions. But although it has only 5 percent of the world's population, the United States produces 25 percent of the world's greenhouse emissions, and without its participation there will be little progress.[5]

We must intensify the search for nonpolluting, inexpensive energy sources, and rethink the relative risks and benefits of nuclear versus fossil fuel energy. We must expand nonnuclear, nonfossil energy sources such as biomass,[6] the sun, wind, and hydropower, as well as continue the search for inexpensive, safe, inexhaustible, and non-polluting energy that nuclear fusion just might provide.

Ozone Layer Damage. A different kind of threat is occurring in the ozone layer, which is the layer in the upper atmosphere (ten to twenty miles in altitude) that is formed naturally by photochemical reactions with solar ultraviolet radiation. The ozone layer is nature's sunscreen. It is made up of three atoms of oxygen which, if brought down to sea level, would form a layer only about three millimeters thick. According to the UN World Meteorological Organization, the ozone layer has been depleted by 2.5 percent per decade since the late 1970s. Every 1 percent drop in ozone means that roughly 1.3 percent to 1.5 percent more ultraviolet light reaches Earth's surface.

The hole in the ozone layer was first discovered in 1985. It is now widely accepted that the use of chlorofluorocarbons (CFCs) as coolants in refrigerators, air conditioners, and foam packaging releases chlorine atoms into the atmosphere and destroys ozone molecules. The breach in this protective layer admits higher levels of ultraviolet radiation (UVB) to the earth's surface, causing cataracts, skin cancer, premature skin aging, and weakening of the immune system in humans, as well as activating retroviruses such as herpes. Biologists also worry that damage to plants and animals could quickly affect the entire food chain, and that eventually life itself, with the exception of marine life, could not exist.

The use of CFCs was banned in the United States, and following the Montreal Protocol of 1987 Europe began to reduce their use as well. DuPont, which is the world's largest producer of CFCs, pledged to phase out all production of the product worldwide, with a goal of 95 percent cutback by the year 2000. CFC-producing nations agreed to eliminate CFCs by 1995. Data have shown a slowdown in the rates of growth of CFCs, constituting a modest success story of international collaboration. But even if all the UN member nations were to comply with agreements, it would take until 2050 to 2070 for the ozone layer to completely recover. Moreover, the hole in the ozone is widening again in 2007 because of the use of refrigerant in India and China.[7]

Air Pollution. The industrial revolution resulted in a 500-fold increase in coal use. Consequently, asthma appears to be increasing rapidly around the world, especially among children and older persons. In the UK it took a dreadful London smog in December 1952 that killed about four thousand people to stimulate effective action. In the United States, the Clean Air Act of 1970 has made an enormous difference in air quality. Yet, although particulate air pollution has declined somewhat, we still need to understand better its harmful components and their mechanisms of action and reduce it much further.

Water Pollution. Sanitation and clean water helped bring us the Longevity Revolution. Ironically, according to the World Commission on Water for the 21st Century, today more than one billion people worldwide lack access to clean drinking water. The problem is most severe in Asia, the world's most polluted and populated region. By about 2025, half the world's population is projected to experience a severe shortage of water for drinking and irrigation, with agriculture using up 70 percent of the world's fresh water. Rice cultivation, which is critical to the nutrition of millions of people, requires great quantities of water.

Many factors are responsible for water pollution, including the rising number of motor vehicles and electrical utilities that emit sulfur dioxide and nitrogen oxides, resulting in the acid rain that contaminates rivers, lakes, and other water supplies, drifting to the East Coast from coal-burning factories of Midwestern and southern states.

Pesticides and herbicides also threaten the water supply. Cancer and heart disease, as well as other life-threatening illnesses, have been linked to polluted drinking water. For instance, higher levels of cancer, particularly leukemia and lymphomas, are reported in Iowa, Nebraska, and Illinois. Farmers and farm workers are at greatest risk of exposure because of pesticides that filter into the drinking water.

We need to create nontoxic, biological approaches to insect and disease control in agriculture. An example is the introduction of ladybugs as predators into farm areas of the United States to reduce the number of aphids that feed on grain crops.

DISEASE

The World Health Organization (WHO) reports that cholera exists in many parts of Africa, Bangladesh, China, India, Indonesia, Malaysia, the Philippines, Thailand, and Vietnam, as well as in the former Soviet Union, where the decline in living standards, sanitary conditions, and health care have led to a dramatic increase.

Until recently, cholera was rarely seen in cities, but with growing urbanization and poverty, it has become a threat. There is no effective vaccine. Untreated victims may die within hours of dehydration. However, with prompt replacement of body fluids and salts, along with the administration of antibiotics, most patients can now be saved and epidemics are no longer marked by high death rates. The issue is whether public health and medical responses can be mobilized quickly enough in epidemics.

The World Health Organization sponsored an intensive, worldwide campaign to spray the *Anopheles* mosquito breeding areas with DDT. Two major problems developed. Mosquitoes began to develop resistance to DDT, requiring the introduction of new insecticides like Dieldrin and Lindane, and these insecticides posed a significant ecological hazard to other living species. Developed nations passed a ban on the use of these insecticides, which led to a resurgence of the disease. However, by 2001, DDT was being used once again in some areas, such as South Africa, and in some twenty other nations where cases of malaria were increasing.

Malaria is now out of control in many parts of the world, notably sub-Saharan Africa and other tropical regions. Research on genetically engineered malaria vaccines have progressed to the testing stage in humans. There is a worldwide race to find a vaccine against the most common and lethal form, Plasmodium *falciparum,* which causes about 95 percent of deaths from this disease. In any case, bed nets to protect against mosquitoes should be more universally available.

Tuberculosis is making a comeback in the United States and Western Europe, especially in the inner cities, largely as a result of AIDS. Persons with AIDS are unusually susceptible to tuberculosis, which is highly contagious and can result in an epidemic or "coepidemic." Tuberculosis declined steadily in the 1960s and 1970s, leading some to predict its eradication by the end of the century. However, the downward trend stopped and the number of cases rose again in the 1980s, prompted by AIDS, homelessness, drug and alcohol use, crowded living conditions, and malnutrition among the inner city poor. In the United States 22,400 cases were reported in 1988, and eighteen hundred deaths. Although the disease can usually be cured with medication, 50 percent of untreated patients die.

Worldwide, nearly a third of the population (1.75 billion people) is infected with the bacterium, most in the latent, carrier stage. About 10 percent will develop a life-threatening form of the disease, especially when the immune system is compromised. There are eight million new cases of tuberculosis each year; over 95 percent in developing countries. Two to three million die each year, which is one-quarter of all the preventable deaths in the world.

Young children and older persons[8] are the most vulnerable because of their sensitive immune systems. (Older persons account for 30 percent of all active cases and 60 percent of all deaths from tuberculosis.) Moreover, multiple antibiotic-resistant strains[9] throughout the world pose a grave threat.[10]

AIDS has taken more lives than the Black Plague. In 2007, an estimated sixty-five million people were living with HIV and an estimated death toll of over twenty-five million has been attributed to AIDS since its discovery in 1981.[11]

Much is already known about how the disease is transmitted and operates in the body. Whether or not a vaccine or cure is discovered soon, we know that infection can be prevented by safe sex and circumcision. It remains to be seen whether this epidemic will follow the pattern of great killers of the past and eventually run its course and decline in virulence, or whether the AIDS virus is different altogether, threatening the human race in a new way.

Since it is transmitted from mother to child, AIDS directly interferes with the perpetuation of the species. The risk of transmitting the virus from mother to child ranges from 25 to 30 percent in developing countries and 15 to 30 percent in developed countries.

By 2001, there were twenty-six million cases of AIDS in Africa, although malaria was still the number one killer. AIDS had reduced average life expectancy by fifteen years in sub-Saharan Africa, with more than 70 percent of all estimated cases of AIDS in the world in that region.

The UN has planned a multibillion dollar global AIDS fund to which nations and foundations would contribute. The fund, which seeks $7 to $10 billion per year, will be modeled in part on the health initiatives backed by the Bill and Melinda Gates Foundation, such as the Global Alliance for Vaccines and Immunizations (GAVI), to which the Gates Foundation gave $750 million in 2001.

Deforestation

Deforestation, especially the increasing destruction of the tropical rain forests,[12] affects the climate and the supply or source of potentially new medicines. Vegetation absorbs some of the developed world's carbon dioxide. In addition, burning fossil fuels (coal, oils, and natural gas) shrouds villages, towns, and cities with air pollution and spews sulfuric acid on forests, resulting in stunted forest growth and the eventual death of trees and entire forests. Forests are also cut to create pastures for cattle to provide hamburger for the worldwide expansion of fast food restaurants. This is a very destructive way to provide human nutrition, both for ecological and health reasons. Desertification is the extreme.[13]

EXTINCTION OF SPECIES

Ironically, as the human species has gained life expectancy, ancient forms of plants and animals are disappearing at an alarming rate. This depredation is due to war, population growth, pollution, farming, fishing, logging, and other factors. We are losing genes, habitats, and ecosystems. Scientists estimate that in less than fifteen years, 10 to 20 percent of the known species may become extinct. Edward O. Wilson, Harvard University naturalist,[14] estimates that twenty-seven thousand species per year, seventy-four per day, three per hour, are being destroyed. Many wild and primitive varieties of plants and animals show disease resistance, adaptability to adverse environmental conditions, and other characteristics that could be used to improve crops and livestock. Modern medicine has already been the beneficiary of nature's largesse, in that more than one-fourth of all drugs owe their existence to plant substances. For example, cyclosporine A, a powerful autoimmune suppressant that plays a large part in the management of many illnesses, has greatly improved the success rate of organ transplants. It is from spores in a soil sample in Norway.

The UN Food and Agriculture Organization proposed a new global system for collecting and storing endangered genetic resources of plants, including rootstocks, seeds, and tissues.[15] However, the growth of genetic engineering is making such access a geopolitical and geoeconomic issue, and controversy rages over who will maintain control over access to such genetic resources. In fact, the United States declined to participate, preferring a less extensive storage system, administered by a commission largely supported by the United States.

Developing countries contain large numbers of endangered plant and animal species. These countries should be funded to save species, acquire gene technology, and profit from their own natural resources. Seed companies from industrialized countries have profited from transplanting the genetic traits of wild and primitive plant varieties from developing nations into hybrid varieties.

Invertebrates. Invertebrates, among the dominant species of the world, along with microbes, are critically important to the planet and to humanity. Insects are essential to the decomposition of mammalian excreta that becomes fertilizer. Mindless destruction by insecticides can endanger ecosystems and, thereby, the food chain. Without invertebrates, humans would barely survive a few months.

U.S. food producers have come to depend upon the application of pesticides and fungicides to crops. The level of pesticides that infants and children receive is of special concern because, although standards regarding safe levels of consumption have been established, they relate to a 160-pound adult male. A child may receive up to 35 percent of his entire lifetime dose of carcinogens by his fifth birthday. Pesticides may also interfere with the fertility of some animals, including humans.

Radiation

The cold war left the world with uninhabitable land, the leavings of uranium, weapons-grade plutonium, and radioactive waste that will remain toxic for millennia.

Where can the leavings of the superpower struggle as well as the radioactive waste of nuclear energy plants be safely disposed? And what of nuclear energy plants located over earthquake faults? One terrifying example of the destructive potential of this type of energy is the 1986 Chernobyl nuclear power plant explosion, which released ten times the radiation given off by the atomic bomb dropped on Hiroshima in 1945. Chernobyl was closed nearly ten years after the disaster, but the safety of nuclear plants is a realistic worry for all nations. In truth, there are no scientifically established safe doses of radiation.

The only serious obstacle to constructing a bomb is the limited availability of purified fissionable fuels. Nuclear threats continue, even at the individual level.[16]

Antibiotic Overuse

Longevity has been greatly aided by antibiotics. However, antibiotic overuse has bred drug-resistant bacteria, and pathogenic bacteria have enough variations of themselves to resist at least one of the over one hundred antibiotics now available. There has been a sharp increase in antibiotic-resistant strains of bacteria that cause pneumonia, meningitis, and other diseases. Dr. Stuart Levy of Tufts University has written that "antibiotic usage has stimulated evolutionary changes unparalleled in recorded biologic history."[17]

Pandemic Flu

Flu epidemics illustrate the adaptability of pathogens. Worldwide flu pandemics occur every ten to twenty years or so, with the emergence of an influenza virus that strikes a population that has not had recent exposure and lacks immunity. Animals are the source of pandemics that generally arise in Asia and involve viruses that have undergone an exchange of genes. For example, the growth of Asian fish farms may increase the risk, particularly when fish farms use fresh fertilizer from pigs and poultry, giving influenza viruses of pigs, ducks, and humans the opportunity to mingle and exchange genes. Many experts believe we face an inevitable worldwide pandemic of killer flu[18] in particular when and if the avian flu virus mutates so that human-to-human transmission occurs. Many feel it is a matter of when, not if, and that millions could die. To date we do not have systems in place for rapid vaccine production, care, and treatment.

Overpopulation and Economic Development

Since the beginning of recorded time, humankind has been making efforts to control reproduction. Given population momentum, were the rate to decline to 2.1 over the next half century, the world's population would still reach eleven billion before leveling off near the end of the twenty-first century. If the rate were 2.5, there would be twenty-eight billion by 2150! (See Chapter 17.)

Especially worrisome are young men between fourteen and thirty years of age, in great numbers in the developing world, who are frequently unemployed and given to violence and thus are a source of political instability.

Until the world enjoys comprehensive family planning, there will be excesses of population relative to resources. Asia, South America, and Africa will experience awesome poverty, although, tragically, AIDS could sharply lower population levels, especially in Africa and Asia. Where mortality rates continue to decline, the population of older persons will increase. A segment of this population will include the frail and the demented, who could pose so great a burden that curbs on longevity might be sought. Thus, longevity itself becomes a threat to longevity in the absence of new population and resource policies.

Lifestyle

What we do to ourselves is a threat to longevity. Alcohol, smoking, the sedentary life, and poor diet are principal issues, as discussed in Chapter 11. There is growing evidence that humans thrive best on vegetarian or near-vegetarian diets so long as they consume adequate protein, vitamins, and minerals. Even though humans have long loved to eat meat, our Stone Age ancestors are thought to have consumed five times the fiber, one-fourth the salt, and half the fat of a modern American diet. The fat content of wild animal meat is less than that of domesticated animals; in fact, humans have eaten large amounts of fatty domesticated meat only since the Industrial Revolution, and particularly since the beginning of the twentieth century. Tobacco is probably the single largest cause of preventable death around the globe.

Modern Malthusianism

What about the Malthusian theory that humankind will outrun its own food supply? Is this a description of an inevitable biological reality or is it subject ultimately to human control? As a matter of fact, food supplies are plentiful in nations that reproduce below or at zero population growth, although even some densely populated nations, such as Bangladesh, have become self-sufficient in food.

Generally, there is not only a scarcity of food, but it is inequitably distributed throughout the world. The planet's food surpluses could vanish within months if persistent bad weather, crop failures, new blights,[19] or wars occurred in one of the world's important farming regions. The availability of modern varieties of grain could lead to neglect of primitive and native varieties that have valuable genetic traits, such as resistance to drought and tolerance to toxic soil salts. Moreover, modern varieties tend to need pesticides and fertilizers, which pollute the world's waters.

Homo Sapiens as Part of Nature

The environmental movement has been extraordinarily important in bringing attention to the degradation of the environment and the need to maintain a livable biosphere. However, certain assumptions require fresh examination. A kind of romantic, sanctimonious sentimentality, even hysteria, has been built into the environmental movement, with little consideration given to the reality of people living in the developing world—for example, those who must depend upon fossil fuels for their survival. There is the mistaken idea that a human being is somehow artificial, or at war with that which is natural. It must be stressed that we are part of nature and have altered nature as no other animal form has ever done.[20]

We need to build a constructive alliance between nature and economic development, ensuring mutual adaptability.

Reprise

Since infectious diseases remain a major cause of death throughout the world, we have not even completed the first phase of the Longevity Revolution. Poverty and malnutrition, the synergistic roots of so much disease, remain omnipresent.

How can the rich nations of the world stand by when so much is already possible? And when they themselves risk danger? And when economic realities necessitate the expansion of markets for their products and services?

According to the Lederberg-Shope-Oakes report[21] prepared for the Institute of Medicine, "Despite the appearance of security, however, there is only a thin veneer protecting humankind from potentially devastating infectious disease epidemics. Alone or in combination, economic collapse, war, and natural disasters, among other social disruptions, have caused (and could again cause) the breakdown of public health measures and the emergence or re-emergence of a number of deadly diseases. We certainly cannot aspire to decontaminate the Earth with respect to all potential microbial rivals."

The problem, of course, is managing short-term needs and greed on the one hand and taking the long-term view on the other. In addition, it is morally repugnant, politically irresponsible, and economically impossible for the developing world to carry out appropriate environmental practices to the disadvantage of its own peoples. In fact, it is the developed world that accounts for the greater amount of degradation and pollution—so-called eco-imperialism.[22]

THE MICROBIAL WORLD

Joshua Lederberg, research geneticist and Nobel Prize winner, wrote, "The human occupation of Planet Earth has left us with a conceit that our species is at the very top of the food chain, predatory on all others, rapidly destroying their habitats, and diminishing the genetic diversity of domesticated species. Carnivores that once terrorized human bands have been all but exterminated. Rodents abound, and threaten some food stores; but they can be contained with concrete and warfarin.

Insects nibble our agricultural surpluses; at their worst, swarms of lo-
custs will be regional not global pests. Barring geno-suicide, the
human dominion is challenged only by the pathogenic microbes, for
whom we remain the prey; they the predator."[23,24]

Can we secure a truce with the microbial world? Out of self-interest,[25]
as well as on humanitarian grounds, the industrialized world should
allocate billions to help eliminate disease in the developing world. For
instance, the WHO urges the industrialized nations to put up $100
million each year to buy drugs to overcome tuberculosis. About $135
million could eliminate polio in five years, according to WHO and
UNICEF experts. Much more should be spent in a worldwide immunization initiative. Expanding upon the Lederberg report, among the
most urgent steps we must take is a global policy to deal with the world
of microbes through the use of:

- Intensified global surveillance of diseases, microbes, and vectors.
  (Imagine if the deadly Ebola virus were to mutate and become contagious by air droplet!) We need rapid-strike epidemiological and
  treatment forces just as we do a permanent rapid deployment of
  military forces for peacekeeping. Doctors Without Borders is a
  model.

- Continued use of satellite and computer communication, such as
  ProMED (Program for Monitoring Emerging Disease), for surveillance. This is privately funded through the Federation of American
  Scientists and has twenty-five thousand subscribers who can inform
  each other instantaneously about any curious outbreaks in the
  ninety countries they represent.

- Nationwide computer systems to track infectious disease outbreaks.

- Continual training and research support for virology and microbi-
  ology, with emphasis upon exploration of modern biological concepts and techniques.

- Expansion of the field of veterinary medicine.

- Better diagnosis and less casual use of antibiotics.

- Better management of vaccine production and supply. In 2004, one vaccine manufacturer (Chiron) could not produce the flu vaccine. Governmental subsidies for vaccine manufacturers are essential, since market forces cannot succeed here.

- Development of methods for rapid production of flu vaccines that do not require the use of egg, to which many people are allergic.

- Specific studies that elucidate the mechanisms by which a microbe crosses from one species to another.

- Specific studies on enhancement of immunity and reduction of virulence.

- Immunization programs—worldwide. In 1974, the WHO created the Expanded Programme for immunization to ensure that every child in the world was vaccinated against six major diseases—polio, measles, smallpox, diphtheria, pertussis, and tetanus. By 1990, 80 percent of children had been vaccinated, but this figure, unfortunately, has not risen since.

- Continued search for single-dose inoculation that would protect all infants from all major childhood diseases, such as a supervaccine which would provide protection over the entire life course.

- New antibiotic and vaccine development utilizing molecular biology. For example, by decoding the DNA of the TB bacillus, it will be possible to defuse the pathogen's genetic defense acquired in the course of evolution to overcome the human immune system.

- New biological approaches to the control of disease vectors such as mosquitoes to prevent malaria.

- Public and professional education about environmental and pathogenic threats to health and longevity.

- More effective protection of food supply by FDA and the Department of Agriculture. At best, the FDA is only able to inspect less than 1 percent of imports.

- A wash hands program. Many public bathrooms in the developed world already have sensor-operated faucets and toilets that are touch-free.

PUBLIC HEALTH POLICY

The power of the surgeon general of the Public Health Service has been significantly reduced. This office should have a meaningful budget, and the physician holding this office should be able to speak directly to the American people unencumbered by politics.

If necessary, this first physician of the United States should be able to close down an airport or other public places where there is danger of contagion. He or she should lead in public health education and should be America's top doctor as commander of the six thousand members of the service's commissioned corps. He or she should be in office for seven years and not beholden to the presidential administration in power.

There should be comparable positions of authority in nations and regions around the world. In addition, the World Health Organization (WHO) should be greatly strengthened by added support of member nations to greatly expand its budget.

The WHO and Public Health Service along with the NIH and the

CDC should be empowered (and receive generous funding) to prepare for a flu pandemic[26] or other biological emergencies (such as bioterrorism) that will enable them to create the infrastructure for early scientific detection, vaccine production, quarantine, care, and treatment.

Support is also needed for the WHO's plan in collaboration with various governments, foundations, and anti-TB organizations to create a global drug facility to buy and supply drugs to nations and NGOs, free if need be. (As of March 2001, only Canada contributed to this facility.)

The political merger of the anti-nuclear, environmental, and health movements should be protective of our hard-won longevity. The industrialized world, the United States, the European Union, Japan, and Oceania, with the support of the UN, should take leadership in preserving the twenty-first-century gains in longevity. Achieving a balance between the individual nations (and their claims of sovereignty) and global realities constitutes the battle that has raged over the power of the League of Nations and, later, the United Nations. This issue will be debated, and I fear wars will continue to be fought over it, for a long time, perhaps too long to deal with the threats to life and longevity. Our failed husbandry of Earth, our aggressive nature and incapacity to resolve conflicts peaceably, our failures at governance of ourselves, and our ignorance and intolerance have the real potential of creating "shortgevity." We still see political instability and economic perturbations that at worst end in poverty, starvation, and hunger for millions.

I have particularly emphasized the evolutionary struggle of various predator pathogens upon humanity, but the ultimate issue is quite different. It is within the cultural evolution of homo sapiens to overcome the penchant for misdirected aggression, the inability to solve conflicts, rapaciousness, and greed. Thus, the ultimate threat to humanity and to longevity is humanity itself.

PART VI: IMAGINING LONGEVITY

CHAPTER 20

THE GOOD LIFE:
QUALITY OF LIFE IN THE ERA OF LONGEVITY

Throughout this book one idea has been stressed: longevity is desirable if accompanied by a life of high quality. But what makes for such a good life?[1] Most of us want love, meaningful work, safety and security, energy and health, and, to varying degrees, power, fame, freedom (control and choice), and wealth, and we want to live in a society that supports these goals.

How can we measure quality of life?[2] There is no simple answer. It is an amorphous concept, constantly changing with the historical period and one's culture, personal background, stage of life, and socioeconomic status. A person's definition of quality of life is, and should be, highly individualized and subjective.

Danger lies in a sector of society attempting to quantify, qualify, or create a cut-and-dried definition of what constitutes quality of life. Neither governments nor physicians should be given the authority to determine for an individual the point at which quality of life is so diminished as to make life itself undesirable or unnecessary.

For example, care must be taken when making assumptions about how older people perceive their lives. A popularly held view is that old people who seem no longer able to enjoy life may not wish to live. To an objective eye, their lives may seem so limited as to appear pointless, but to the individual living that life, each remaining day, week, or year is often precious. Longevity itself may entail quality of life and provide choices and opportunities for happiness. In fact, evidence exists that old people may prefer a longer life with illness to a shorter life of higher quality.[3]

Can a higher quality of life in and of itself extend the length of life? Having goals, passion, positive emotions, purpose, and structure in one's life have been associated with longevity. Length of life and quality of life are certainly intertwined. Medicines and surgeries have done much to advance life's quality, and outpatient cataract surgery and hip replacements are excellent examples of that.

On the other hand, aggressive surgical and chemotherapeutic interventions to treat cancer and other diseases that extend life may incapacitate a patient. Public concern has grown over the quality of life one may expect with diminished capabilities. From treatments for angina to depression, cancer, and stroke, people are ambivalent. They want to do everything possible to survive illness and trauma but often become upset when the outcome compromises the quality of their lives. For example, an open and full discussion with the patient and family about the use of a ventilator for breathing is very important, for it can make the last days of one's life miserable.

According to the National Center for Health Statistics, disability has increased for each decade in the last thirty years of life, with the outcome being a growth of medicated survivors in the later years. The fastest-growing group, those who are eighty-five or older, are susceptible to new pathologies from lifelong health habits, chronic diseases, and exposure to environmental factors, such as occupational hazards.

However, contrary to the common wisdom that the longer one lives the more quality may be compromised, disability rates may now be falling proportionately, despite the growing numbers of persons of advanced age.[4]

To help quantify the relationship between health and quality of life, in 1971 D.F. Sullivan of the National Center for Health Statistics introduced the concepts of disability-free life expectancy and active life expectancy. Still later, the concept of the health span was put forth by the U.S. Surgeon General, and others use the expression "quality time." All of these terms have been defined using different criteria, some overlapping, but all useful. In addition, a variety of measurements have been developed in connection with the basic first tier of quality of life, such as physician Sidney Katz's activities of daily living (such as the ability to walk and take care of oneself)[5] and social scientists Robert Havighurst and Bernice Neugarten's idea of life satisfaction, which takes into account the pleasures of everyday life and of holding a positive self-image.[6] Moreover, a variety of rating systems have been created with regard to specific disease states, such as cardiac disability and depression. Psychiatrists Thomas Holmes and Richard Rahe have assigned values to various stresses,[7] and Bruce McEwen measures "allostatic load," which evaluates stress and its accumulated effects over a lifetime.[8] The management of daily hazards is essential.

Perhaps the classic and simplest equivalent of quality of life is happiness, which is usually regarded as being a result of one's own doing and of one's circumstances, and mistakenly that it can be bought. However, some American psychologists regard happiness as largely determined by genes, similar to the set-point concept of weight control (which posits that whether gaining or losing weight, there is a tendency to return to a fixed weight). It has been suggested that wealth, education, marriage, and family have only marginal and transient effects upon levels of happiness (about 8 percent). These conclusions were deduced from the National Health and Nutrition Examination Survey, which followed some six thousand men and women over ten years.[9] Consistency in mood[10] was observed during the study. Other studies comparing identical and fraternal twins have shown similar results.

Losses and trauma can have profound effects on an individual and lead to clinical depression despite the set-point. On the other hand, Dr. David T. Lykken, a behavioral geneticist, proposes that "a 'steady diet' of simple pleasures"[11] will keep one above one's set-point.[12]

Very special issues affect women that can enhance or diminish their quality of life. The first of these is caregiving, which primarily burdens women and impacts heavily on their quality of life. This is true of societies around the world. (It must be recognized, however, that giving care offers satisfaction as well.)

The second issue is poverty. Women around the world are more likely than men to experience economic deprivation throughout life and in old age in particular.

On the other hand, studies have shown that single women are the happiest, followed by married men. Married women follow; least happy are single men.

HOW WE SPEND OUR TIME

Daniel Kahneman of Princeton University was awarded the Nobel Prize in economic sciences in 2002 for his work integrating psychological research into the economic sciences. He offers the Day Reconstruction Method, which incorporates psychological and social components, to measure well-being. A National Well-Being Account, akin to GDP, could provide a noneconomic measure of well-being of people at all ages. Kahneman has said that "time is the ultimate currency."[13]

Time is the essential commodity to ensure quality of life, in intimacy and in health care, where it is most expensive, even more than technology. Reducing the use of technology could be rewarded by paying the doctor and nurse for their time, rather than relying so much on instruments. But, unfortunately, the medical payment structure does the reverse.

Time is the irreplaceable resource to be spent most wisely, which is so difficult to do.

How do we balance immediate gratification with long-term pleasures (saving for a new home, a vacation, etc.)?

Life in Europe, including social protections and leisure (shorter workweeks and more vacation time), supports the good life. It is said that Europeans work to live while Americans live to work. This is an oversimplification with a grain of truth.[14] Americans generally have a work ethic, but less so what might be labeled a quality of life ethic (see Figure 20.1).

The United States is one of the few industrialized nations that has not established maximum working hours or annual leave entitlements through legislation.[15] Working hours continue to rise in America.

Environments and Quality of Life

Suburbs without sidewalks and urban sprawl are not conducive to quality of life for older persons, the disabled, or persons of any age. Long commutes to work undermine family life. Adjusting our environment to the requirements of different age groups is one way society can enhance quality of life. For example, underfoot treads found in train stations in Japan and other countries help guide people with limited vision safely on and off trains.

Speed is the result of the industrial age that brought us so-called efficiency, such as the accelerated line speed in factories. The velocity of a culture, as well as the frequency, intensity, and multiplicity of stimuli, can be especially harmful for older people because of their slowed reaction time and disabling conditions.

We are made weary by speed and efficacy—24/7, the Internet, and e-mail, cell phones, and overtime.

Figure 20.1—America the Industrious. Average Annual Vacation Days.*

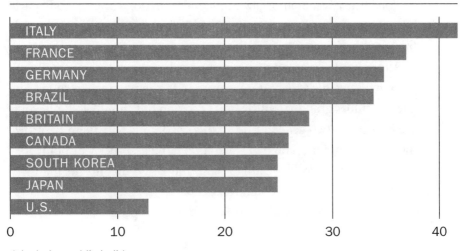

* Includes public holidays.

World Tourism Organization (2002).

Active Pursuit of Quality of Life

The process is an active one, a matter of acquiring the skills, knowledge, and goals to pursue a life of deep personal interest and developing lifelong health habits that prevent illness and promote personal well-being and active engagement in one's immediate community and larger society. To carry out this active process requires freedom of choice; hopefully, informed choice.

Indeed, we each have a responsibility to actively promote our own quality of life and make constructive contributions to those around us. As Bernard Berenson wrote, there are "'life-enhancing experiences' that help make 'life a work of art.'"[16] This brilliant art historian found that his aesthetic enjoyment grew sharper and more profound with age. One's quality of life ultimately reflects one's passion, purpose, and spirit, which are not easy to objectify and measure, however cleverly we construct our scales, and, ultimately, upon informed choices. Here Schopenhauer was very clear that with aging comes the peeling away of illusions.

Legacy

Philanthropies, memorialization, and donation of one's body parts have become popular in recent times. Not only do these acts bring fulfillment; they suggest an afterlife.

All forms of legacy should be encouraged in our culture, as should the growth of an *ethic to the future,* a troth to the generations to come. The latter includes grandparentage and adoption but should go far beyond the immediate two generations. In this era of longevity there are growing numbers of four- and five-generation families. Brock Chisholm, former head of the World Health Organization, said that in true maturity one thinks beyond two generations.

Americans are more philanthropic than Europeans, in part because our tax laws encourage us to be.

Would that philanthropy were driven less by the desire to mitigate or escape estate taxes and instead focused predominantly upon the greatest needs of society,[17] such as education, public health, medical research, science, the arts, music, the humanities, housing, the eradication of poverty,[18] the protection of the vulnerable (children, the disabled, and old persons), and the preservation and beautification of the environment.

Many philanthropists are motivated to control their legacy and manipulate their image. The well-to-do should be taught philanthropy, the needed skills, decision making, how to follow up, and how to obtain and evaluate results. This is being done in many families.

Organ Donors. Donating organs, tissues, cells, and blood is one way to contribute, dramatically and measurably, to the longevity and quality of life of other human beings. A donor not only contributes to the longevity of another; in a sense the donor's life is extended beyond death.

For a time, older persons were not considered appropriate candidates for transplants because they were likely to have diseases. But the old can and do donate organs and tissues to the young and, by 2005, disease, substance abuse, and obesity, to say nothing of age, were, in practice, no longer a basis for automatically rejecting donor organs. Even so, there will never be enough vital human organs, such as the heart and lung, to meet the need.

Until the United States adopts a universal automatic donor (presumed consent) policy, Americans need to be educated by physicians, other health care workers, social workers, and clergy so that they appreciate the profound contribution that organ, blood, cornea, and skin donors make. They need to be given support so they can work through conflicting feelings regarding the use of their body parts and those of their loved ones, and are encouraged to voluntarily sign donor cards.

Hospitals, physicians, and other health care providers need to solicit donation cards far ahead of time and not wait until a person is dying. Over forty states do have laws requiring hospitals to solicit organ donations from families with dying and dead members. There is a similar law regarding hospitals that receive Medicare and Medicaid funds. But so far these laws have had a minimal effect. The time of death is tense and grievous, and the window of opportunity for a transplant is limited after a patient dies. Prospective donors need to be encouraged to discuss their wishes with family, for family permission at the time of death is a greater determinant of whether organs will be donated than is a signed donor card. Families are barriers to donation more than is doctors' failure to ask, and less than half of all eligible donor families consent to organ donation. If a person cannot donate body parts, the entire body can be donated to medical schools for teaching and research, regardless of age and the presence of disease.

As Andrew Carnegie bluntly stated, "He who dies rich dies disgraced." Everyone should be a philanthropist in the broadest sense, giving one's body as well as one's time and capabilities during life as well as a portion of one's estate and possessions to the common good, with a clear sense of a future and of those who will occupy it. These actions will enrich one's quality of life immeasurably.

The Role of Society

Of course, we await the hoped-for evolution of society to a higher plane. For society, too, has its responsibility to support medical research and cultural institutions in order to improve life's quality and to enrich society's concern for its members. The solution to Alzheimer's disease would make an extraordinary contribution to quality of life. More research should be directed against the nonlethal conditions such as arthritis and sinusitis, so common in old age, as well as conditions impairing vision, hearing, and brain function. The last third of the twentieth century in particular was characterized by major new technological advances in medicine, such as hip and knee replacements, cataract operations with lens implants, new antidepressants, blood-pressure-lowering drugs, and Viagra, all of which have improved quality of life. Sleep, too, as well as relaxation and freedom from stress, contributes to quality of life. Sleep problems can be the result and the cause of poor health. Unfortunately, the oncoming generations may suffer a greater incidence of diabetes, hypertension, and coronary heart disease because of the national obesity epidemic and the sedentary life.

The UN Human Development Index (HDI) was devised in 1990 by Amartya Sen and a Pakistani economist, the late Mahbub ul Haq, to measure the quality of life among countries. It includes life expectancy at birth, educational attainment, and income per capita. It ranks 173 countries.[19] Its annual reports also alert the world to special issues, such as climate change and the water crisis.

Statistics Canada has been developing Canada's first national index of well-being, And the small Himalayan kingdom of Bhutan has also been focusing on the happiness and well-being of its people. Should the United States?

Attitudes Toward Death

In the NIMH studies in the 1950s and 1960s of healthy community residents, some 15 percent of older people were consciously frightened of death,[20] and for them just being alive was defined as a good quality of life. Thirty percent tended to avoid or deny death and the remaining 55 percent realistically accepted it. On the other hand, there are those who had a distinct fear of longevity, especially of having to live a long time in a debilitated state, which is a dominant theme in Western civilization. It is significant, therefore, that when the Research on Aging Act was passed by Congress, creating the National Institute on Aging, the language of the act made clear that the goal was the preservation and continuation of the vigorous, healthy, middle period of life, not the pursuit of an extended life for its own sake.

Albert Camus[21] wrote that the idea of suicide provokes the ultimate philosophical question, the value of one's life. Homicide and suicide are among the ten most frequent causes of death in the industrialized world. What does this say about quality of life in our time?

Fortunately, palliative care, including end of life care, which is growing in America, focuses on quality of life.

The Life Review

How do I review my life and sort out the past? How do I make new adjustments in my view of myself? How do I sum up my life? What are my last wishes and instructions? How will I be remembered?

The life review is a common, perhaps universal, experience that occurs in the later years or when one is facing death. It is intensely personal, often poignant, especially experienced in isolation. It is characterized by a return of memories, both good and bad. These memories may be intense and include the sounds and smells of childhood. When confronting one's past conflicts (usually associated with guilt and shame) and regrets, there are efforts toward atonement and resolution as well as reconciliation with individuals with whom one may have become alienated. When there is successful resolution of conflicts and one comes to terms with one's life, it usually involves forgiveness of others and of oneself, and is rewarded by a growing sense of serenity.

QUALITY-OF-LIFE INDICATORS

Borrowing from many sources, I have compiled a comprehensive list of various aspects of quality of life (see Table 20.1) that apply throughout life but become more acute as one grows older. This list is neither exhaustive nor arranged in order of importance.

Some aspects are obvious and basic, such as physical, financial, personal, and social well-being. Others are more complex, such as possession of a sense of purpose, moral well-being, and, perhaps, spirituality. Coming to terms with one's life (for example, through a life review) and the reality of life satisfaction is especially complex.

I believe the first basic tier of indices of quality of life includes possession of intellectual capacity; the ability to perform activities of daily living; freedom from pain and suffering; preservation of the senses and sensuality; a social support system; an adequate financial base; mastery over one's life (independence, autonomy, and *choice*); a purpose outside of oneself that offers a sense of usefulness; and some degree of happiness and morale. And love, even when occasioned by heartbreak through conflict or loss. Quality of life also comprises freedom, legal protections, and human rights, and some degree of leisure.

Far from being immutable, even these basic elements in the first tier are relative and their importance varies widely according to personality and circumstances. Certain personality characteristics can be adaptive or transcend the most profound physical and mental pathology. This explains why some individuals with severe dementia, with paralysis, or who are living a limited life in a nursing home still enjoy their lives and consider them to be of relatively high quality. Personality characteristics help us understand how Stephen Hawking continues to make important contributions to the field of physics despite his debilitating and progressive neurological disorder, and why the family of an end-stage dementia patient may insist on preserving that person's quality of life in spite of inevitable deterioration.

Moving into the second tier and beyond, subjectivity and relativity become even more pronounced. For one person, a good quality of life may mean having a rich network of friendships. For another it may mean having the freedom to be alone and enjoy music in solitude. We must be very careful in our judgments about this elusive concept, for it goes to the core of a person's being.

The great Greek lawmaker Solon said, as paraphrased by the philosopher Mortimer Adler, "Only when your life is over can someone else commenting on your life declare that you *had* lived a good life and could be described as a person who had achieved happiness." From this perspective, longevity is irrelevant. Some could have remarkable short lives surpassing long, unremarkable ones. Others could profit from longevity, given more time to carry out one's ambitions, enjoy one's grandchildren, and engage in leisure.

Robert Fogel writes, "What is the good life?" From the time of Socrates until the beginning of the twentieth century, that was only an issue for the rich. Now it is becoming an issue for ordinary people. If you worked three thousand hours a year, you didn't have a lot of time to wonder what to do with your life."[22]

Ultimately, the good life depends upon the freedom to make choices, even reinvent oneself and pursue the unknown, with new adventures. Ironically, once choices and, therefore, commitments are made, however, freedom of choice diminishes. Less time is left to make new choices and commitments. Irresolution can be a fatal flaw. Following William James, not making a decision constitutes a decision in itself. Even so, the good life remains an active work in progress, evolving until one's life is completed.

CHAPTER 21

IMAGINING LONGEVITY:
CHALLENGES OF THE FUTURE

Man seeks to become immortal, but seldom considers the consequences. From alchemy to modern science, from incantations to experiments, the search continues. Largely missing has been a simultaneous quest to understand what extended longevity would mean for individual people and cogent arguments in support of any advantages to society at large.

What passes for wisdom and judgment in older persons may be no more than cunning, guile, caution, and enough experience in life's various struggles to survive and find ways to avoid precipitous action.

Religious, spiritual, and medical rituals elaborately mask the dominating fear of death. Death, after all, is the master tailor of thought, and its garments come in many styles. It shapes our stories. It motivates our actions, noble and ignoble. Death invokes high adventure. It beckons equally the foolhardy and courageous to go to war and to engage in acts of daredevil defiance. It stimulates creativity. But it also prompts the search for longevity.

And now we have come to the possibility of humankind gaining some further degree of control over its life expectancy: through new conquests of disease, the regulation of aging per se,[1] or by achieving some hubristic and unlikely control of evolutionary destiny by breaking the species barrier to the length of life.[2] (See Table 21.1 to observe the impact of immortality on the growth of population.)

We have no known way to stop or reverse the basic aging processes in humans, but the future is not an automatic extension of the present. And, possibly, means now exist to slow aging in humans.

On the other hand, we have the potential to make good use of the additional time given us because of advances in public health and medical science. Would we make even better use of even more time? Is there a reason to finance my proposed Apollo-level research project on aging and longevity to buy us more time? Are we mature enough to be measured, for example, by our ability to resolve conflicts and to advance material, cultural, and intellectual resources that truly enrich *all* lives?

This book has focused upon several necessary and appropriate questions: How can increased longevity be financed? How might the generations better cohabit the same new sphere of increasing longevity? How might we better organize the medical and social machinery needed to maintain added life? How could the riches of the earth be properly used with relatively fewer newborns and more Methuselahs?

But questions remain whether human history so far justifies the extension of life beyond the maximum time individuals and societies already enjoy. Is it likely that we will be transformed by added longevity? Would our energies and resources be better served by dedicating ourselves to humanity's transformation?

Science and scholarship, public welfare, public health, and wealth have brought many benefactions, among them the advances in health that are modest but notable successes against the firm reality of death and a partial victory over the short and brutish life.

Some might say there is not much one can do but laugh at our ultimate helplessness before death and our curious, often fatuous and destructive behavior. To celebrate even a soupçon of longevity seems awkward in the face of human profligacy, inflated expectations, and ineffable sorrow.

Old age has been appraised over the last several decades as a promising, newly robust, healthier stage of life in the course of social reinvention and the establishment of a new vocabulary. Some studies suggest older people are less likely than younger people to react aggressively when problems arise in their relationships.[3] According to studies conducted by Laura Carstensen,[4] old people are better able to regulate their emotions than younger adults, the modern statement of "mellowing with age." Faced with disasters such as 9/11 and Hurricane Katrina, older people have proven remarkably resilient. Can we build upon these studies and anecdotal observations? If so, can we rededicate this, life's final chapter, to finding solutions to the problems that confront humanity?

Of all the great goals that might be imagined—tolerance, the conquest of poverty, the establishment of constructs for the peaceful resolution of conflicts, true concern for future generations, the quieting of our interior demons, human rights for all, and the mastery of forgiveness, among others—the longevity quest seems far less important. There are momentous matters of scale that we have rarely managed: scurrying about in our daily efforts to survive in times of misery, shouldering through the crowded marketplace of products, services, and ideas, self-centered and willing to crush any opposition by the "other" with malice and mendacity and seemingly wired to kill.[5] Our present state is still all too close to the Hobbesian jungle, and so our possible reinvention is only a desperate hope, a hope that we might be ennobled by longevity. Both our perfectibility and our social progress could come with increasing longevity, argued William Godwin and the Marquis de Cordorcet, Enlightenment philosophers and prolongevists who were roundly criticized by the Reverend Thomas Malthus. "Human nature modified," was Elie Metchnikoff's hope.

How possibly can older persons, world-weary, subjected to life's cruelties and corrosive ageism, often facing poverty, illness, and disability, pioneer in redefinition and take on the great challenging issues? How can old people adapt to what might be labeled *responsible aging*, one road to take to deserve the right to a longer life?

Given life's painful realities and appreciating the world's perils, older people usually hold back from speaking their minds, after years of practiced compliance, to which is added fear, self-doubt, inner conflicts, lack of education, and ignorance. No wonder the rigidity, docility, timidity, and the disillusion of many but not all in old age.

Arguably, the Longevity Revolution came so fast and is so new that there has not been time to practice optimism in getting things done nor even to simply reflect, rebel, and rebuild. Of course, there are older persons who do show leadership. Some speak out courageously. Yet more is lost by those who fear change than by those who seek it. Inaction and atrophy are the ultimate negative forces and a whimpering climax to life cowed by insecurity. This passivity of the old is not new. It reflects generations of caution, apathy, neglect, and subtle oppression. How might older persons be awakened and become engaged responsibly in the present, building a worthy epitaph to the future? How might they be mobilized to found a movement dedicated to various forms of social engagement? This is the true challenge.

Older persons would do well to heed the words of Cicero in *De Senectute*: "Old age will only be respected when it fights for itself, maintains its rights, avoids dependence on anyone, and asserts control over its own to the last breath."

LONGEVITY AND MORALITY

It is reasonable that humanity struggles to defeat illness and delay death. It is not surprising that some scientists seek to discover longevity genes today just as the alchemists in the age of Paracelsus endeavored to convert lead into gold, and the short Hobbesian life into longevity. But I ask again, is longer life a foolish goal to aspire to when we continue to abuse and kill one another, ourselves, and our environment? The ancient and modern history of human evil is painful—slavery, human sacrifice, rape, murder, war: it is endless.

Given the destructive character of humanity, would success advancing longevity simply extend the evil that human beings are capable of, giving them more time to wreak harm? The twentieth century, and so far the twenty-first, does not suggest a very optimistic answer to this dark and bleak question.

If we were given more time, would maturation of human character be possible? George Bernard Shaw believed humans needed centuries for civilizations to mature and for individuals to become truly civilized.

The great medical historian and scholar Henry Sigerist wrote that classic Greece amassed "an artistic and intellectual capital from which the Western world has been drawing for over 2000 years."[6] Moreover, we must draw upon the equally extraordinary "artistic and intellectual capital" amassed in the Eastern world. Ayurveda, "knowledge of long life" or "wisdom of long life," is but one example.

We also need new perspectives for psychotherapy as well as a philosophy of longevity, building upon the work of Martin Buber and Victor Frankl, among others, who sought meanings *in* life and new possibilities as well as traditional and new means of *redemption* through acts of expiation, atonement, and forgiveness.

Most authors who write of old age focus upon understanding aging, the inevitable crises of illness, loss, dying, and death. The prospect of an enriched longevity is little touched upon. One notable exception is the extraordinary novelist and Nobelist Kenzaburo Oë, who writes of modern Japan and informs us how we can live creatively with disability.

We need guidance in the new world of longevity, all the more so if we should succeed further in prolonging life. Aside from the thoughts of a handful of scholars and thinkers whose wisdom has come down to us through the ages,[7] little in Western literature exists to contribute to a systematic philosophy of the conduct of life in the later years. Since longevity was comparatively rare in the past, this is not so surprising.

In Gabriel Garcia Marquez's *Love in the Time of Cholera* we see the fulfillment of love made possible by longevity. Fifty-one years, nine months, and four days after its initiation, the love of Marquez's protagonist Florentino for Fermina Daza is fulfilled at last. Marquez notes that love, like cholera, reveals similar clinical symptoms such as dizziness, nausea, and fever, and both can result in death.

With older persons in greater numbers and proportions, their impact is increasingly felt in society, and modern artists are beginning to deal with great age, longevity, and death. Truly great films about old age have been made in Europe, Japan, and America. The films of Ingmar Bergman, Vittorio de Sica, Akira Kurosawa, and Billy Wilder come to mind.

Both the philosophy of childhood and the philosophy of old age have been neglected. We need guides that help us live out our lives, enjoy longevity, and contribute responsibility to others in the later years. We need a book on the philosophy of old age (distinct from longevity). For example, in *The Philosophy of Childhood,* Gareth B. Matthews wrote that his aim was "to convince my students that philosophy is a natural activity, quite as natural as making music and playing games." Matthews points out that Descartes taught students to do philosophy by "starting over, to ask fundamental questions as a child may." Thus, the naive questions of childhood are an important part of philosophy. As Matthews says, "The philosophic thinking of children may be quite independent of cognitive maturation." However, we rarely listen to children. In fact, we knock out a considerable amount of philosophic perspective and imagination from the minds of children. We shirk their questions. We superimpose complexity on the elementary truths of life. Perhaps, in old age, there can be a return to the more elemental, powerful forces, the freedom that comes from not having to obey a boss, a social structure, or even the scholarship of a lifetime. Perhaps it is old people who can point out the guideposts for others to follow in this new land of old age.

Around the 563 BC, Siddhartha Gautama, the Buddha, at twenty-nine years of age, discovered misery in the world, witnessing first a decrepit old man "wasted by age." This was a turning point in the rich Prince Gautama's protected life. Next he saw a sick man and learned that all men are subject to sickness. Then he saw a dead body. Finally, he saw a man who was serene despite old age, sickness, and death. Having experienced these four signs, Siddhartha decided to search for a solution to suffering. He sought enlightenment.

There is no reason to oppose increasing longevity. We already have increasing control over our own longevity, through our behavior, and *science will find new ways to extend life* and its quality. Evidence suggests that morbidity can be further compressed, and society can adapt to the growing numbers of older persons.[8]

What is at issue is quality of life, especially the interrelationships of population with societal and natural resources. Until we can do better, it is probably just as well if we do not have a breakthrough in longevity. Or, perhaps were we to have a breakthrough, would we move faster in making adjustments?

Or were we to further prolong life, would it reduce incentive? Would it destroy the motivation of scientists, artists, and others, freed from the threat of early death driving them on? Would the increased numbers and proportions of older persons provoke economic stagnation?

Robert Hutchings Goddard dreamed of interplanetary travel at the turn of the twentieth century and became the inventor of modern rocketry and author of the age of space flight. The twentieth century touched the frontier of space. Now in the twenty-first century we may be poised at the frontier of biological time—the prospect of germline engineering, the means by which the human species would direct its own evolution and extend its life span. James Watson, cowinner of the 1962 Nobel Prize for discovering the structure of DNA, favors such an approach. Enthusiasts over the future of cell, tissue, and organ replacement imagine successive, comprehensive reconstitutions of the body. Replacement or regenerative medicine would push death back, presumably indefinitely.

One must not doubt the possibility of the unexpected in science and the uneven evolution of knowledge.

.   .   .

The ancient Sumerian epic of Gilgamesh, which dates from an estimated fifteen hundred years before the legendary Homer and is older than the Bible, has been told and retold. It is a story of the revolt against death by the great King Gilgamesh and the death of his friend Enkidu. Gilgamesh sought to regain his friend's life, comprehend death, and secure immortality for himself. This became his quest. He did find a plant, the means to immortality. But Gilgamesh was arrogant, "blind with love of self and with rage." He left the plant unguarded. A serpent smelled its sweet fragrance and devoured it. The serpent shed its skin, which became symbolic of regeneration and eternal life. When Gilgamesh rose from a pool he found the plant was gone and the discarded skin of a serpent lay nearby. He sat down on the ground and wept.

APPENDIX: TABLES

TABLE 1
Share of Population in the United States
2000

	Number (million)	Share of Total Population
Total All Ages	282.92	100
Children (1995–2003)	19.22	6.8
Generation Y (1981–1995)	61.32	21.8
Generation X (1965–1980)	59.74	20.9
Baby Boom (1946–1964)	83.19	29.5
World War II + (–1945)	59.45	21.0

Source: Population Divisions, U.S. Census Bureau, 2000. *Intercensal Estimates of the United States Civilization Population by Age and Sex 2000.* Washington, DC: Census Bureau. Accessed on March 9, 2006.

TABLE 1.1
The New Longevity: Life Expectancy from Age 65

MALE		FEMALE	
Rank/Country	**Additional Years**	**Rank/Country**	**Additional Years**
1. Hong Kong	17.1	1. Japan	22.0
1. Japan	17.1	2. France	20.9
3. Israel	16.6	3. Hong Kong	20.7
3. Switzerland	16.6	4. Switzerland	20.5
5. France	16.4	5. Italy	20.4
5. Greece	16.4	6. Spain	20.3
7. Australia	16.3	7. Canada	20.1
7. Canada	16.3	8. Australia	20.0
7. Spain	16.3	8. Sweden	20.0
7. Sweden	16.3	10. Norway	19.6
11. Italy	16.1	11. New Zealand	19.5
11. New Zealand	16.1	12. Austria	19.3
13. United States	16.0	13. Belgium	19.3
14. Norway	15.7	14. United States	19.2
15. Austria	15.6	15. Finland	19.1
16. England and Wales	15.5	16. Germany	19.0
17. Germany	15.3	17. Israel	18.9

TABLE 1.1, cont.
The New Longevity: Life Expectancy from Age 65

MALE		FEMALE	
Rank/Country	Additional Years	Rank/Country	Additional Years
18. Belgium	15.2	18. Netherlands	18.8
18. Singapore	15.2	19. England and Wales	18.7
20. Chile	15.1	19. Greece	18.7
21. Finland	14.9	21. Northern Ireland	18.5
21. Northern Ireland	14.9	22. Chile	18.4
23. Denmark	14.8	23. Denmark	18.1
24. Netherlands	14.7	24. Portugal	17.9
25. Portugal	14.3	25. Ireland	17.7
26. Ireland	14.2	25. Singapore	17.7
26. Scotland	14.2	27. Scotland	17.4
28. Czech R	13.4	28. Poland	17.0
28. Poland	13.4	29. Czech R	16.9
30. Romania	13.0	30. Slovakia	16.6
31. Slovakia	12.9	31. Hungary	16.0
32. Bulgaria	12.5	32. Romania	15.5
33. Hungary	12.2	32. Russian Federation	15.5
34. Russian Federation	11.6	34. Bulgaria	15.1

Centers for Disease Control and Prevention, Health, United States, 2003.

TABLE 1.2
Years of Life Remaining at 50

1900–1902	21.3
1909–1911	21.0
1919–1921	22.5
1929–1931	21.1
1939–1941	23.0
1949–1951	24.4
1959–1961	25.3
1969–1971	25.9
1979–1981	27.9
1989–1991	29.0
2000	30.0

Source: National Vital Statistics Report, Vol. 51, No. 3, December 19, 2002, Table 11.

TABLE 1.3
10 Countries Where Men Outlive Women

Country	Male	Female
1) Botswana	37	36
2) Kenya	37	36
3) Zambia - ?*	39	39
4) Niger	42	41
5) Zimbabwe	50	49
6) Pakistan - ?	62	62
7) Bangladesh - ?	63	63
8) Maldives	66	64
9) Tonga - ?	71	71
10) Qatar	75	74

Source: World Health Organization. Annex table 1: Basic indicators for all WHO Member States. The World Health Report 2005. Geneva: World Health Organization, 2005.

*I have placed a ? after the countries where the differences in life expectancy are close and the reliability of data are uncertain.

TABLE 2.1
"Silver Industries" or The Mature Market

Businesses involved directly with the older population (50 and over) and with longevity issues:

Financial Services

❏ insurance (life, disability, long-term care, other insurance, and annuity products),

❏ banking (including trust departments, estate management),

❏ financial management (for savings and investment [e.g., mutual funds], retirement and estate planning).

Legal Services

❏ legal services (for wills, living wills and advance directives, durable powers of attorney, trusts, and other documents related to estates).

Health-Care Services

❏ health-care services,

❏ long-term care (adult day care, high-tech home care, low-tech home care, nursing homes, hospitals, rehabilitation centers, health maintenance organizations and variants of group medical practice combined with hospital and other institutional care),

❏ pharmaceuticals, nutritional products, medical devices,

❏ medical equipment (e.g., wheelchairs) and supplies,

❏ personal care and transportation (homemakers, chore personnel, drivers, meal preparation).

TABLE 2.1, cont.
"Silver Industries" or The Mature Market

Businesses involved directly with the older population (50 and over) and with longevity issues:

Housing

___

❏ housing for retired persons (e.g., Marriott, Hyatt and other for-profit companies and nonprofit organizations that build or manage continuing care retirement communities [CCRCs], congregate housing facilities and assisted living; also companies that design and build houses),

❏ household products, especially those that promote independent living (including prosthetic, assistive devices and robotics for such tasks as dressing, toileting, picking up objects, moving about the household, bathing, and personal safety, telephone, and other electronic equipment for the safety and support of persons with physical and mental disabilities).

Other

___

❏ clothing (special fits), cosmetics,

❏ travel, hospitality, and tourism, recreation and entertainment,

❏ information and communication (computers, cellular phones),

❏ education (such as Elderhostel),

❏ spending on children and grandchildren,

❏ reminiscence and remembrance (books, videos),

❏ end-of-life services (funerals, cemeteries, memorials).

Note: There are sixty-nine million Americans over 50 who can be divided into three markets: youngest, 50–64, in their peak years, young retirees, 65–79, older retirees, 80 and above.

TABLE 3.1
"Dependency Ratio"
Comparison of Numbers of Persons Under 18 and Over 65
(The "Dependents") to the Middle Generations (The "Workers")

Year	Number of Persons Under Age 18 per 100 Persons Age 18 to 64	Number of Persons Age 65+ per 100 Persons Age 18 to 64	Total Number of "Dependents" per 100 "Workers"
Estimates			
1900	72.6	7.3	79.9
1910	65.7	7.5	73.2
1920	64.0	8.0	72.0
1930	58.6	9.1	67.7
1940	48.8	10.9	59.7
1950	51.1	13.4	64.5
1960	65.3	16.9	82.2
1970	61.1	17.6	78.7
1980	46.2	18.7	64.9
1990	41.7	20.3	62.0
1995	42.7	20.9	63.6
2000	41.3	20.4	61.7
Projections			
2005	39.8	20.2	60.0
2010	38.3	21.1	59.4
2020	39.8	27.7	67.5
2030	42.1	35.6	77.7
2040	42.1	36.5	78.6
2050	39.0	33.4	72.4

Source: U.S. Bureau of the Census (January 2000). Resident Population Estimates of the United States by Age and Sex: April 1, 1990, to July 1, 1999, Current Population Reports, and Population Projections of the United States by Age, Sex, Race, and Hispanic Origin: 2001 to 2050. Washington, DC; U.S. Government Printing Office.

TABLE 3.2
Population of Countries
With at Least 10 Percent of Their Population Age 65 and Over, 2003

Region or Country	Total Population	Total 65 and Over	% 65 and Over
Italy	57,998,353	10,893,973	18.8
Japan	127,214,499	23,720,030	18.6
Greece	10,625,945	1,947,336	18.3
Germany	82,398,326	14,643,067	17.8
Spain	40,217,413	7,075,743	17.6
Sweden	8,970,306	1,545,515	17.2
Belgium	10,330,824	1,777,398	17.2
Bulgaria	7,588,399	1,293,949	17.1
Portugal	10,479,955	1,749,225	16.7
France	60,180,529	9,801,524	16.3
Croatia	4,497,779	723,788	16.1
Estonia	1,350,722	217,199	16.1
Austria	8,162,656	1,282,955	15.7
United Kingdom	60,094,648	9,429,087	15.7
Finland	5,204,405	805,215	15.5
Latvia	2,322,943	358,400	15.4
Switzerland	7,408,319	1,131,164	15.3
Ukraine	48,055,439	7,212,722	15.0
Georgia	4,710,921	706,380	15.0

TABLE 3.2, cont.
Population of Countries
With at Least 10 Percent of Their Population Age 65 and Over, 2003

Region or Country	Total Population	Total 65 and Over	% 65 and Over
Denmark	5,394,138	802,456	14.9
Norway	4,555,400	676,160	14.8
Hungary	10,057,745	1,492,216	14.8
Slovenia	2,011,604	298,344	14.8
Serbia and Montenegro	10,823,280	1,592,794	14.7
Lithuania	3,620,094	530,425	14.7
Luxembourg	456,764	65,985	14.4
Belarus	10,322,151	1,478,835	14.3
Romania	22,380,273	3,169,849	14.2
Czech Republic	10,251,087	1,432,188	14.0
Netherlands	16,223,248	2,241,317	13.8
Russia	144,457,596	19,203,848	13.3
Malta	395,178	51,969	13.2
Uruguay	3,381,606	442,733	13.1
Canada	32,207,113	4,167,291	12.9
Poland	38,622,660	4,924,081	12.7
Australia	19,731,984	2,502,665	12.7
United States	290,342,554	35,878,341	12.4
Hong Kong S.A.R.	6,809,738	836,153	12.3

TABLE 3.2, cont.
Population of Countries
With at Least 10 Percent of Their Population Age 65 and Over, 2003

Region or Country	Total Population	Total 65 and Over	% 65 and Over
Puerto Rico	3,878,679	461,501	11.9
Iceland	291,064	34,055	11.7
Slovakia	5,416,406	630,190	11.6
New Zealand	3,951,307	457,805	11.6
Ireland	3,924,023	447,070	11.4
Cyprus	771,657	85,629	11.1
Macedonia	2,063,122	217,965	10.6
Argentina	38,740,807	4,042,311	10.4
Martinique	425,966	43,818	10.3
Armenia	3,001,712	306,182	10.2
Moldova	4,439,502	452,797	10.2
Bosnia and Herzegovina	3,989,018	401,929	10.1

Source: U.S. Census Bureau, International Data Base, 2004.

TABLE 4.1
2006 Official Poverty Data
Realities for All People Over 65 Per Individual

Poverty Line for 65+   Percentage of the 65+ population below the poverty line 9.42%:

$9,669/Year or Less

About $185/Week or Less

About $26/Day or Less

Poverty: Percentage of the 65+ population below 125% of poverty line: 16.65%:

$12,086/Year or Less

About $232/Week or Less

About $33/Day or Less

Poverty: Percentage of the 65+ population below 150% the poverty line: 22.35%

$14,503/Year or Less

About $279/Week or Less

About $39/Day or Less

Source: U.S. Census Bureau 2006 "Official Poverty Tables," accessed August 2007 from http://pubdb3.census.gov/macro/032007/pov/toc.htm. The Census Bureau 2006 Official Poverty Tables are derived from the 2007 Current Population Survey (March Supplement).

TABLE 4.2
"The Social Security Vanishing Act"
What the 2006 Cost-of-Living Adjustment Is Up Against

Social Security

Cost-of-Living adjustment: up 4.1%

Average monthly benefit in 2005: $963

Average monthly benefit in 2006: $1,002

Medicare

Part B: Monthly premium: $88.50 (up $10.30 from 2005)

Part D: (Drug Benefit): Average monthly premium: $32

Inflation (9 months since Jan. 1, 2005): up 5.1%

Energy Prices: up 42.5%

Food Prices: up 2.1%

Medical Care Costs: up 4%

Source: AARP Bulletin, December 2005.

TABLE 5.1
Divorce Rate for Selected Countries

Cases per 1,000 People

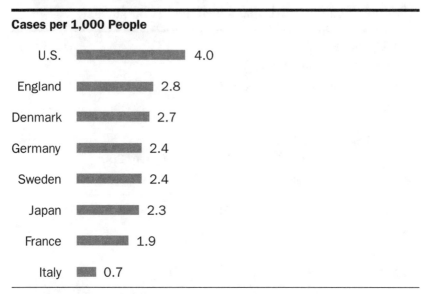

U.S.	4.0
England	2.8
Denmark	2.7
Germany	2.4
Sweden	2.4
Japan	2.3
France	1.9
Italy	0.7

Note: France, Germany, and Italy data is for 2000; Denmark and Sweden, 2001; all others, 2002.

Sources: Japanese Ministry of Health, Labor and Welfare; Council of Europe; Associated Press.

TABLE 6.1
Leading Causes of Death in the United States for People 65 and Older in 2001

Cause of Death	No. of Deaths	Percentage
Heart Disease	582,730	32%
Malignant Neoplasms	390,214	22%
Cerebrovascular Diseases	144,486	8%
Chronic Lower Respiratory Diseases	106,904	6%
Influenza and Pneumonia	55,518	3%
Diabetes Mellitus	53,707	3%
Alzheimer's Disease	53,245	3%
Nephritis, Nephrotic Syndrome, and Nephrosis	33,121	2%
Unintentional Injuries	32,694	2%
Septicemia	25,418	1%
Other	320,383	18%
Total	**1,798,420**	**100%**

Source: Centers for Disease Control and Prevention, National Center for Health Statistics, National Vital Statistics System, 2003.

TABLE 6.2
Leading Causes of Death in the United States in 2000

Cause of Death	No. of Deaths	Death Rate Per 100,000 Population
Heart Disease	710,760	258.2
Malignant Neoplasms	553,091	200.9
Cerebrovascular Diseases	167,661	60.9
Chronic Lower Respiratory Diseases	122,009	44.3
Unintentional Injuries	97,900	35.6
Diabetes Mellitus	69,301	25.2
Influenza and Pneumonia	65,313	23.7
Alzheimer's Disease	49,558	18.0
Nephritis, Nephrotic Syndrome, and Nephrosis	37,351	13.5
Septicemia	31,224	11.3
Other	499,283	181.4
Total	**2,403,351**	**873.1**

Source: Centers for Disease Control and Prevention, National Center for Health Statistics, National Vital Statistics System, 2003.

TABLE 6.3
2001 Healthy Life Expectancy (HALE) at birth in all WHO Member States

Country	Years	Country	Years
Japan	73.6	Belgium	69.7
San Marino	72.2	United Kingdom	69.6
Sweden	71.8	Malta	69.2
Australia	71.6	Singapore	68.7
France	71.3	Slovenia	67.7
Monaco	71.3	United States	67.6
Iceland	71.2	China	63.2
Austria	71.0	India	51.4
Italy	71.0	Swaziland	33.9
Andorra	70.9	Rwanda	33.8
Luxembourg	70.6	Afghanistan	33.4
New Zealand	70.3	Zimbabwe	31.3
Germany	70.2	Malawi	29.8
Greece	70.2	Angola	28.7
Denmark	70.1	Sierra Leone	26.5
Canada	69.9		

Healthy life expectancy estimates published here are not directly comparable to those published in *The World Health Report 2001* because of improvements in survey methodology and the use of new epidemiological data for some diseases and revisions of life tables for 2000 for many Member States to take new data into account. The futures reported in the table along with the data collection and estimation methods have been largely developed by the World Health Organization (WHO) and do not necessarily reflect official statistics.

Source: *The World Health Report 2002,* Annex Table 4. Further development in collaboration with Member States is under way for improved data collection and estimation methods. Figures not endorsed by Member State as official statistics.

TABLE 6.4
Actual Causes of Death in the United States in 1990 and 2000

Actual Cause	No. (%) in 1990*	No. (%) in 2000**
Tobacco	400,000 (19)	435,000 (18.1)
Poor Diet and Physical Activity	300,000 (14)	400,000 (16.6)
Alcohol Consumption	100,000 (5)	85,000 (3.5)
Microbial	90,000 (4)	75,000 (3.1)
Toxic Agents	60,000 (3)	55,000 (2.3)
Motor Vehicle	25,000 (1)	43,000 (1.8)
Firearms	35,000 (2)	29,000 (1.2)
Sexual Behavior	30,000 (1)	20,000 (0.8)
Illicit Drug Use	20,000 (>1)	17,000 (0.7)
Total	**1,060,000 (50)**	**1,159,000 (48.2)**

* McGinnis, J. Michael, and Foege, William H., Actual causes of death in the United States. *JAMA* (1993) *270*: 2207–2212.

** Mokdad, Ali H., Marks, James S., Stroup, Donna F., and Gerberding, Julie L., 2004.

TABLE 6.5
Chronic Conditions by Age and Gender (1995)

Age	Male	Female
50–64	Hypertension	Arthritis
	Hearing Impairment/Deafness	Hypertension
	Arthritis	COPD
	Heart Disease	Heart Disease
	Diabetes	Hearing Impairment/Deafness
	Chronic Obstructive Pulmonary Disease (COPD)	Diabetes
	Visual Impairment/Blindness	Visual Impairment/Blindness
65–74	Arthritis	Arthritis
	Hypertension	Hypertension
	Hearing Impairment/Deafness	Heart Disease
	Heart Disease	Hearing Impairment/Deafness
	Diabetes	Cataract
	COPD	Diabetes
	Cataract	COPD
75+	Hearing Impairment/Deafness	Arthritis
	Arthritis	Hypertension
	Hypertension	Hearing Impairment/Deafness
	Heart Disease	Heart Disease
	Cataract	Cataract
	Visual Impairment/Blindness	Diabetes
	Diabetes	COPD

Source: National Center for Health Statistics, National Health Interview Survey, unpublished data, August 2001.

TABLE 6.6
Disorders of Longevity

❏ Fetal, Infant, and Childhood Origins—Genetic, Environmental, and Behavioral (such as low birth weight)

❏ Polygenic Disorders, such as coronary heart disease

❏ Medicated and Surgical Survival and Success of Neonatology

❏ Lifelong Exposure to Environmental and Behavioral Factors

❏ Poor Lifestyle and "Wear and Tear"

❏ Accumulation of Deleterious Genetic Components Through Natural Selection

❏ Gompertz/Aging as Risk Factor/Declining "Defense" and "Repair" Mechanisms

❏ Prions (misfolded proteins) incubate for long periods of time

❏ Conversion of some acute diseases to chronic diseases, (such as AIDS and some cancers)

❏ Late-Life Adverse Events, such as pneumonia

TABLE 7.1
Expectation of Years of Life Remaining at Single Years of Age, United States: 2000

Age	Total Population	Male	Female
0	76.9	74.1	79.5
5	72.5	69.8	75.1
10	67.6	64.9	70.1
15	62.6	59.9	65.2
20	57.8	55.2	60.3
25	53.1	50.6	55.4
30	48.3	45.9	50.6
35	43.6	41.3	45.8
40	38.9	36.7	41.0
45	34.4	32.2	36.3
50	30.0	27.9	31.8
55	25.7	23.8	27.4
60	21.6	19.9	23.1
65	17.9	16.3	19.2
70	14.4	13.0	15.5
75	11.3	10.1	12.1
80	8.6	7.6	9.1
85	6.3	5.6	6.7
90	4.7	4.1	4.8
95	3.5	3.1	3.5
100	2.6	2.4	2.7

Source: National Vital Statistics Reports, Vol. 51, No. 3.

TABLE 11.1
Guralnik's Chair Test

Fold Your Arms Across Chest

Stand Up from Sitting Position

Do These Squats 5 Times ASAP

Scoring: The Amount of Time it Takes to Do 5 Squats

❑ If Chair stand time is 16.70 sec or more: 1 Point

❑ If Chair stand time is 13.70 to 16.69 sec: 2 Points

❑ If Chair stand time is 11.20 to 13.69 sec: 3 Points

❑ If Chair stand time is 11.19 sec or less: 4 Points

Key Predictor of Frailty is the Strength of the Quadriceps

Data derived from a study at the National Institute on Aging and participants were 71 and older.

The lower the score the more an individual should do squats every other day; that is, stand up from the sitting position with one's back held against the wall.

TABLE 12.1
Suicide Rates per 100,000 by Age, United States, 2000

Older People (Ages 65+)	15.6 per 100,000 persons
Young People (Ages 15–24)	9.9 per 100,000 persons
All Ages	11 per 100,000 persons

Source: Office of Statistics and Programming, National Center for Injury Prevention and Control, Centers for Disease Control and Prevention (CDC) Data Source: NCHS Vital Statistics System for Number of Deaths, Bureau of Census for Population Estimates. 2005. Suicide Injury Deaths and Rates per 100,000. Atlanta: Centers for Disease Control and Prevention.

TABLE 12.2
Health Care Expenditures per Capita in Selected Countries, 2002

USA	$5,267*
Canada	$2,931
Germany	$2,817
U.K.	$2,161

*The U.S. has the highest expenditures but the lowest life expectancy and highest infant mortality compared with these three countries.

TABLE 12.3
The Quality of Medical Services Varies Widely

❏ 3% of hospital patients hurt by medical error.

❏ 1 in 300 patients die from such mistakes.

❏ 90,000 people die from such mistakes.

❏ 180,000 elderly outpatients die or are harmed by drug toxicity.

❏ 7,000 patients die from drug errors each year.

❏ 554 errors in four months were found at one six-bed intensive care unit—
147 were potentially serious or life-threatening.

❏ 55% of recommended care actually gets administered.

❏ $2,000 annual cost to employers per insured worker, due to poor quality care.

Compiled by *Forbes* magazine. Sources: Lucian Leape, *New England Journal of Medicine,* CDC; *Forbes*; *Journal of the American Medical Association*; Institute of Medicine, Quality and Safety in Health Care; RAND; Midwest Business Group on Health.

2006 Mercer Health & Benefits LLC, all rights reserved.

Presentation by Alexander Domaszewicz, Health Care Consumerism and Account-Based Strategies Symposium, Envisioning the Future, MetLife/ILC, Washington, DC, April 3, 2006.

TABLE 13.1
Impact of Retirement of the Baby Boomers

❏ About 50% of 1.6 million federal workers are eligible to retire by 2008.

❏ Nuclear power industry expects 28% of 58,000 workers to retire within 5 years; another 18% of young employees leave.

❏ By 2010, more than 40% of RNs will be over age 50; 50% of nurses will retire within 15 years.

❏ Record low number of students in petroleum engineering programs even as 60% of employees retire by 2010.

❏ Retirement of executive and managerial talent.

Source: David W. DeLong, *Lost Knowledge,* 2004

TABLE 14.1
Lack of Public Knowledge About Social Security

The results:

❏ Only half of all adults know that Society Security guarantees payment for life.

❏ Only a quarter of adults know that Social Security guarantees protection against inflation.

❏ Only about half of all adults know that Social Security provides life and disability coverage for spouses or children of workers who die or are disabled.

❏ Only about one adult in every six knows that Social Security has lower administrative costs than private pension and retirement plans.

❏ Almost half of adults do not know that Social Security has never failed to pay benefits.

Source: Harris Poll with International Longevity Center (2005).

TABLE 14.2
Sources of Income

Income of people age 65 and older in the U.S. in 2000:

❑ 40 percent from Social Security

❑ 18 percent from pensions

❑ 23 percent from continuing employment

❑ 18 percent from assets

❑ 1 percent from other sources

Sources: *Income of the Population 55 or Older, 2000,* Social Security Administration, Office of Research, Evaluation and Statistics, February 2002.

http://www.ssa.gov/policy/docs/statcomps/income_pop55/2000/sect1.html#t1_1

Hungerford, Thomas, Rassette, Matthew, Iams, Howard, and Koenig, Melissa, Trends in the economic status of the elderly, 1976–2000, *Social Security Bulletin* 2001–2002, 64(3):12–22.

http://www.ssa.gov/policy/docs/ssb/v64n3/v64n3p12.html#mn1

TABLE 14.3
Social Security as Economic Stimulus

Social Security helps the U.S. economy, which is two-thirds consumer-driven and stimulates local economies. Examples include:

❑ Florida enjoys $55 billion annually in income from Social Security beneficiaries—more than tourism or agriculture.

❑ Other Sunbelt states such as Arizona and California enjoy similar benefits.

❑ Morganton, NC, calculates that each retired couple living in their community generates as much economic benefit as 3.7 manufacturing jobs.

Source: "North Carolina Community Seeks Active Retirees," press release, John Cantrell, Councilman of Morganton, NC, 2004.

TABLE 16.1
Presidential Electoral Votes and the 65+ Votes

State	Electoral Votes	Population 65 and Over
California	54	3.6 Million
New York	33	2.4 Million
Texas	32	2 Million
Florida	25	2.8 Million
Pennsylvania	23	1.9 Million
Illinois	22	1.5 Million
Ohio	21	1.5 Million
Michigan	18	1.2 Million
New Jersey	15	1.1 Million
North Carolina	14	.97 Million
Total	**257**	**17.5 Million**

TABLE 16.2
Amount Spent on Federal Lobbying in 2004, in Millions of Dollars

Health Care	$325
Communications, Technology	283
Finance, Insurance	279
Business, Retail, Services	165
Transportation	162
Energy, Natural Resources	157
Defense	93
Single-Issue Groups*	90
Manufacturing	88
Agriculture	79

Sources: *Political Money Line*; Secretary of the Senate; Clerk of the House.

* Includes abortion, environment, and gun control groups.

TABLE 17.1
Most Populated Nations
2050

India*	1.53	Billion
China	1.48	Billion
United States	349	Million
Pakistan	345	Million
Indonesia	312	Million
Nigeria	244	Million
Brazil	244	Million

* India overtakes China between 2045 and 2050. China's infant mortality rate is better, 38 deaths per 1,000 live births, compared with 70 in India.

TABLE 17.2
Shrinkage of Europe and Japan
in Millions

	2001		**2050**
Europe	726.3	➔	603.3
Japan	127.3	➔	109.2

Source: UN Population Fund.

TABLE 17.3
Countries or Areas with Highest Infant Mortality

Death/1,000 Live Births

Highest		Lowest	
Sierra Leone	170	Japan	4
Afghanistan	151	Singapore	5
Malawi	138	Norway	5
East Timor	135	Germany	5

TABLE 18.1
Leading Infectious Causes of Death Worldwide, 2001

Cause	Rank	Estimated Number of Deaths
Respiratory Infections	1	3,871,000
HIV/AIDS	2	2,866,000
Diarrheal Diseases	3	2,001,000
Tuberculosis	4	1,644,000
Malaria	5	1,124,000
Measles	6	745,000
Pertussis	7	285,000
Tetanus	8	282,000
Meningitis	9	173,000
Syphilis	10	167,000

Source: WHO, 2002.

TABLE 18.2
Increasing Burden of Diseases and Injuries
Change in rank order of DALYs* for the 15 leading causes

1999 Disease or Injury

1. Acute lower respiratory infections.

2. HIV/AIDS

3. Perinatal Conditions

4. Diarrheal Disease

5. Unipolar Major Depression

6. Ischemic Heart Disease

7. Cerebrovascular Disease

8. Malaria

9. Road Traffic Injuries

10. Chronic Obstructive Pulmonary Disease

11. Congenital Abnormalities

12. Tuberculosis

13. Falls

14. Measles

15. Anemias

DALY* = Disability-Adjusted Life Year
Source: WHO, Evidence, Information and Policy, 2000.

TABLE 18.3
Life Expectancy in 50 Places

2004 estimates for life expectancy at birth. Life expectancy in the 50 places with the longest-lived populations (* indicates tie):

Rank/Country	Years	Rank/Country	Years
1. Andora	83.5	19. Spain	79.3
2. Macau	82	20. Norway	79.2
3. *San Marino	81.5	21. Israel	79.1
3. *Singapore	81.5	22. Jersey	79.09
5. Hong Kong	81.3	23. Faroe Islands	79.05
6. Japan	81	24. Aruba	78.98
7. Switzerland	80.31	25. Greece	78.94
8. Sweden	80.3	26. Martinique	78.88
9. Australia	80.2	27. Austria	78.87
10. Iceland	80.18	28. Virgin Islands	78.7
11. Guernsey	80.17	29. *Malta	78.68
12. Canada	79.9	29. *Netherlands	78.68
13. Cayman Islands	79.8	31. Luxembourg	78.58
14. Italy	79.5	32. Germany	78.54
15. Gibraltar	79.52	33. Montserrat	78.53
16. France	79.44	34. New Zealand	78.49
17. Monaco	79.42	35. Belgium	78.44
18. Liechtenstein	79.4	36. Saint Pierre and Miquelon	78.28

TABLE 18.3, cont.
Life Expectancy in 50 Places

2004 estimates for life expectancy at birth. Life expectancy in the 50 places with the longest-lived populations (* indicates tie):

Rank/Country	Years	Rank/Country	Years
37. United Kingdom	78.27	44. Saint Helena	77.5
38. Finland	78.24	45. Puerto Rico	77.49
39. Isle of Man	78.16	46. Cyprus	77.46
40. Guam	78.12	47. Denmark	77.44
41. Jordan	78	**48. United States**	**77.43**
42. Guadeloupe	77.7	49. Ireland	77.36
43. Bermuda	77.6	50. Portugal	77.35

Source: *CIA World Factbook 2004.*

TABLE 20.1
Quality-of-Life Indicators

Physical Well-Being

Energy and Function

Sexuality

Quality Health Care and Health

Freedom from Pain

Preservation of Senses (e.g., vision and hearing)

Adequate Rest and Sleep

Mobility

Financial and Material Well-Being

Financial Security and Independence

Income from a Variety of Sources

Property

Employment

Social Mobility

Personal Well-Being

Mental Health and Happiness

Self-esteem/Dignity

Identity, Continuing Growth/Re-invention of the self through life

Body Image/Appearance

Memory

Control Over One's Life/Independence

Morale

Freedom from Excessive Stress

Adaptiveness

Choice/Opportunity

Education

Love

TABLE 20.1, cont.
Quality-of-Life Indicators

Social Well-Being

Having Family, Friendship, Neighborhood, Social Network, and Support System

Internet Access

Environmental Well-Being Clean Air and Water

Public Order and Safety (Crime-Free)

Purposeful Well-Being

Contribution to Others/Altruism, Philanthropy

Productive Aging or Engagement

Knowledge (Truth)

Aesthetic Well-Being

Exposure to Theatre, Music, Arts, Humanities (Beauty)

Leisure time

Learning

Joyfulness

Pleasures—Big and Small

Merriment—Exhilaration

Food—Recreation—Travel

Adventure—Excitement

Moral Well-Being

Clear Conscience (Goodness)

Spirituality

Beyond Self

Personal Beliefs

Serenity—Tranquility

Meditation

TABLE 20.1, cont.
Quality-of-Life Indicators

Enjoyment of One's Lifetime

Feeling that "It has been good to be alive"

Life Satisfaction

Reminiscence and Life Review

Accomplishments

Full Life

Creativity

Serenity

Present-ness

Elementality, Simplicity

Freedom from Preoccupation with Past and Future

End of Life

Quality of Dying

Sense of Control

Quality of Care (Control of Pain and Suffering)

TABLE 21.1
Population Growth Rates With and Without Immortality

Year	Birth Rate	Death Rate	Growth Rate	Population Doubling Time (Years)
1000*	~70	~69.5	~0.1	~800 – 1,000
1900	50	30	2.0	35
1950	45	15	3.0	26
2000	15	10	0.5	140
Immortality	~15	0	1.5	~53
Immortality**	~10	~0.1	~0.9	~80

* The birthrate and death rate in the year 1,000 cannot be known with certainty. These numbers are used to illustrate that vital rates were extremely high by comparison to today, and that the birthrate throughout most of human history hovered, on average, just above the death rate.

** Birthrates would likely decline if immortality was achieved. The estimated birthrate of 10 per thousand is speculated and perhaps even an overestimate. A death rate of zero is impossible to achieve in the real world where accidents, homicide, and suicide are present. The difference between the vital rates under the more realistic demographic conditions that might occur in the presence of immortality would lead to a growth rate of less than one percent and a population doubling time of approximately 80 years.

I created two immortality scenarios—one with a birthrate of 15 per thousand and a death rate of 0. The second was with a birthrate of 10 per thousand and a death rate slightly above zero (a more likely scenario if immortality is ever achieved). Let me provide a brief explanation. One of the major complaints about the idea of immortality is that it would create runaway population growth. In the long-term, say, past 100 years or so, this would indeed be a problem. However, if we achieved immortality today, the population growth rate would actually be smaller than it was in the post–World War II era, and actually, not that much lower than it is today. Remember, the growth rate is defined by the birthrate minus the death rate. If we are immortal, then the growth rate is defined only by the birthrate, which is already at exceedingly low levels. Thus, true immortality would not have that much of an effect on the growth rate of the population.

S. Jay Olshansky
April 25, 2007.

NOTES

Preface

1. As the cultural historian Jacques Barzun has said, "Revolution is a process, not an event."
2. Baby boomers in other nations have similar hopes. The "population bulge" that they represent will severely test the capacity of all nations and of individuals themselves to provide for a decent old age unless plans are laid *now*.
3. Butler, Robert N. Medicine's responsibility to the longevity revolution, *Medical Times* (1986) *114*: 25–26.

Chapter 1: What Is the Longevity Revolution?

1. When the American Revolution began in 1775, only 2 percent of the population was over age sixty-five (compared to nearly 13 percent today). Average life expectancy for both sexes was thirty-five years.
2. http://www.utexas.edu/depts/classics/documents/Life.html.
3. Of course, similar events occurred in Asia. Increasingly, more is being discovered about the development of China.
4. What was the Industrial Revolution due to after all? Ultimately, fossil energy. What did it bring? Clean water, nighttime illumination, and winter warmth, but also pollution and global warming.
5. Robert W. Fogel and Donna L. Costa, "A Theory of Technophysio Evolution, with Some Implications for Forecasting Population, Health Care Costs, and Pension Costs," *Demography* (1997) *34*: 49–66.
6. Taller people have always lived longer, according to a study of bones dating from the ninth century. But the little people of Krk, the largest of Croatia's Adriatic islands, are an exception.

7. See Robert W. Fogel, "Can We Afford Longevity?" In *Longevity and Quality of Life,* Robert N. Butler and Claude Jasmin (eds.) (New York: Kluwer Academic/Plenum Publishers, 2000), 47–59.

8. René Dubos, *Mirage of Health: Utopias, Progress and Biological Change* (New York: Harper, 1959).

9. Bacterial infection (usually streptococcal) of the uterus and genital tract causing severe pain. In the eighteenth and nineteenth centuries, it affected an average of six to nine women in every thousand deliveries, killing two to three with peritonitis or septicemia. It caused half of all deaths related to childbirth.

10. American physician and author Oliver Wendell Holmes also called attention to the contagiousness of puerperal fever in 1843.

11. Defined by Friedrich Engels, collaborator of Karl Marx, as "the class of modern capitalists, owners of the means of social production and employers of wage labor."

12. See Donald Sassoon, *One Hundred Years of Socialism: The West European Left in the 20th Century* (New York: The New Press, 1996).

13. In 2001, the emphasis of the George W. Bush administration on faith-based charities has raised troubling issues connected with maintaining a firewall between church and state, deeply embedded in the American Constitution and tradition. The idea that the government's responsibility to the poor should be turned over to religious charities has been urged by Marvin Olasky, author of *The Tragedy of American Compassion* (Washington, DC: Regnery). The Mormon Church has built a notably successful private welfare system that expects recipients to give some service in return for avoiding the "evils of a dole."

14. Actually, two social insurance programs were created, a federal-state program of unemployment compensation and a federal program of old age national insurance, *not* an investment program.

15. See Oliver Wendell Holmes, American physician and poet, "The Deacon's Masterpiece: The Wonderful One Hoss Shay" in *The Complete Poetical Works of Oliver Wendell Holmes* (Boston: Houghton Mifflin, 1895). This poem is suggestive of what James F. Fries at Stanford has called the "compression of morbidity," which many have begun to experience: less intensive and shorter duration of illness and disability before death.

16. In addition, legal and illegal immigration, usually of younger persons, decreased the share of the population over sixty-five.

17. Aubrey de Grey, "A Cure for Aging," *The Speculist,* August 6, 2003, www.speculist.com/archives/000056.html.

18. John Harris, "Intimations of Immortality—The Ethics and Justice of Life Extending Therapies," Hatch Lecture, International Longevity Center, 2002.

19. R.M. Suzman, D.P. Willis, and K.G. Manton (eds.), *The Oldest Old* (New York: Oxford University Press, 1993).

20. Kenneth Manton, XiLiang Gu, and Vicki L. Lamb, "Change in Chronic Disability from 1982 to 2004/2005 as Measured by Long-term Changes in Function and Health in the U.S. Elderly Population," *Proceedings of the National Academy of Science* (2006) *103*: 18374–18379.

Chapter 2: The New Longevity Is the Biggest Challenge

1. In "The Birth of a Revolutionary Class," *The New York Times Magazine,* May 15, 1996: 46–47.

2. Catherine Hoffman, Dorothy P. Rice, and Hia Y. Sung, "Persons with Chronic Conditions: Their Prevalence and Costs," *Journal of the American Medical Association* (1996) *276*: 1473–1479.

3. Ibid., 1477.

4. http://www.cdc.gov/nccdphp/overview.htm.

5. http://www.partnershipforsolutions.com/dms/files/chronicbook2002.pdf, *Chronic Conditions: Making the Case for Ongoing Care* (2002).

6. http://meps.ahrq.gov/papers/rf5_99–0006/RF5.htm#Fig2.

7. http://www.aahsa.org/public/nursbkg.htm.

8. http://www.cms.hhs.gov/statistics/nhe/historical/t7.asp.

9. *Health Care Financing Review,* summer 2004, vol. 25, no. 4.

10. www.cms.hhs.gov/statistics/nhe/projections-2003/t1.asp.

11. www.cms.hhs.gov/statistics/nhe/projections-2003/t13.asp.

12. www.cms.hhs.gov/statistics/nhe/projections-2003/t10.asp.

13. Health is herein defined as a state of physical and mental well-being and freedom from disease such that only minimal health services are required and full function is present. Longevity is herein defined as both average life expectancy and any increase thereof.

14. David Cutler and Mark McClellan, "Is Technological Change in Medicine Worth It?" *Health Affairs* (2001) *20*: 11–29.

15. William D. Nordhaus, *The Health of Nations: The Contribution of Improved Health to Living Standards*, Working paper 8818, National Bureau of Economic Research, 2003.

16. James P. Smith, "Healthy Bodies and Thick Wallets: The Dual Relation

between Health and Economic Status," *Journal of Economic Perspectives* (1999) *13*: 145–166.

17. Kelly Greene, "Florida Frets It Doesn't Have Enough Elderly," *The Wall Street Journal* October 15, 2002; Kevin Sack, "More Retirees Discover Small Towns," *New York Times,* May 25, 1997. Also see International Longevity Center, *Arkansas: A Good Place to Grow Old?* 2005.

18. Robert William Fogel, *The Escape from Hunger and Premature Death, 1700–2100: Europe, America and the Third World* (Cambridge, UK: Cambridge University Press, 2004).

19. Ibid., 77.

20. There are competing challenges. For example, we have not proactively adapted to serious environmental changes such as acid rain, global warming, etc.

Chapter 3: "The Greedy Geezer"

1. Humphrey Taylor, "A Remarkable Lack of Intergenerational Conflict: How Should Government Spending Be Divided between Young and Old?" in *Longevity and Quality of Life: Opportunity and Challenges,* Robert N. Butler and Claude Jasmin (eds.), Kluwer Academic Press/Plenum Publishers, 2000, 209–217. Similar findings have been reported in France. Claudine Attias-Donfut, "Are We Moving toward a War between the Generations?" in Butler and Jasmin, 197–205.

2. Ann Robertson, "The Politics of Alzheimer's Disease: A Case Study in Apocalyptic Demography," in Meredith Minkler and Carroll L. Estes (eds.) *Critical Perspectives on Aging* (Amityville, NY, Baywood, 1991), 135–150.

3. "The Coming Conflict as We Soak the Young to Enrich the Old," *The Washington Post,* Jan. 5, 1986.

4. Free Press, 1996.

5. In 1999, one in six of American children were counted poor; in 2003, one in five (17.6 percent) were, the highest poverty rate in ten years. The United States has more poor children than any other industrialized nation. This is due in part to the American emphasis on individual responsibility and its opposition to redistribution of wealth. The official poverty line is $18,810 for a family of four, stringent indeed (2003). In August 2004, the median hourly wage in the United States was $15.52 or $25,179 per year (www.census.gov).

6. Latest data from Merton C. Bernstein, professor of law, Washington University, June 25, 1996.

7. Of course, children also receive Social Security. About four million did in 2003.

8. Linda K. George, "Family Finances and Aging," Chapter 8 in *Life in an Older America,* Robert N. Butler, Larry K. Grossman, and Mia A. Oberlink (eds.) (New York: The Century Foundation Press, 1999).

9. However, property taxes prompt intergenerational conflicts when older people do not vote for school taxes. This is often because they can't easily pay property taxes on fixed incomes. The states (not the federal government) should contribute more to public education and offer income-related property tax rebates. Still, older citizens should not oppose school taxes on ideological grounds. They benefit from the universality of Social Security and Medicare; in turn, they should contribute to the young since the young through taxes contribute to (or will contribute to) them. This reflects the intergenerational social contract.

10. Eric Kingson, Barbara A. Hirshorn, and John M. Cornman, *Ties That Bind: The Interdependence of Generations* (Cabin John, MD: Seven Locks Press, 1986).

11. U.S. Census Bureau, October 2003, *Grandparents Living with Grandchildren, 2000*, Census 2000 brief.

12. U.S. Census Bureau, 2003, *Income, Poverty and Health Insurance Coverage in the United States: 2003.* Washington, DC: GPO.

13. Craig S. Smith, "Old Folks in Beijing Dance in the Streets, Annoying the Young," *Wall Street Journal,* Aug. 5, 1996.

14. According to the *July 2004 Employment Report,* issued by the Bureau of Labor Statistics, in addition to the 8.2 million Americans who are officially unemployed, there are 75.7 million, sixteen years of age and older, who are not in the labor force and 24.2 million part-time workers.

15. The average cost of raising an American child born in 1999 through age seventeen has been estimated to be about $160,000 by the Department of Agriculture. If adjusted for inflation, the cost could rise to $237,000. Those figures only cover basic needs such as food, housing, clothing, health care, and transportation.

16. Leonard Silk and Mark Silk with Robert Heilbroner, Jonas Pontusion, and Bernard Wasow, *Making Capitalism Work* (New York: New York University Press, 1996). See also Robert H. Binstock, "Older Persons and Health Care Costs," Chapter 4 in *Life in an Older America,* Robert N. Butler, Larry K. Grossman, Mia A. Oberlink (eds.) (New York: The Century Foundation Press, 1999).

Chapter 4: Ageism

1. The necessary revolt of the young against their parents and their parents' generation is not ageism, but is essential to social evolution, and is usually functional and constructive.

2. Terms such as old guard, fuddy-duddy, and superannuated are other examples of disparagement. Homosexual men refer to older homosexuals as trolls who inhabit bars called wrinkle rooms.

3. See Robert Binstock, "Older Persons and Health Care Costs," Chapter 4 in *Life in an Older America,* Robert N. Butler, Lawrence K. Grossman, and Mia R. Oberlink (eds.) (New York: The Century Foundation, 1999), 75–99.

4. Viagra is considered a recreational drug in Germany and its use is not covered by the sickness funds. In the United States, some managed care companies allow three tablets each month. Also see Robert N. Butler and Myrna I. Lewis, *The New Love and Sex After Sixty* (New York: Ballantine Books, 2002) (4th ed.).

5. Interview with Carl Bernstein, "Age and Race Fears Seen in Housing Opposition," *Washington Post,* March 7, 1969; "The Effects of Medical and Health Progress on the Social and Economic Aspects of the Life Cycle" (paper delivered at the National Institute of Industrial Gerontology, Washington, D.C., March 13, 1969), 1–9; "Ageism: Another Form of Bigotry," *The Gerontologist* (1969) *9:* 243–246. I decided to spell ageism as the English spell ageing on the grounds that the absence of the *e* could lead to frequent mispronunciation. One remarkable example of ageism: Once I was not so jokingly called a necrophiliac because of my interest in the care of older persons.

6. Barbara Myerhoff, *Number Our Days* (New York: Dutton, 1978).

7. *Emergency Preparedness for Older Persons.* International Longevity Center–USA, 2003.

8. George H. Mead, *Mind, Self, and Society* (Chicago: University of Chicago Press, 1934).

9. See Becca R. Levy, "Eradication of Ageism Requires Addressing the Enemy Within," *The Gerontologist* (2001) *41:* 578–579.

10. See International Longevity Center, *Ageism in America,* 2006.

11. In my original formulation, ageism or prejudice with respect to age was observed to operate in both directions; that is, the old can be prejudicial toward the young as well as the young toward the old.

12. Asia is usually cited for its reverence for age, but modernity is bringing about change. See *The Tokyo Story* (1953) directed by Yasujiro Ozu. It is the

story of an older couple's visit to their children, their disappointments, their sadness. Also see Charlotte Muller and Catherine Silver, *A Comparative Study of Values and Value Transmission between Japan and the U.S.* (International Longevity Center: USA and Japan, March 1995). There was little difference.

13. On the other hand, the Beatles song "When I'm Sixty-Four" is sympathetic to older persons.

14. Harold Sheppard, "The New Ageism and the 'International Tension' Issue" (International Exchange Center on Gerontology, University of South Florida, 1988).

15. International Longevity Center, *Old and Poor in America*, 2001.

16. U.S. Census Bureau, April 2003, *The Older Population in the United States: March 2002*, http://www.census.gov.prod/2003pubs/p20-546.pdf.

17. Near-poor is defined as 125 percent or 150 percent of the poverty rate.

18. The amount of underfunding in corporate pension plans currently totals $450 billion, and the amount in government pension plans is $300 billion. Pending federal legislation to shore up underfunded pension plans proposes eliminating core retirement protections, such as giving employers the power to reduce worker's pensions and take away certain pension benefits that older employees have already earned. This would establish a dangerous precedent. Statistics here are from *Ageism in America,* Report of the International Longevity Center, 2006.

19. We have seen all of the most famous "isms" in medicine, sadly. We saw racism in the Tuskegee study when African Americans were not told that they had syphilis and were not treated for it. We saw sexism in the Texas contraceptive trial in which women were placed on placebos without being told in order to test birth control pills. Ageism was seen in the Brooklyn Jewish Chronic Disease Study in which demented older patients were given live cancer cells without consent. Out of these shocking experiments the field of bioethics grew.

20. See Cato 6. "A Piece of My Mind: Dirtball," *Journal of the American Medical Association* (1982) *247*: 3059–3060; Samuel Shem, *The House of God* (New York: Richard Marek, 1978). This novel was written by a young doctor, Dr. Stephen Bergman, describing his internship, and published under a fictitious name.

21. Robert Pear, *New York Times,* Feb. 18, 2002, A11.

22. Fifty-four percent of nursing homes fail to meet minimum standards, yet only 0.5 percent of them nationwide are cited and penalized for patterns of widespread problems that cause harm to residents.

23. Polly Newcomb and Paul P. Carbone, "Cancer Treatment and Age: Patient Perspectives," *Journal of the National Cancer Institute* (1993) *85*: 1581.

24. See International Longevity Center, *Clinical Trials and Older Persons: The Need for Greater Representation,* 2003.

25. In 2001, Dr. Harold F. Shipman, fifty-five, who practiced in Hyde, England, was discovered to have been "perhaps the world's most prolific serial killer," having ended the lives of older people, particularly older women, usually widows. The number: about three hundred, possibly 345! Most were over age seventy-five. Kurt Eichenwald, "True English Murder Mystery: Town's Trusted Doctor Did It," *New York Times,* May 13, 2001. There are also contemporary examples of gruesome murders of helpless older persons in nursing homes and hospitals by nurses and nurse's aides. The film *I Can't Sleep* is based on a famous murder case in France where two "granny killers" killed over twenty old women before they were caught in 1987.

26. Robert N. Butler, "The Facade of Chronological Age: An Interpretive Summary of the Multidisciplinary Studies of the Aged," *American Journal of Psychiatry* (1963) *119*: 721–728, 1963, reprinted in *Middle Age and Aging: A Reader in Social Psychology,* Bernice L. Neugarten (ed.) (Chicago: University of Chicago Press, 1968).

27. EEOC enforces Title VII of the Civil Rights Act of 1964, which prohibits employment discrimination on the basis of race, color, religion, sex, or national origin; the Age Discrimination in Employment Act; the Equal Pay Act; Title I of the Americans with Disabilities Act, which prohibits employment discrimination against individuals with disabilities in the private sector and state and local governments; prohibitions against discrimination affecting individuals with disabilities in the federal government; and sections of the Civil Rights Act of 1991. Further information about the commission is available on its Web site at www.eeoc.gov.

28. See discussion by Erdman Palmore, "Ageism in Gerontological Language," *The Gerontologist* (2000) *40*: 645, 2000 and in *Ageism: Negative and Positive* (New York: Springer, 1999). Palmore has worked to develop an ageism survey instrument. Also see Palmore, "The Future of Ageism," Issue Brief, International Longevity Center, 2004.

29. See: AARP, "Virtually Invisible: The Image of Midlife and Older Women on Prime Time T.V.," 2000.

30. Betsy McKay and Suzanne Vranica, "How a Coke Ad Campaign Fell Flat with Viewers," *The Wall Street Journal,* March 19, 2001.

31. Robert N. Butler, *The Human Rights of Older Persons* (International Longev-

ity Center, 2002). Also, "Declaration of the Rights of Older Persons," *The Geron-tologist* (2002) *42*: 152–153.

32. Bernanos is the author of *Diary of a Country Priest.*

Chapter 5: The Changing Family and Longevity

1. Only 3 percent of all mammals and birds are estimated to congregate in family structures.

2. U.S. Census Bureau. *America's Families and Living Arrangements: 2003.* Washington, DC: November 2004. Surprising to some, age fifteen is selected by the Census Bureau in marriage statistics.

3. Ibid.

4. I do not intend here to describe the uncertain prehistoric origin of family structure such as the evolution of monogamy, its economic influences, and issues of patriarchy and matriarchy. Those interested, for example, can study what is known of the ancient, biblical, Greek, Roman, and Medieval family in Europe and the Confucian family of China and Japan to see the patriarchal figure in action. I do wish to observe the family's continuing evolution around a central, enduring core.

5. U.S. Census Bureau, 2003, *2003 American Community Survey Summary Tables,* Washington DC; Census Bureau, September 8, 2004, Facts for features #CB04-FF.15–2: Grandparents Day 2004, http://www.census.gov/Press-Release/www/releases/archives/facts_for_features_special_editions/002319 .html.

6. Uhlenberg, Peter. *Implications of Increasing Divorce for the Elderly.* Paper presented at the United Nations International Conference on Aging Populations in the Context of the Family. Ketakyushu, Japan.

7. *Social Structure and the Family: Generational Relations* (Englewood Cliffs, NJ: Prentice-Hall, 1965).

8. An exploratory transcultural study of the transmission of values across the generations, conducted by the International Longevity Centers (U.S./Japan) found, among many things, that Americans are as respectful and supportive of their older members as the Japanese, contradicting the usual view that the Japanese are vastly superior in this regard; also see Charlotte Muller and Catherine Silver, *A Comparative Study of Values and Value Transmission between Japan and the U.S.* (ILC-USA and ILC-Japan), March 1995.

9. James N. Morgan, "The Role of Time in the Measurement of Transfers and Economic Well-being," in Marilyn Moon (ed.), *Economic transfers in the*

United States: Studies in Income and Wealth 49: 936; 199–238. Chicago: University of Chicago Press, 1984.

10. Margaret Blenkner, "Social Work and Family Relationships in Later Life with Some Thoughts on Filial Maturity," in Ethel Shanas and Gordon F. Streib (eds.) *Social Structure and the Family: Generational Relations* (Englewood Cliffs, NJ: Prentice-Hall, 1965).

11. Low birth weight, inadequate dietary folic acid, fetal alcohol syndrome, crack, and AIDS are among the problems of babies of low income families.

12. The Alan Guttmacher Institute, *Induced Abortion in the United States* (New York: The Alan Guttmacher Institute, 2005).

13. It is said that Americans have sex early, marry late, and divorce often. Indeed, adult Americans are estimated to average seven sexual partners over a lifetime.

14. U.S. Census Bureau. January 24, 2003. *Survey of Income and Program Participation* (SIPP), 1996 Panel, Wave 10 (last revised Jan. 11, 2005).

Chapter 6: The Disorders of Longevity

1. Lest we forget, TB, malaria, AIDS, hepatitis B, respiratory and diarrheal diseases, and infectious diseases in acute and chronic forms still dominate much of the developing world. William H. McNeill wrote, mistakenly, in 1983, "One of the things that separate us from our ancestors and make our contemporary experience profoundly different from that of other ages is the disappearance of epidemic diseases as a serious factor in human life." *The New York Review,* July 21, 1983.

2. J.P. Barker (ed.), *Fetal and Infant Origins of Adult Disease* (London: British Medical Journal Publishing Group, 1992). The Barker hypothesis refers to diseases in later years whose origins occurred in fetal life and early infancy. Low birth weight (a function of an adverse intrauterine environment) appears to account for later heart disease and many other problems. One could also consider as an extension of the Barker hypothesis the opposite, i.e., early favorable health conditions, such as absence of lead, advance later productivity and intellectual development. The Bogalusa Heart Study at Louisiana State University Medical Center is the longest, most detailed study of children in the world. The focus is on understanding the early natural history of coronary artery disease and essential hypertension. Since 1972, the Bogalusa Heart Study has been responsible for conducting cardiovascular risk factor research both in the community and in the laboratory. Observations show that the

major causes of adult heart diseases, atherosclerosis, coronary heart disease, and essential hypertension begin in childhood. Documented changes occur between five and eight years of age.

3. The role of inflammation in aging and disease is receiving special attention. Caleb E. Finch and Eileen M. Crimmins, "Inflammatory Exposure and Historical Changes in Human Life Spans," *Science* (2004) *305*: 1736–1739.

4. Type 1 diabetes typically occurs in childhood and is attributed to a virus. Type 2 diabetes is called postmaturity or late-life diabetes and is linked to aging and obesity but also can occur in childhood.

5. Robert N. Butler, Huber Warner, T. Franklin Williams, et al., "The Aging Factor in Health and Disease," *Aging Clinical and Experimental Research* (2004) *16*: 104–112.

6. Spinal stenosis, a narrowing of the space around the spinal cord, can be painful and debilitating but is treatable—not always successfully—by pain medication and physical therapy, and, when unrelenting, by surgery. It usually occurs when people are in their sixties, seventies, eighties, and beyond.

7. Unfortunately, however, AIDS is a massive problem in the developing world and, for a variety of reasons, including the cost of antiretroviral drugs, is a disaster.

8. Of course, people over fifty-five also get AIDS and must practice safe sex.

9. James E. Birren, Robert N. Butler, Samuel W. Greenhouse, Louis Sokoloff, and Marian R. Yarrow (eds.), *Human Aging I: Biological and Behavioral Study,* 1963 Public Health Service Publication No. 986, Washington, DC: GPO. (Reprinted 1971 and 1974.)

10. E.g., cells of skin and GI tract.

11. E.g., in general, nerve and muscle cells.

12. Benjamin Gompertz, "On the Nature of the Function Expressive of the Law of Human Mortality and on a New Mode of Determining Life Contingencies," *Philosophical Transactions of the Royal Society of London* (1825) *115*: 513–585.

13. Peter Brian Medawar, *An Unsolved Problem of Biology* (London: Lewis, 1952).

14. It is important to realize that so-called senility is not an inevitable consequence of aging but an outdated term that is applied to mental and even physical deterioration in general.

15. Thirteen percent of African Americans and 10.2 percent of Hispanics have diabetes, compared to about 6.5 percent of whites.

16. Kenneth Manton, Eric Stallard, and Larry Corder, "Education-specific Estimates of Life Expectancy and Age-specific Disability in the U.S. Elderly Population: 1982 to 1991, 1992," *Journal of Aging and Health 9*: 419–450.

17. Focuses upon people sixty-five and older.

18. James F. Fries, "Aging, Natural Death and the Compression of Morbidity," *New England Journal of Medicine* (1982) *303*: 130–135; Ernest M. Gruenberg, "The Failures of Success," *Milbank Memorial Fund Quarterly/Health and Society* (1977) *55*: 3–24.

19. It should be noted that activities of daily living may be impaired as a result of poor physical fitness due to deconditioning.

20. Jay S. Olshansky, Dan Perry, Richard A. Miller, and Robert N. Butler. "In Pursuit of the Longevity Dividend," *The Scientist* (2006) *20*: 28–36.

21. For fun and to make points about human limitations, several of us offered ideas for the reconstruction of the body.

22. See Institute of Medicine, "To Err Is Human: Building a Safer Health System," 1999.

23. Jay S. Olshansky, Bruce Carnes, and Robert N. Butler. "If Humans Were Built to Last," *Scientific American* (2001) *284*: 50–55.

24. Linda P. Fried, M.D., M.P.H., is director of the Division of Geriatric Medicine and Gerontology at Johns Hopkins.

25. Jack M. Guralnik is senior investigator and chief, Epidemiology and Demography Section, National Institute on Aging.

26. A.M. Minino and B.L. Smith, "Deaths: Preliminary Data for 2000," *National Vital Statistics Report 49*(12) (Hyattsville, MD: National Center for Health Statistics, 2001). (Most recent table as of August 2004.)

27. *National Vital Statistics Reports* 52(3), September 18, 2003.

28. Nonetheless, 30 percent of stroke victims are under sixty-five.

29. Carl Zimmer, "Do chronic diseases have an infectious root?" *Science* (2001) *293*: 1974–1977.

30. Since detection of the bacterium is easy it may be possible to eradicate stomach cancer by antibiotics. Timothy Wang has reported that in some countries 90 percent of the population has been infected by the age of nine; in the United States, 30 to 40 percent of the entire population. Stomach cancer is the seventh cause of cancer deaths in the United States.

31. Lynn Payer, *Medicine and Culture: Varieties of Treatment in the United States, England, West Germany and France* (New York: Holt, 1988).

32. Larissa K.F. Temple, Robin S. McLeod, Steven Gallinger, and James G. Wright, "Defining Disease in the Genomics Era," *Science* (2001) *293*: 807–808.

33. Paula Kiberstis and Leslie Roberts, "It's Not Just The Genes," *Science* (2002) *296*: 685.

34. Neil A. Holtzman and Theresa M. Marteau, "Will Genetics Revolutionize Medicine?" *New England Journal of Medicine* (2002) *343*: 141–144; Paul

Lichtenstein, et al., "Environmental and Heritable Factors in the Causation of Cancer," *New England Journal of Medicine* (2000) *343*: 78–85.

35. Kristin K. Barker, *The Fibromyalgia Story: Medical Authority and Women's World of Pain* (Philadelphia: Temple University Press, 2005).

36. Erdman Palmore, "Gerontophobia versus Ageism," *The Gerontologist* (1972) *12*: 213.

37. R.N. Butler, "The Life Review: An Interpretation of Reminiscence in the Aged," *Psychiatry* (1963) *26*: 65–76.

38. Bruce S. McEwen, "Allostasis and Allostatic Load: Implications for Neuropsychopharmacology," *Neuropsychopharmacology* (2002) *22*: 108–124; see http://www.macses.ucsf.edu/Research/Allostatic/notebook/allostatic.html.

39. See McEwen, 2000.

40. Bruce A. Carnes, Larry R. Holden, S. Jay Olshansky, Tarynn M. Witten, and Jacob S. Siegel, "Mortality Partitions and their Relevance to Research on Senescence," *Biogerontology* (2006) *7*: 183–198.

41. This would be the counterpart of Vaupel's finding that the shape of the Gompertz curve improves in the 1990s.

42. Personal comments, 2003.

43. Robert N. Butler, Huber Warner, T. Franklin Williams, et al., "The Aging Factor in Health and Disease," *Aging Clinical and Experimental Research* (2004) *16*: 104–112.

44. Leonard Hayflick, "Has Anyone Ever Died of Old Age?" (New York: International Longevity Center, 2003).

45. Robert Kohn, "Cause of Death in Very Old People," *Journal of the American Medical Association* (1982) *247*: 2793–2797.

46. Such special populations as the Framingham heart study and Baltimore Longitudinal Study of Aging.

47. Thomas T. Perls, Laura Alpert, and Ruth C. Fretts, "Middle-Aged Mothers Live Longer," *Nature* 1997, Sept. 11; *389*(6647): 133.

Chapter 7: Human Interest in Longevity

1. Developed with the National Council on the Aging and the International Longevity Center.

2. Federal Law: Employee Retirement Income Security Act of 1974. The latest such poll I could find.

3. Gerald J. Gruman, *A History of Ideas about the Prolongation of Life* (New York: Springer, 2003).

4. It has been proven that peoples in these places do *not* live longer, and no elixir to extend life exists.

5. S. Jay Olshansky, Bruce A. Carnes, and Christine Cassel, "In Search of Methuselah: Estimating the Upper Limits to Human Longevity," *Science* (1990) *250*: 634–640; Olshansky, Carnes, and Aline Désesquelles, "Prospects for Human Longevity," *Science* (2001) *291*: 1491–1492.

6. Tracing its roots to 1898, the NIH currently has more than 18,500 employees (including nearly six thousand scientists) and funds about fifty thousand competitive grants to more than 212,000 researchers at over 2,800 universities, medical schools, and other research institutions.

7. Not everyone agrees. See Daniel Callahan, *What Price Better Health? Hazards of the Research Imperative* (Berkeley: University of California and NY Milbank Memorial Fund, 2003).

8. See, especially, Hamilton, Moses III, E. Ray Dorsey, David H.M. Matheson, and Samuel O. Their, "Financial Anatomy of Biomedical Research," *Journal of the American Medical Association* (2005) *294*: 1333–1342.

9. In 2003 the W.M. Keck Foundation gave a $40 million gift to the National Academies (the National Academy of Sciences, the National Academy of Engineering, and the Institute of Medicine) to run a grant program for interdisciplinary research.

10. Bureau of Labor Statistics, Department of Labor. Pharmaceutical and medicine manufacturing, Bulletin 2541. *The 2004–2005 Career Guide to Industries.* Washington, DC, 2005.

11. Biotechnology Industry Organization, 2005. http://www.bio.org.

12. Commissioned by the Funding First initiative of the Mary Woodard Lasker Charitable Trust.

13. Small, identifiable segments of DNA that might lie near a gene.

14. The order with which specific diseases become hot national topics or popular charities is not equivalent to their incidence and prevalence in society. For example, heart disease remains the number one disease, not Alzheimer's disease or breast cancer or AIDS. Lung cancer is much more common than breast cancer. (There are 158,000 annual fatalities from lung cancer and 44,300 from breast cancer.) Some diseases, of course, have a particular impact. Cancer of the breast and prostate affects intimacy, sexuality, and a sense of self. Even so, although prostate cancer kills nearly as many American men as breast cancer kills women, prostate cancer receives less money for research and inspires less activism.

15. To give only a few examples, the estimated total cost for heart disease is $125.8 billion, cancer $96.1 billion, and stroke $30 billion (Research! America 1991

figures). The Dana Alliance for Brain Initiatives counts up the cost of neurological and psychiatric disease to $634 billion per year, affecting one hundred and eighty million people. Advocates like to compare other expenditures, e.g., Americans spend $40 billion on gambling and $45 billion on tobacco but only $33 billion on medical research. The annual health costs of smoking is $150 billion, according to one study. It is essential that costs of illness be carefully calculated and overlaps be avoided where possible or acknowledged and explained. The anguish and cost of diseases is obvious enough and it is unwise to unnecessarily expose medical science to criticism. Epidemiologists and health economists who are estimating the extent and cost of illness should not be such advocates as to lose their objectivity. Note: the excellent work of Dorothy Rice, Cost of Illness, Issue Brief, International Longevity Center, 2004.

16. One extremely painful example of special interest to gerontology is progeria for children who age rapidly and do not survive their teens. The National Progeria Foundation in association with the National Institute on Aging has supported work that has resulted in identification of a relevant gene. A therapy is being tested.

17. There is the International Alliance of Patients' Organizations (IAPO) headquartered in the United Kingdom.

18. Albert and Mary established the Lasker Awards in Medicine, often called the American Nobel Prize.

19. There have been many successful trans-NIH coordinating efforts.

20. Some older persons, especially professionals and the well-to-do, are creating their own community arrangements to provide services such as home repairs and home health services to shepherd them through the final chapter of their lives. These grass-roots efforts include Beacon Hill Village in Boston. These arrangements are set up as nonprofit corporations and, so far, are only possible for the well-to-do.

21. New York: Harcourt, Brace.

22. India has more than three thousand towns and cities, but only eight have comprehensive water treatment plants.

23. See Brookings Institution, *A Fifth of America*, 2006, which focuses on sixty-four counties where suburbs were built after WWII.

24. International Longevity Center, "Arkansas, A Great Place to Live?" 2005.

25. Jane Jacobs, *The Death and Life of Great American Cities* (New York: Random House, 1961).

26. Formerly the American Association for Retired Persons.

27. If 80 percent of the housing units in a community have at least one occupant fifty-five or older, the community can be certified as "adult."

28. The idea of trading one's car for rides began in Portland, Maine, with the Independent Transportation Network.
29. Dr. William H. Thomas, a geriatrician, introduced the Eden Alternative and the social model of the Green House.
30. Includes the Florence Gould Foundation as well as the Japanese and the French governments.
31. Dr. Victor Rodwin is codirector of WCP, a joint project of the International Longevity Center and the Robert F. Wagner School of Public Service at New York University; see Victor Rodwin and Michael Gusmano, *Older in World Cities* (Nashville, TN: Vanderbilt University Press, 2006).
32. Will Durant, The Life of Greece Part II, *The Story of Civilization* (New York: Simon & Schuster, 1939).
33. DARPA is the Defense Advanced Research Projects Agency.

Chapter 8: Alzheimer's

1. Evidence varies concerning the incidence of Alzheimer's disease. See D.A. Evans and others, "Prevalence of Alzheimer's Disease in a Community Population of Older Persons. Higher Than Previously Reported," *Journal of the American Medical Association* (1989) *262*: 2551–2556. Irizarry, Michael C. and Bradley T. Hyman, Chapter 5 in *Principles of Neuroepidemiology,* T. Batchelor and M.E. Cudkowicz (eds.) (Boston: Butterworth-Heineman, 2001).
2. Personal communication, Robert Katzman, neurologist and pioneer student of Alzheimer's disease, Dec. 14, 1994: "If one uses an average duration of life of Alzheimer's patients of nine years today and the conservative prevalence estimate of about 2.5 million individuals with Alzheimer's disease, over 270,000 patients would die each year. But since most Alzheimer's disease patients are older persons and hence at high risk of dying even if not demented, how many of these deaths can be attributed to Alzheimer's disease?"
3. Senility is the popular term which covers the various dementing diseases of old age, the most common of which is Alzheimer's disease. Some clinicians believe there are two or more forms of Alzheimer's disease.
4. National Institute on Aging Task Force, "Senility reconsidered: treatment possibilities for mental impairment in the elderly," *Journal of the American Medical Association* (1980) *244*: 261–263.
5. Robert Katzman and Robert Terry in association with Kathryn Bick of the Neurology Institute catalyzed a very important conference on senile de-

mentia and related conditions in 1977. The NIA and NIMH joined in sponsorship of this conference. See Robert Katzman, Robert D. Terry, and Kathryn L. Bick (eds.), *Alzheimer's Disease: Senile Dementia and Related Disorders* (New York: Raven Press, 1977).

6. The hippocampus helps to form memories after receiving input from elsewhere in the brain. Hippocampal damage makes it difficult to acquire new knowledge.

7. Dennis J. Selkoe, "Alzheimer's Disease is a Synaptic Failure," *Science* (2000) *298*: 789–791.

8. V.R. Kral, "Psychiatric Observations under Severe Chronic Stress," *American Journal of Psychiatry* (1951) *108*: 185–192.

9. Ronald D. Petersen and others, "Mild Cognitive Impairment: Clinical Characterization and Outcome," *Archives of Neurology* (1999) *56*: 303–308. Although I use the singular for simplication, I believe Alzheimer's disease is more than one disease, as many students of this condition do.

10. Compare the threshold at which osteopenia (bone thinning with age and poor conditioning) becomes osteoporosis (a disorder of longevity). In 1981, Robert Terry reported that there is no loss of neurons in the cerebral cortex during aging, but subsequently he showed a loss of synapses in normal aging. Terry's work has been confirmed.

11. In 1993, Roses and other Duke University researchers discovered the first possible genetic risk factor for late-onset familial Alzheimer's disease, the gene for apolipoprotein E on chromosome 19, specifically the variant or allele Apo E–4.

12. Down syndrome is *not* a genetic disease but an acquired chromosomal abnormality. Persons with Down syndrome have an extra copy of chromosome 21 that may contain a gene or genes that may be predisposed to cancers and Alzheimer's disease. The onset of Alzheimer's disease in Down syndrome generally occurs between forty-five and fifty-five. Down syndrome patients are now living longer. The CDC reported in 2002 that between 1983 and 1997, the average life expectancy of Down syndrome patients advanced from twenty-five to forty-nine. (During the same time period, American average life expectancy grew from seventy-three to seventy-six.) Presumably, this will increase the number of Alzheimer's patients.

13. About 70 percent of early onset cases are due to the chromosome 14 defect, 25 percent due to chromosome 1, and 5 percent to chromosome 21. But there could still be an undiscovered FAD gene. The gene on chromosome 14 is called presenilin–1 (PS1) and the one on chromosome 1 presenilin–2

(PS2). The three rare genes account for the 2 percent to 7 percent of early onset cases.

14. We now have more information about the genes in Alzheimer's disease than about other complex diseases such as heart disease, cancer, and stroke.

15. The work of Eric Kandel, the Nobelist who studied the sea slug, aplysia, has found that learning results in protein formation.

16. See Darab K. Dastur and others, "Effects of Aging on Cerebral Circulation and Metabolism in Man," Chapter 6 in *Human Aging I: A Biological and Behavioral Study*, James E. Birren et al., (eds.), DHEW Publication (ADM), 1971 (reprinted 1974), 74–122. Also see Jack C. de la Torre and V. Hachinski, "Cerebrovascular Pathology in Alzheimer's disease," *Annals of the New York Academy of Sciences* (New York: 1999).

17. R.D. Terry and H.M. Wisniewski, "Structural and Chemical Changes of the Aged Human Brain." In *Genesis and Treatment of Psychologic Disorders in the Elderly*. Aging, Volume 2, S. Gershon and A. Raskin (eds.) (New York: Raven Press, 1975), 127–141.

18. S. Ramón y Cajal, *Trab. Lab. Invest. Biol.* VIII, 2 (1910), F. Reinoso-Suárez (trans.), in *Neuroplasticity: A New Therapeutic Tool in the CNS Pathology*, R.L. Masland, A Portera-Sanchez, and G. Toffano (eds.) (Padua, Italy: Liviana Press, 1987), 31–37.

19. Jonas Frisen and Arturo Alvarez-Buylia, working independently.

20. David R. Kornack and Pasko Rakic, "Cell Proliferation without Neurogenesis in Adult Primate Neocortex," *Science* (2001) *294*: 2127–2130. This study was conducted on monkeys.

21. But they may be related.

22. See Stanley J. Rapoport, "Functional Brain Imaging to Identify Affected Subjects Genetically at Risk for Alzheimer's Disease," *Proceedings of the National Academy of Sciences* (2000) *97*: 5696–5698.

23. B.L. Plassman, K.M. Langa, G.G. Fisher, S.G. Heeringa, D.R. Weir, M.B. Ofstedal, J.R. Burke, M.D. Hurd, G.G. Potter, W.L. Rodgers, D.C. Steffens, R.J. Willis, and R.B. Wallace, "Prevalence of Dementia in the United States: The Aging, Demographics, and Memory Study," *Neuroepidemiology* (2007 Oct. 29) *29(1–2)*: 125–132.

24. *New York Times*, June 1, 1993.

25. For care and on productivity, respectively.

26. Timothy A. Salthouse, a neuroscientist, analyzed studies meant to show that mental exercise arrests mental decline. He found no evidence.

27. Aggression makes the life of the caregiver extremely difficult. The intercon-

nections of brain and behavior are a focus of investigation, including the role of testosterone in the male.

28. Hayden Bosworth, B. Schaie, K. Warner, and Sherrey L. Willis, "Cognitive and Sociodemographic Risk Factors for Mortality in the Seattle Longitudinal Study," *Journal of Gerontology* (1999) *548*: 273–282.

29. It is essential to accelerate drug discovery research for Alzheimer's Disease. The Lauder family created the Institute for the Study of Aging (ISOA) and the Alzheimer's Drug Discovery Foundation (ADDF) to do so, under the direction of Dr. Howard Fillit.

Chapter 9: The "Biology of Extended Time"

1. Caleb E. Finch, *Longevity, Senescence and the Genome* (Chicago: University of Chicago Press, 1990).

2. Ibid.

3. Atlantic salmon are not semelparous and may reproduce as many as six times.

4. Similarly, in caloric restriction, reproduction is delayed and longevity increases; that is, death is deferred.

5. Annibalel A. Puca and others, "A Genome-wide Scan for Linkage to Human Exceptional Longevity Identifies a Locus on Chromosome 4," *Proceedings of the National Academy of Sciences* (2001) *98*: 10505–10508.

6. Caleb E. Finch and Gary Ruvkun, "The Genetics of Aging," *Annual Review: Genomics and Human Genetics* (2001) *2*: 435–462.

7. Agnes M. Herskind and others, "The Inheritability of Human Longevity: A Population-based Study of 2,872 Danish Twin Pairs Born 1870–1900," *Human Genetics* (1996) *97*: 319–323.

8. The law of mortality extends across species.

9. Of course, this could be a survival effect, that is, the least fit die earlier.

10. C. Finch, *Longevity, Senescence and the Genome.*

11. The existence of more than one structural type in the same species.

12. The relationship between body size and longevity vary among species.

13. On the other hand, Jonathan Weiner (*The Beak of the Finch*, Knopf, 1994) believes Charles Darwin underestimated the power and speed of natural selection. Weiner concluded, from studies of the beak of the finch, that evolution could be remarkably rapid. These were studies in nature, not evolution in a test tube or cage. Moreover, pathogenic bacteria, developing resistance, also demonstrate rapid evolution. Creation is a continuing process.

Moreover, many biologists had thought that species remained unchanged and that there were no new species since the beginning of the world. By the late 1700s, biologists began to question these beliefs. After all, farmers produced new varieties of plants and animals by selective breeding.

14. Donald Johanson and Blake Edgar, *From Lucy to Language* (New York: Peter A. Neuraumont, 1996).

15. Today called chromosomes, genes, DNA.

16. Finch, *Longevity*, Chapters 4 and 10. Also see Finch, "Variations in Senescence and Longevity Include the Possibility of Negligible Senescence," *Journal of Gerontology* (1998) *53A*: B235–B239.

17. *Pleo* derives from the Greek and means more.

18. Females survive males in most but *not* all species. Men outlive women in ten countries.

19. Annachiara De Sandre-Giovannoli and others, "Lamin A Truncation in Hutchinson-Gilford Progeria," *Science* (2003) *300*: 2055.

20. They focused on these nations because of their high rates of intermarriage.

21. Hundreds of years of living in geographically isolated areas such as the Pale of Settlement, a place now in three countries, Russia, Poland, and Ukraine, has resulted in a uniquely homogeneous genetic makeup of this population. (Couzin, J., "Aging research: is long life in the blood?" *Science* [2003], *302*: 373–375.)

22. Nir Barzilai and others, "Unique Lipoprotein Phenotype and Genotype Associated with Exceptional Longevity," *Journal of the American Medical Association* (2003) *290*: 2030–2040.

23. Bard J. Geesaman and others, "Haplotype-based Identification of a Microsomal Transfer Protein Marker Associated with the Human Lifespan," *Proceedings of the National Academy of Sciences of the United States of America* (2003) *100*: 14115–14120.

24. International Longevity Center, *Longevity Genes: From Primitive Organisms to Humans* (New York, 2004).

Chapter 10: Cycles, Clocks, and Power Plants

1. Huber Warner, Robert N. Butler, Edward Schneider, and Richard Sprott (eds.). *Modern Biological Theories of Aging*, New York: Raven Press, 1987.

2. Usually enzymes are proteins made up of the same amino acids that constitute life. Telomerase is made up of protein with RNA which serves as a template of the six recurring genetic letters in humans—TTAGGG—which

make up telomeres. Elizabeth Blackburn, Carol W. Greider, and Jack W. Szostak predicted them and discovered telomerase.

3. Calvin B. Harley, A. Bruce Futcher, and Carol W. Greider, "Telomeres Shorten During Ageing of Human Fibroblasts," *Nature* (1990) *345*: 458–460.

4. Nam W. Kim and others, "Specific Association of Human Telomerase Activity with Immortal Cells and Cancer," *Science* (1994) *266*: 2011–2015.

5. Not everyone considers *in vitro* aging to be relevant to *in vivo* cellular aging or longevity. But a study reported in 2006 suggests relevance (Utz Verbig, Mark Ferreira, Laura Condel, Dee Carey, and John M. Sedivy, "Cellular senescence in aging primates," *Science* [2006] *311*: 1257).

6. The first truly immortal human cell culture is named after Henrietta Lacks, whose cancerous cervical tissue has been cultivated since 1952. They are called HeLa cells.

7. Vincent J. Cristofalo and others, "The Relationship between Donor Age and the Replicative Lifespan of Human Cells in Culture: A Reevaluation," *Proceedings of the National Academy of Sciences* (1998) *95*: 10614–10619.

8. Junko Oshima, Judith Campisi, T. Charles, A. Tannock, and George M. Martin, "Regulation of c-*fos* in Senescing Werner Syndrome Fibroblasts Differs from that Observed in Senescing Fibroblasts from Normal Donors," *Journal of Cellular Physiology* (1995) *162*: 277–283.

9. Denham Harman, "Aging: A Theory Based on Free Radical and Radiation Chemistry," *Journal of Gerontology* (1956) *11*: 298–300.

10. Known as a superoxide anion.

11. There have been a number of trials in animals and humans of such antioxidants as vitamin E, C, and A, alpha-lipoic acid, 1-carnitine, CoQ_{10} and DHEA. Coenzyme Q_{10} (CoQ) is a cofactor for energy metabolism in the mitochondria, often taken as a diet supplement.

12. There is some evidence of dangers in taking beta carotene.

13. One Finnish study did *not* find that beta carotene and Vitamin E prevented heart disease or cancer. However, the average subject in the study had smoked a pack of cigarettes daily for thirty-six years! It is not reasonable to assume that a short-term intervention provided in this study would reverse or halt pathological processes known to take place over decades. Any protective effects of these substances would most likely be limited to the phase of initiation of the pathological process such as cancer.

14. Green tea contains vitamin C and half the caffeine of coffee. It also has antibiotic properties. Twenty percent of the tea drunk in the world is green

tea, consumed in Asia. Polyphenols, found in green tea, have been shown to lower cholesterol and inhibit enzymes that produce cancer-causing substances in animals. Black tea also contains antioxidants called flavonoids, which are also in red wine. Garlic, licorice, soy, cranberry juice, and carrots are also believed to contain unusual amounts of antioxidants.

15. Elissa Epel and Elizabeth H. Blackburn, "Accelerated Telomere Shortening in Response to Life Stress," *Proceedings of the National Academy of Sciences* (2004) *101*: 49, 17312–17315.

16. Meaning maintained throughout evolution.

17. C. Wallace Douglas writes about "ox-phos" diseases, which could include Alzheimer's disease.

18. Samuel E. Schriner and others, "Extension of Murine Lifespan by Overexpression of Catalase Targeted to Mitochondria," *Science* (2005) *308*: 1909; Mark K. Shigenaga, Tory M. Hagen, and Bruce N. Ames, "Oxidative Damage and Mitochondrial Decay in Aging," *Proceedings of the National Academy of Sciences* (1994) *91*: 10771–10778.

19. People have voluntarily put themselves on a caloric restricted diet. They call themselves "CRONIES" (see www.caloricrestriction.org and www.cron web.org). Luige Fontana and John Holloszy of Washington University have begun to study them.

20. T. Konrad Howitz and others, "Small Molecule Activators of Sirtuins Extend *Saccharomyces cerevisiae* Lifespan," (2003) *Nature 425*: 191–196.

21. Orthologs are genes in different species that evolved from a common ancestral gene.

22. Lenny Guarente and colleagues at the Massachusetts Institute of Technology found the gene silences DNA by removing acetyl groups causing chromatin to bunch up closer to the DNA, blocking access to the cell's gene-transcription machinery.

23. Worms and people took different evolutionary pathways some six hundred million years ago. Nonetheless, they retain certain genetic mechanisms in common, including the insulin-signaling control of metabolism. Disrupting this pathway makes the worms live longer.

24. Lenny Guarente, *Ageless Quest: One Scientist's Search for Genes That Prolong Youth* (Cold Spring Harbor, NY: Cold Spring Harbor Laboratory Press, 2002). Also see David A. Sinclair and Lenny P. Guarente, "Unlocking the Secrets of Longevity Genes," *Scientific American* (2006) *294*: 48–57.

25. One would have to drink capacious amounts of pinot noir to gain the favorable effects of resveratrol.

26. Max Rubner proposed the disputed "rate of living" theory in 1900. Das

problem der Lebensdauer und seine Beziehungen zum Wachstum under Ehrnarung. Munich: Oldenburg, 1908, 150–204.

27. International Longevity Center, *Biomarkers of Aging: From Primitive Organisms to Man,* 2001. It should be understood that there are markers that assay the likelihood of further longevity including disease markers (e.g., high cholesterol) and fitness level. There are also functional markers such as activities of daily living.

28. A genome is the genetic composition of a single species. Genomics is the name that has been given to the study of living things in terms of their full DNA sequences or genomes. *Sequencing* refers to working out the order of the biomedical building blocks, or base pairs, that make up an organism's DNA. The human genome is thought to have perhaps thirty thousand or forty thousand genes.

29. As Nancy Wexler has said, "After the completion of the Human Genome Project, everyone will have a preexistent condition."

30. "Longevity of symphony conductors," *Statistical Bulletin of MetLife Insurance Co.* (1980) *61*: 2–4.

31. James S. House, Cynthia Robbins, and Helen L. Metzner, "The Association of Social Relationships and Activities with Mortality: Prospective Evidence from the Tecumseh Community Health Study," *American Journal of Epidemiology* (1982) *116*: 123–140.

32. Although we do not have scientific studies, probably unmarried cohabitating couples do as well as married couples.

33. See Judith Campisi, "Suppressing Cancer: The Importance of Being Senescent," *Science* (2005) *309*: 886–887.

34. The study, *Aging in Today's Environment,* called for the development of "gerontotoxicology."

35. There are some beginnings of national populations laboratories; e.g., there is the National Longitudinal Survey of Youth, which has studied a representative sample of 12,686 high school students that graduated between 1980 and 1982. The National Hispanic Community Health Study is a $61 million project, in which sixteen thousand people eighteen to seventy-four were studied for seven years at scientific sites.

36. Radcliff College's Murray Research Center is the largest repository of longitudinal studies.

37. I have been a volunteer in the Baltimore Longitudinal Study (BLSA) since 1978. When I was NIA director, Myrna Lewis insisted that I add women to the BLSA, to which I readily agreed. In addition to the scientific importance of this step, it was also fortunate for NIA when the NIH became a target of

criticism by the Congressional Women's Caucus at the end of the 1980s. It is very difficult and expensive to study nutrition in humans. For example, it is hard to secure accurate longitudinal dietary histories, conduct necessary nutritional balance studies in clinical research centers and consider relevant molecular and physiological phenomena, observe cultural and individual taste, and specify genetic components, and to do all this and more over the entire life course. Yet such a grand experiment would save countless lives and enhance quality of life.

38. Arthur Schopenhauer, "The Ages of Life." In *Counsels and Maxims*, T. Bailey Saunders (tr.). London: Swan Sonnenschein & Co., 1890.

39. Daniel Rudman and others, "Effects of Human Growth Hormone in Men over 60 Years Old," *The New England Journal of Medicine* (1990) *323*: 1–6.

40. Thomas Perls, Neal R. Reisman, and S. Jay Olshansky, "Provision or Distribution of Growth Hormone for 'antiaging'," *Journal of the American Medical Association* (2005) *294*: 2086–2090.

41. Such approaches could be useful against wasting from diseases such as AIDS.

42. Testosterone, estrogen, growth hormone, DHEA, and melatonin are among the hormones that tend to decline with aging.

43. In 1996, Maxine Papadakis, professor of medicine at the University of California, San Francisco, studied fifty-two men, seventy and older, with varied growth hormone levels, and found similar changes in body composition as Rudman's but no improvements in muscle strength, endurance, and mental abilities. She found untoward side effects, such as swelling in ankles and lower legs.

44. Maria A. Fiatarone and others, "High Intensity Strength Training in Nonagenarians: Effects on Skeletal Muscle," *JAMA* (1990) *263*: 22.

45. Irwin Rosenberg, director of the Tufts Center on Nutrition, also contributed to this concept and work.

46. I borrow freely here from work with colleagues.

47. S. Jay Olshansky, Daniel Perry, Richard A. Miller, and Robert N. Butler, "In Pursuit of the Longevity Dividend," *The Scientist* (2006) *20*: 28–36.

Chapter 11: Toward a Prescription for Longevity and Quality of Life

1. Statement written in 1900. By Nobelist Elie Metchnikoff, *The Prolongation of Life*, republished by International Longevity Center, 2004.

2. *Advancing Health Literacy: A Framework for Understanding and Action.* The

National Assessment of Adult Literacy (NAAL) estimates that only 12 percent of the adult U.S. population is "proficient" at understanding health information. Institute of Health Medicine, *Health Literacy: A Prescription to End Confusion,* Washington, DC: National Academies Press, 2004.

3. Anna Maria Herskind and others, "The Heritability of Human Longevity: A Population-based Study of 2,872 Danish Twin Pairs Born 1870–1900," *Human Genetics* (1996) *97*: 319–323.

4. An example of how misleading reports by both investigators and the media can be: that a diet or a medicine cuts the death rate from a disease by 50 percent can mean one death in one million compared to two deaths in a million!

5. The first meta-analysis was undertaken by Karl Pearson, a statistician, in 1904, to overcome the limits of small sample sizes.

6. "Crossover" refers to switching groups who have received the drug or hormone with the control or placebo group in the course of study.

7. Men require more calories than women. Calorie intake depends on many factors, principally activity level.

8. Some baby boomers have shown an interest in two hybrids: neutrichemicals and cosmiceuticals, hybrids of foods and drugs and cosmetics and drugs, respectively.

9. Thomas A. Wadden and others, "Randomized Trial of Lifestyle Modification and Pharmacotherapy for Obesity," *New England Journal of Medicine* (2005) *353*: 2111–2120.

10. See *Food Marketing to Children and Youth: Threat or Opportunity?* Institute of Medicine (IOM) report on childhood obesity, 2005.

11. S. Jay Olshansky and others, "Potential Decline in Life Expectancy in the United States in the 21st Century," *New England Journal of Medicine* (2005) *352*: 1138–1145.

12. See Alice Waters, "Eating for Credit," *New York Times,* Feb. 24, 2006.

13. A 20-ounce bottle of cola has sixty-seven grams of sugar and 250 empty calories. Some see colas as liquid energy, contributing mightily to childhood obesity.

14. A pound of fat is equal to 3,500 calories, so a daily reduction of five hundred calories should result in a pound per week of fat loss.

15. According to a study by the National Institute of Child Health and Human Development (NICHD).

16. William G. Bowen and Sarah A. Levin. *Reclaiming the Game: College Sports and Educational Values.* Princeton, NJ: Princeton University Press, 2003.

17. The sessions are to be based on the Department of Health and Human

Services Public Health Service Guideline, "Treating Tobacco Use and Dependence: A Clinical Practice Guideline," (PHS2000 Guideline), available online at www.surgeongeneral.gov/default.htm.

18. One of the biggest and most successful federal lawsuits in history accused cigarette makers of fifty years of fraud, deceptive advertising, and dangerous marketing practices. Yet, at the close of the nine-month trial, the Justice Department reduced the penalties it was seeking from the tobacco industry to $10 billion from $130 billion (2005)!

19. "Health Risks from Exposure to Low Levels of Ionizing Radiation: BEIR VII Phase 2," http://www.books.nap.edu/catalog/11340.html.

20. Jason Lazarou, Bruce H. Pomeranz, and Paul N. Corey. "Incidence of Adverse Drug Reactions in Hospitalized Patients: A Meta-analysis of Prospective Studies," *JAMA* (1998) *279*: 1200–1205.

21. This includes an estimated five thousand to six thousand deaths each year from intestinal bleeding from such commonly used medicines as ibuprofen and aspirin, according to a paper in the *American Journal of Gastroenterology*.

22. Cholesterol-lowering statins might help prevent bone fractures and mitigate the effects of the flu.

23. James Nyberg and Robert N. Butler, *Post-marketing Drug Surveillance* (New York: International Longevity Center, 2004). France has a network of "pharmaco-vigilance" centers to identify adverse drug reactions. In 1938, Congress passed the Food, Drug and Cosmetic Act, which required pharmaceutical manufacturers to prove drug safety for the first time.

24. Herbert Schunkert, "Pharmacotherapy for Prehypertension—Mission Accomplished?" *The New England Journal of Medicine* (2006) *354*: 1742–1744.

25. National Highway Traffic Safety Administration, U.S. Department of Transportation, *The Economic Impact of Motor Vehicular Crashes 2000*, May 2002.

26. That DNA is inherited in chunks is the principle central to what has been named the HapMap. A chunk is called a haplotype. The HapMap describes the common pattern of human genetic variation.

27. New York: Routledge, 2005.

28. The International Longevity Center, *Putting the Brain Back into Medicine*, 2006.

29. Spiegel, Karen, Rachael Leproult, and Eve Van Cauter, "Impact of Sleep Deficit on Metabolism and Endocrine Function," *The Lancet* (1999) *354*: 1435–1439.

30. S. Jay Olshansky and others, "Potential Decline in Life Expectancy in the

United States in the 21st Century," *New England Journal of Medicine* (2005) *352*: 1138–1145.

31. International Longevity Center, *Is There An "Anti-Aging" Medicine?* New York: International Longevity Center, 2001; Robert N. Butler, Michael Fossel, Mitchell Harman, et al., "Is There an "Anti-Aging" Medicine?" *Journal of Gerontology: Biological Sciences* (2002) *57*: B333–B338; General Accounting Office, *Anti-Aging Products Pose Potential for Physical and Economic Harm.* Report to Chairman, Special Committee on Aging, U.S. Senate (Washington, DC: GAO, 2001, GAO–01–1129).

32. In 2001, people spent $17.8 billion in the United States on dietary supplements, $4.2 billion of it for herbs and other botanical remedies. In 1994, there were four hundred unregulated products on the U.S. market. In 2004, there were twenty-nine thousand.

33. See "Dangerous Supplements," *Consumer Reports* (2004) *69*: 12–17. Note especially, "Twelve supplements you should avoid," p. 15.

Chapter 12: Redesigning Health Care for an Older America

1. ACE units, hospital units dedicated to all acute care of the elderly, have been developed all over the United States.

2. About 14 percent of Medicare beneficiaries have congestive heart failure among their chronic conditions, and these beneficiaries account for 43 percent of Medicare spending. About 18 percent of Medicare beneficiaries have diabetes, yet account for 32 percent of Medicare spending.

3. This was one of Maggie Kuhn's signature lines stated in many, many speeches.

4. Linda T. Kohn, Janet M. Corrigan, and Molla S. Donaldson (eds.), "Committee on Quality of Health Care in America." In *To Err Is Human: Building a Safer Health System* (Institute of Medicine, 2000).

5. In our family, a medical family, whenever a relative is in the hospital, a family member or friend has to be in attendance day and night to protect the patient, checking drug dosages, for example. Hospitals should encourage the participation of families in the care of the family members, including those in the intensive care unit (ICU).

6. There were 144 allopathic and osteopathic medical schools in 2005.

7. Today, one-year clinical fellowships in geriatrics are financed through the federal government, mostly using Medicare Graduate Medical Education funds and the Veterans Administration. The National Institute on Aging provides research fellowships. The Health Resources and Services Administration

has provided the Geriatric Academic Career Award (GACA). Advanced fellowships have been provided by private philanthropy, especially by the Brookdale, John A. Hartford, and Donald W. Reynolds Foundations, the Commonwealth Fund, Atlantic Philanthropies, and through the Alliance for Aging Research and the American Federation for Aging Research (AFAR).

8. Unfortunately, there has been a decline in the number of primary care physicians in the United States, which further complicates the care of our older population. See Thomas Bodenheimer, "Primary Care—Will It Survive? Perspective," *The New England Journal of Medicine* (2007) *355*: 861–864.

9. See, for example, David Casarett, Jennifer Kapo, and Arthur Caplan, "Appropriate Use of Artificial Nutrition and Hydration—Fundamental Principles and Recommendations," *The New England Journal of Medicine* (2005) *353*: 2607–2612.

10. Health care workers are recruited from the developing world, such as the Philippines and Ghana, to the UK, Canada, Australia, and the United States, depriving the developing countries of needed health personnel.

11. Medicare payments to doctors were cut by 5.4 percent in 2004 and remain under threat.

12. They could merge as one specialty.

13. The HELP committee of the House of Representatives refers to Health, Education, Labor and Pension.

14. Of a $2.4 trillion federal budget!

15. Now folded into Medicare Payment Advisory Commission.

16. *AARP Bulletin*, May 2007. Also note Joseph Violante, national legislative director for the Disabled American Veterans (DAV), who calls the VA medical system "excellent."

17. Cynthia X. Pan, M.D., Emily Chai, M.D., Jeff Farber, M.D. ILC Report: *Myths of the High Medical Cost of Old Age and Dying.* International Longevity Center, 2007.

18. I have built upon the work of an ILC task force reported in *Redesigning Health Care for an Older America: Seven Guiding Principles* (International Longevity Center, 2004).

19. In 2000, $257 billion is the estimate of the national economic value of informal, family caregiving. This is twice the amount paid today for home health and nursing home care, which in 2000 totaled $126 billion. Peter Arno, an economist, unpublished update of the previous estimates. See *Economic value of informal caregiving: 2000.* Presentation before the American Association for Geriatric Psychiatry, Orlando, FL, 2002; and Peter S. Arno and C. Memmott, "The Economic Value of Informal Caregiving," *Health Affairs* (1999) *18*: 182–188.

20. International Longevity Center, *Caregiving in America,* 2006.

21. International Longevity Center, *Myths of the High Medical Cost of Old Age and Dying,* 2007.

22. International Longevity Center, *Clinical Trials and Older Persons: The Need for Greater Representation* (New York: International Longevity Center, 2003).

23. The per capita cost of health care in America on average is $5,267 per year (in 2002) compared to $2,931 in Canada and $2,736 in France.

24. "Premiums and Profits," *Consumer Reports,* September 2007, p. 20.

25. So-called Medicare Advantage is costly to taxpayers but profitable to the health insurer. Members of these plans are supposed to receive extra benefits, in addition to standard fee-for-service Medicare. Medicare Advantage costs 12 percent more than government-administrated Medicare benefits. See, "Extra Payments from Medicare = Extra Profits for Insurers," *Weekly Medicare Consumer Advocacy Update 7*: Aug. 16, 2007.

26. Institute of Medicine, *Insuring America's Health: Principles and Recommendations.* (Washington, DC: National Academies Press, 2004).

27. In the meantime, State Child's Health Insurance Program (SCHIP) should be expanded.

Chapter 13: Live Longer, Work Longer

1. This number will crest between 2020 and 2030.

2. Nickk M. Vanston, "Maintaining Prosperity in an Aging Society," in *Longevity and Quality of Life,* Robert N. Butler and Claude Jasmin (eds.) (New York: Kluwer Academic/Plenum, 2000), 252–262.

3. See Butler, "The Relation of Extended Life to Extended Employment Since the Passage of Social Security in 1935." *Milbank Memorial Fund Quarterly/Health and Society* (1983) *61*: 420–429.

4. Disability evaluation is a complex issue but at present it is inadequate and painfully slow. There is a two-year backup in evaluation! Many people with disabilities want to remain independent and can work but can lose disability payments if they do so. Substantial and more generous earnings should be allowed before losing benefits, utilizing a sliding scale.

5. See the standard economics textbook by Paul Samuelson and William D. Nordhaus, *Economics* (McGraw-Hill, 2004).

6. Butler, Testimony to Subcommittee on Retirement and the Individual, Senate Special Committee on Aging. July 15, 1969. Also see: Butler, *Why Survive? Being Old in America* (1975) reprinted by the Johns Hopkins Press, 2002.

7. A viatical settlement is the sale of a life insurance policy by the owner before it matures. This offers the seller immediate cash. The buyer's price is discounted from the face value policy, in excess of the premiums paid and the current surrender value. This arrangement usually applies to individuals with less than two years of life expectancy in countries without state-subsidized health care or where health costs are very high (such as the United States).

8. In 1982 I introduced the notion of productive aging, which in retrospect seems to me to overemphasize monetization and commoditization. I now believe that productive engagement is a more felicitous term. See Butler, *Why Survive?* (especially Chapter 4, "The Right to Work"); Butler and Herbert P. Gleason (eds.), *Productive Aging: Enhancing Vitality in Later Life* (New York: Springer, 1985); Butler, Mia Oberlink, and Malvin Schechter (eds.), *The Promise of Productive Aging: From Biology to Social Policy* (New York: Springer, 1990); Butler and Schechter, "Productive Aging." In *Encyclopedia of Aging,* 2nd ed. (New York: Springer, 1995), 763–764.

9. The unpaid work of women is said to be the equivalent of GDP that is measured!

10. Barbara Butrica, Richard W. Johnson, Karen E. Smith, and Eugene C. Steuerle, "The Implicit Tax on Work at Older Ages," Urban Institute, http://www.urban.org/url.cfm?ID=1001021, (PDF) October, 2006.

11. The Occupational Safety and Health Administration (OSHA) should be further empowered to use ergonomics and impose stiff penalties for failures, e.g., to protect the backs of nursing home employees by providing full body slings for lifting patients. According to OSHA in 2005, more than 6,800 workplace-related deaths occurred as well as 4.2 million injuries and illnesses. Nonetheless, since 2001 the fatality rate has fallen 7 percent and the injury rate 19 percent. When the ADA was passed, only 25 percent of the fifteen million disabled of working age (altogether there are fifty-four million disabled Americans) were employed. Two-thirds of those who were not working wished to do so. Like older employees, studies show superior job performance by the disabled as well as excellent safety and attendance records. Yet, despite ADA, the unemployment rate for working-age persons with disabilities remains high and incomes low.

12. Linda Greenhouse, "Victory for Employers," *New York Times,* Jan. 9, 2002.

13. As noted earlier, disability rates among older persons have declined since 1982. See Kenneth G. Manton, Larry Corder, and Eric Stallard, "Chronic Disability Trends in the Elderly. United States Populations: 1982–1994." *Proceedings of the National Academy of Sciences* (1997) 94: 2593–2598. They used data from the National Long-term Care Survey.

14. There are 140 million workers in the United States.

15. Arthur Kramer, International Longevity Center, *Maintaining and Achieving Cognitive Vitality,* 2002. Howard M. Fillit and Butler (eds.), *Cognitive Decline: Strategies for Prevention* (London: Greenwich Medical Media, 1997).

16. Marion C. Diamond and others, "Rat Cortical Morphology Following Crowded-enriched Living Conditions," *Exp. Neurol.* (1987) 96: 241–247.

17. Butler, *Why Survive?*

18. U.S. Bureau of Labor Statistics.

19. Independent Sector, *Value of Volunteer Time,* http://www.independentsec tor.org, 2004.

20. *Seattle Post-Intelligencer,* June 16, 2004.

21. Originally Sergeant Shriver, the first head of the Peace Corps, discouraged recruitment of older persons, concerned about possible illness and disability as well as legal liability. This contrasts with an advertisement to join the Peace Corps in the *New York Times,* Aug. 10, 2001, that read, "No upper age limit. Seniors are highly regarded and desired."

22. Experience Corps, Civic Ventures, 425 Second Street, Suite 601, San Francisco, CA 94107, (415) 430-0141, www.experiencecorps.org.

23. National Senior Service Corps, Corporation for National Service, 1201 New York Avenue NW, Washington, DC 20525, (202) 606-5000, www.cns.gov. Also of interest to older volunteers is Habitat for Humanity, 121 Habitat Street, Americus, GA 31709-3498, (912) 924-6935, www.habitat.org.

24. Personal communication.

25. See *The Fallacy of the Fixed Lump of Labor,* Issue Brief, International Longevity Center, 2007.

26. Even in Japan, lifetime employment is fading.

27. See Stephen Linder, *The Harried Leisure Class* (New York: Columbia University Press, 1970); Sebastian De Grazia, *Of Time, Work and Leisure* (New York: Twentieth Century Fund, 1962); Mortimer J. Adler, *We Hold These Truths* (New York: Macmillan, 1987).

28. Elie Metchnikoff, *The Prolongation of Life. Optimistic Studies* (New York: G.P. Putnam's Sons, 1908), reissued by Springer and the International Longevity Center, 2004.

29. The unemployed and those who do not seek jobs include those who are very poorly educated, mentally and physically ill, alcoholic or otherwise addicted, and criminal.

Chapter 14: Social Security

1. The numbers and calculations in this chapter reflect conditions and data in general, circa 1997–2005, as available.

2. The United States lags behind Germany, Canada, and other countries that provide better pension benefits and replacement of income.

3. The government must pay T-bond holders before it pays anyone else, including employees, contractors, and even insured bank depositors. Taxpayers provide the ultimate protection.

4. *Life Insurers Fact Book* (American Council of Life Insurance, 2003).

5. Economic growth may get us through Social Security solvency, but we cannot be expected to grow our way out of all our deficits. See presentation at Envisioning the Future conference by David W. Walker, controller-general of the United States, MetLife/ILC, Washington, DC, April 3, 2006.

6. *Wall Street Journal*, June 28, 2004, p. R1.

7. http://www.irs.gov/pub/irs-pdf/p915.pdf; http://www.bankrate.com/brm/itax/tips/20010115a.asp?print=on; http://www.gofso.com/Premium/LE/20_le_ir/fg/fg-SS_Benefits.html.

8. A 1997 estimate, from a memorandum by Willard Witherspoon Jr. entitled *Equivalent Life Insurance in Force under OASDI*, Oct. 19, 1998. Made available by Robert Myers. The corresponding amount of private life insurance in force for 1999 was $13.5 trillion.

9. Average weekly earnings for an individual are $532.02. (Bureau of Labor Statistics, United States Department of Labor. March 2005. http://bls.gob/news.release/empsit.t16.htm.); real median household income is $43,318 per year (Carmen DeNavas-Walt, Bernadette D. Proctor, Robert J. Mills, and U.S. Census Bureau, *Income, Poverty, and Health Insurance Coverage in the United States: 2003.* http://www.census.gov.prod/2004pubs/p60-226.pdf.)

10. The Bush administration argues for converting to "price-indexing," which would significantly reduce the standard of living of future retirees.

11. Essentially all economists agree that the full payroll tax (including the employers' contributions) come out of wages.

12. *Income of the Population 55 or older: 2000* (Social Security Administration, 2000).

13. International Longevity Center, *Old and Poor in America* (New York: International Longevity Center, 2001).

14. About $179/week or about $25/day. See *Federal Register* (2004) *69*: 7336–7338.

15. "Young Workers' Unique Stake in Social Security Reform," a panel at a conference on Social Security and young Americans sponsored by the Cato Institute, America's Future Foundation, and Third Millennium (2002).

16. Humphrey Taylor, "The remarkable lack of intergenerational conflict as to how government spending should be divided between services for young and old." *The Harris Poll #24*, May 20, 1998.

17. (New York: Times Books, 1999). Also see Peter G. Peterson, *Running on Empty: How the Democratic and Republican Parties Are Bankrupting Our Future and What Americans Can Do about It* (New York: Random House, 1999).

18. U.S. Census Bureau, April 2003, *The Older Population in the United States: March 2002*, http://www.census.gov.prod/2003pubs/p20-546.pdf.

19. Peter Whiteford and Gregory Angenent, *The Australian System of Social Protection—an Overview*, Occasional Paper No. 6, 2nd ed. (Canberra, Australia: Department of Family and Community Services, 2002).

20. Arthur M. Schlesinger Jr., *The Age of Roosevelt: The Coming of the New Deal* (Boston: Houghton Mifflin, 1959, 308–309).

21. And, obviously, those who die receive nothing.

22. According to Arthur Levitt, former chair of Securities and Exchange Commission.

23. The Standard & Poor's stock index lost over 40 percent of its value.

24. By 2018, Social Security would draw upon the $1.7 trillion trust fund that resulted from the recommendations of the Greenspan Commission of 1983.

25. Nearly the present life expectation in the United States.

26. Generally regarded as most likely to occur.

27. Between 1996 and 1998, the robust U.S. economy postponed the drop in Social Security solvency three years from 2029 to 2032 and the 75-year deficit from 2.23 percent of taxable payroll to 2.19 percent. By 2002, the Social Security trustees projected a 1.87 percent shortfall.

28. Social Security Advisory Council, 2001 OASDI Trustees Report, http://www.ssa.gov/OACT/TR/TR01/trTOC.html.

29. *Social Security Reform* (rev. ed., 2005). "A Century Foundation Guide to the Issues" (New York: The Century Foundation Press, 2005).

30. FICA, Federal Insurance Contribution Act.

31. There is no limit for the earnings base on which the 1.45 percent Medicare hospital insurance tax is paid.

32. This should be compared with the Chilean system, promoted by the Cato Institute, which requires 10 percent in payroll taxes by workers for retirement,

an additional 3.5 percent for disability, and 7 percent for health insurance, a total of 20.5 percent! Moreover, the entire Chilean population is not covered.

33. From 1978 congressional hearings testimony conducted by David Stockman and Richard Gephardt, both in Congress at the time, noted that only modest additions to worklife significantly improved the financial status of Social Security. (U.S. Select Committee on Population, House of Representatives, *Consequences of Changing U.S. Population: Baby Boom and Bust,* June 1, 2, 1978.)

34. Anthony Webb, *The Dependency Ratio: What Is It? Why Is It Increasing? What Are the Implications?* (New York: Alliance for Health and the Future, 2005).

35. There has been a steady decline in the *total* dependency ratio since 1900.

36. Personal communication, Robert M. Ball, March 30, 1998.

37. The Chamber of Commerce and the National Association of Manufacturers lobby Congress to create individual accounts.

38. There are 141 million American workers.

39. Advocates may prefer the term *ownership society to privatization* or *personal investment accounts,* emphasizing individual choice and personal control, which sound appealing.

40. *Saving Social Security with Stocks: The Promises Don't Add Up,* A Twentieth Century Fund Economic Policy Institute Report (New York, 1997).

41. Once again, special consideration should be given to the payroll tax burden on low-income workers. The lowest decile, for example, might be spared a raise in the payroll tax.

42. Only about 6 percent of U.S. households have incomes above that level. The total removal of the cap would cover two-thirds of the shortfall. Robert M. Ball recommends restoring the maximum taxable earnings base to 90 percent, the level set by Congress in 1983. This figure is adjusted each year in line with inflation (personal communication, December 2004).

43. See Jonathan Gruber and David A. Wise, "Social Security Programs and Retirement around the World: Micro Estimation," NBER Working Paper No. 9407, December 2002.

44. As introduced in 1983, to a degree, by congressional action based upon the Greenspan Commission of 1982.

45. In 2027, when the age for full eligibility for Social Security is sixty-seven, those who retire at sixty-two will receive only 70 percent of their full benefits.

46. Raising the age of full eligibility for Social Security does cut benefits, but if work is available, other benefits linked to productivity accrue to the individual and the nation. In 2003, the National Academy of Social Insurance

estimated that 25 percent of early retirees, those sixty-two to sixty-four, were too frail to do ordinary work.

47. In the United States, the average woman works twenty-seven years and spends twelve years in caregiving.

48. See International Reform Monitor, Social Policy, Labour Market Policy and Industrial Relations (Gutersloh, Germany: Bertellsmann Stiftung, March 2005). Also see Rachel Hennick, *Family Policy in the U.S., Japan, Germany, Italy and France: Parental Leave, Child Benefits/Family Allowances, Child Care, Marriage/Cohabitation, and Divorce*, a briefing paper prepared for the Council on Contemporary Families (New York: Council on Contemporary Families, May 2003).

49. In Japan, pension benefits are indexed to net real wage increases rather than prices. This is calculated minus public pension contributions.

50. Caution in any downward revision of the CPI is mandated because of the high expenses of older persons due to prescription drugs and health care. *Thus any adjustment must focus on the actual living costs of retirees and the disabled, not on the population as a whole.*

51. It has been recommended that up to half of the system's cost eventually be financed by general revenues.

52. *Investment* as an economic term is distinguished from savings and means the purchase of capital goods and structures.

53. Dec. 30, 1996.

54. There is criticism that minorities who do not live as long are disadvantaged by Social Security. However, that is misleading because of survival and disability benefits that offset the reduced life expectancy. The real challenge is to continue to narrow the life expectancy gaps.

55. The values of reduced benefits taken at sixty-two and of higher benefits later are designed to be equivalent, assuming one lives to one's average life expectancy. Suppose one was due $1,000 per month at sixty-five, one would receive $750 per month at 62, and $1,320 per month at 70. Only 20 percent of adult Americans were aware of this change as of 2001.

56. See Gruber and Wise, "Social Security Programs and Retirement around the World," Chicago: University of Chicago Press, 1999.

Chapter 15: The Private Sector

1. As Freddie Mac has advertised, the thirty-year, fixed-rate mortgage invented by government has become wealth-creating (since 1934). Freddie

Mac is a government-sponsored and stockholder-owned corporation that purchases mortgages on the secondary market. By fueling mortgage lenders with liquidity, Freddie Mac encourages homeownership.

2. Select Committee on Population, House of Representatives, *Consequences of Changing U.S. Population: Baby Boom and Bust,* June 1–2, 1978.

3. Survey Research Center, Institute of Social Research of the University of Michigan. *The Health and Retirement Study: A Longitudinal Study of Health, Retirement, and Aging.* Sponsored by the National Institute on Aging, 2005. http://hrsonline.isr.umich.edu.

4. A charitable remainder trust provides tax savings and income to a donor to a charity.

5. Employer Benefit Research Institute (EBRI).

6. By 1999, half of all workers, some seventy-three million, had no employer-based pensions.

7. Part of the Revenue Act of 1978, paragraph (k) of section 401 of the Internal Revenue Code, evolved into a tax-deferred employer-employee matching plan, typically fifty cents from the employer to the employee's dollar. Similar 403(b) plans are used by nonprofit groups. In 1987, federal employees were given a 401(k)-like plan called the thrift savings plan rather than a traditional governmental pension.

8. Economic growth may get us through Social Security solvency, but we cannot be expected to grow our way out of all our deficits. See presentation at a conference called Envisioning the Future by David W. Walker, controller-general of the United States, MetLife/ILC, Washington, DC, April 3, 2006.

9. Statistics from the Department of Labor.

10. The Pension Benefits Guaranty Corporation (PBGC) was created under ERISA in the Department of Labor to monitor pension plans, but has limited resources and personnel.

11. Of course, a future government, faced with deficits, could reduce or end tax-deferred plans.

12. Nearly seventeen thousand Enron pension plan holders will see their benefits paid, thanks to pressure from the federal government's Pension Benefits Guaranty Corporation.

13. Through the 1990s, America enjoyed the longest bull stock market, which began in 1982 and brought the middle class into Wall Street. There has been a 72 percent increase in stock prices. Usually bull markets last for only one or two years. It could be dangerous to privatize Social Security when the memory of the voters, and in this case the baby boomers, is more focused upon a successful bull market and not sensitive to bear markets.

14. Of which 88 percent are still employed.

15. By age thirty-two, a typical U.S. worker has changed jobs nine times.

16. Leslie Wayne, "Pension Changes Raising Concerns," *New York Times* (late edition), Aug. 29, 1994, A1; Sarah Holden and Jack VanDerhei, EBRI Issue Brief No. 238: *Contribution Behavior of 401(k) Plan Participants* (Washington, DC: Employee Benefit Research Institute, October 2001).

17. The failure, as of June 2002, to enact necessary reforms has contributed to the reluctance and lack of confidence of investors to buy stock. Consumer groups, AARP, and Common Cause have fought for reforms. John C. Bogle, founder of the Vanguard Group, and Warren Buffett belong to the Federation of Long-Term Investors, which insists on reforms. Bogle said in *Business Week,* "Our capitalist system is in peril" (June 17, 2002, p. 27).

18. Private companies and companies that file as small business issuers were not required to comply until after December 15, 2005.

19. See Gregory Baer and Gray Ginsler, *The Great Mutual Fund Trap: An Investment Recovery Plan* (New York: Broadway Books, 2004).

20. The United States has excessive corporate and household debt as of 2007.

21. Barbara Ehrenreich, *Nickel and Dimed: On (Not) Getting By in America* (New York: Henry Holt, 2001).

22. www.livingt0100.com; www.webannuities.com.

23. Ellen E. Schulz and Jeff D. Opdyke, "Annuities 101: How to Sell to Senior Citizens," *Wall Street Journal,* July 2, 2002 ("treat them like they're blind 12-year-olds"). Variable annuities are tied to the stock market.

24. Invented about 1970.

25. They do have messy income-tax consequences but are exempt from state and local taxes.

26. Motoko Rich and Eduardo Porter, "Increasingly, the Home Is Paying for Retirement," *New York Times*, Feb. 24, 2006.

27. Jonathan Clements, "Tapping Your House to Fund Retirement: Alternatives to Costly Reverse Mortgages," *The Wall Street Journal,* July 7, 2004.

28. In Europe, the average asset allocation is 56 percent in stocks, 34 percent in bonds and 8 percent in real estate (EFRP data).

29. Adrian Tan, "Securing the Singaporean Nest Egg," *AsiaWise,* May 17, 2001.

30. Indermit Gill, Truman Packard, and Juan Yermo, *Keeping the Promise of Social Security in Latin America* (Washington, DC: World Bank, 2004). In 1999 the World Bank issued a report, "Averting the Old Age Crisis," which called for personal retirement accounts and less dependence upon pay-as-you-go systems. By 2005, the World Bank was less enthusiastic. Its new report "Old Age Income Support in the 21st Century" called for "enhanced focus on

basic income provision for all vulnerable elderly," financed through general tax revenues, not workers' contributions. See *The Economist,* "Pension Reform, Second Thoughts on the Third Age," Feb. 19, 2005, pp. 67–68.

31. Sara E. Rix, *Chile's Experience with the Privatization of Social Security,* Issue Brief No. 23, AARP Public Policy Institute, August 1995.

32. Orszag and Orszag have concluded from studies of individual retirement accounts in OECD countries that individual accounts have entailed significant problems in every developed country that has implemented them. See Michael J. Orszag and Peter R. Orszag, "Individual Accounts: Lessons from International Experience," *Science* (2005) *309*: 250–251.

33. Author of *Stocks for the Long Run* (New York: McGraw-Hill, 2002).

34. Reported by Aaron Lucchetti, *The Wall Street Journal,* Oct. 7, 2002, p. R1.

35. Robert J. Schiller, *Irrational Exuberance* (Princeton, NJ: Princeton University Press, 2000).

36. There was an excellent series examining practices that put Americans' retirement savings at risk in the *Los Angeles Times,* April 23–25, 2006, by Josh Friedman. Many 401(k) accounts are being eroded by unseen fees; annuity providers are targeting the elderly; and teachers unions are steering members into investments with high fees and poor returns.

37. *New York Times,* July 23, 2002.

38. One of the largest pools of investment capital in the world.

39. Gretchen Morgenson and Mary Williams Walsh, "How Consultants Can Retire on your Pension. S.E.C. Turns on an Industry Rife with Conflicts," *New York Times,* Dec. 12, 2004, Business section, p. 1.

40. Kelly Green, "Theft from 401(k)s Is on the Rise," *Wall Street Journal,* March 2, 2005, p. D1.

41. But the reverse is happening. See Deborah Solomon, "Tough Talk of SEC Chief Could Relent," *Wall Street Journal,* Jan. 12, 2005, pp. C1 and C5. Also see Stephen Labaton, "A New Mood in Congress to Forego Corporate Scrutiny," *New York Times,* March 10, 2005, p. C3.

42. Unfortunately, this could encourage small and troubled companies to terminate pension plans. A positive Bush administration effort to strengthen the defined-benefit pension system was weakened by lobbyists. For example, lobbyists want to extend the seven-year catch-up period for corporations to build up their pension liabilities, as the bill allows. In fact, the Pension Protection Act of 2006 has forced companies such as DuPont to scale back their forced DB plan. Also see Mary Williams Walsh, "Major Changes Raise Concerns on Pension Bill," *New York Times,* March 18, 2006.

43. See "The High Cost of Corruption," editorial, *Business Week,* Nov. 29, 2004. Economists don't usually deal with corruption.

44. With many retirees, often more than active employees.

45. That the business sector in the United States plays such a large role in old-age pension and health care came about during World War II when wages were frozen and union leaders negotiated compensatory benefits. As the global economy has grown, these benefits have become burdens, especially to the mature industries that have many retirees.

46. GM provides health care benefits to 1.1 million Americans, including workers, retirees, and their families. That is roughly 0.4 percent of the total U.S. population and explains why GM is the largest private purchaser of prescription drugs. Also see: Mary Williams Walsh and Danny Hakim, "GM and a U.S. Agency See Pensions in Different Lights," *New York Times,* Oct. 3, 2005.

47. In February 2005, G. Richard Wagoner Jr., GM chairman and CEO, called on corporate and government leaders to reform the nation's health care system. He supports some type of national catastrophic reinsurance program, according to the Medicare Rights Center in New York City.

48. William H. Gates and Chuck Collins, "Tax the Wealthy: Why America Needs the Estate Tax," *The American Prospect Online,* June 17, 2002.

Chapter 16: The Politics of Aging and Longevity

1. Capital punishment and war, not to mention child care, are usually left off the list by those who represent the pro-life position.

2. Social democracy is an amalgam of or a democratic compromise between socialism and capitalism that, in varying degrees, we have in Western Europe, the United States, Canada, Japan, and Oceania. Franklin D. Roosevelt's New Deal, on the other hand, has been judged by many to have been reforms that saved capitalism from its excesses. See David M. Kennedy, *Freedom from Fear: The American People in Depression and War* (New York: Oxford University Press, 1999).

3. Mark Twain and Charles Dudley Warner wrote a satirical novel published in 1873 called *The Gilded Age.*

4. The centrist tendencies of both parties have resulted in what is called a "duopoly."

5. Brody Mullins, "Financial Aid-Growing Rule for Lobbyists: Raising Funds for Lawmakers," *Wall Street Journal,* Jan. 27, 2006, p. A1. There is a growing level of corruption in American life beyond politics and including medicine and academe as well as business, finance, the law, and the media. The

latter is all too cozy with political leaders, operating quite differently than did the legendary freelance journalist I.F. Stone. See Frank Rich, "All the President's Flacks," *New York Times*, Dec. 4, 2005.

6. *The American Prospect,* May 2002, A Tipping Point?
7. Charles R. Morris, *The AARP. America's Most Powerful Lobby and the Clash of Generations* (New York: Random House, 1998). Also see Bill Novelli and Boe Workman, *50+: Igniting a Revolution to Reinvent America* (New York: Holtzbrinck, 2007).
8. Peter H. Lindert, *Growing Public: Social Spending and Economic Growth Since the Eighteenth Century* (New York: Cambridge University Press, 2004).
9. William Easterly and Sergio Rebelo, "Fiscal Policy and Economic Growth: An Empirical Investigation," *Journal of Monetary Economics* (1993) *32*: 417–458.
10. Joel Slemrod and Jon Bakija, *Taxing Ourselves: A Citizen's Guide to the Great Debate Over Tax Reform* (Cambridge, MA: MIT Press, 1996). It explores the relationship between the marginal income tax rate, the rate imposed on additional income in a progressive tax system, and productivity.
11. Turkey, Mexico, South Korea, and Ireland.
12. Paul Krugman, "The Memory Hole," *New York Times*, Aug. 6, 2002.
13. See Miles Corak, an economist for Canada's national statistical agency, who edited a Cambridge University Press book on mobility in Europe and North America.
14. Although live tissue can be stored indefinitely, the very process of freezing dead tissue probably results in destroying it, because when the spaces between cells are filled with water and frozen, the ice crystals that form on the outside of the cells sever and tear the host tissue beyond repair.
15. Latin American countries require voting as the responsibility of a citizen of a democracy. Some countries, however, such as Ecuador, say people over sixty should not be obligated!
16. Sunstein, *Designing Democracy.*
17. An idea promoted by Myrna I. Lewis, "A Proactive Approach to Women's Concerns: Women's Longevity Groups and Funds," International Longevity Center, 2005.
18. Abraham Lincoln wrote, "Labor is prior to, and independent of, capital. Capital is only the fruit of labor, and could never have existed if labor had not first existed. Labor is the superior of capital, and deserves much the higher consideration." Quoted in Kevin Phillips, *Wealth and Democracy: A Political History of the American Rich* (New York: Broadway Books, 2002).

Chapter 17: Population Solutions to Longevity

1. At best, these are educated guesses. Moreover, there are major assumptions that China will maintain population control efforts and that there will not be a nuclear world war. Census counts are "squishy." For example, see Daniel Walfish, "National Count Reveals Major Societal Changes," *Science* (2001) *292*: 1823, regarding a China that is "becoming older, better educated and more transient" but with a probable census undercount.

2. Robert W. Kates and others, "Policy Forum: Environment and Development, Sustainability Science," *Science* (2001) *292*: 641–642.

3. The last part of the title, *with Remarks on the Speculations of Mr. Godwin, M. Condorcet, and Other Writers*, questions their optimism about the perfectibility of man. He believed the "poor laws" encouraged fecundity and should be ended. He favored workhouses and not "comfortable asylums."

4. This is the process from high birth and high death rates to low birth and low death rates.

5. As of 2007, AIDS has not significantly affected overall world population growth but had affected Africa profoundly. Nonetheless, Africa still has the world's most rapidly expanding population, having an average of six children per couple.

6. Population Connection is the new name for Negative Population Growth Inc. (NPG), a national nonprofit organization founded in 1972. It appears to be the only organization that calls for a smaller United States, as well as a smaller world population. (Negative Population Growth Inc., P.O. Box 1206, 210 The Plaza, Suite 7G, Teaneck, NJ 07666.)

7. Jeanne Safer, "Women Who Turned Off the Biological Clock," *International Tribune*, Jan. 28, 1996.

8. Culling or shooting of deer and mustangs illustrates population planning that is felt necessary to preserve balance in nature. Persons opposed to family planning policies do not generally question culling.

9. The shorter an organism's life, the greater the number of progeny which can be found in nature, and vice versa. This is also the case with caloric-restricted animals. This is not universal.

10. Charlotte Muller, *Population Growth and Productivity*, Issue Brief (International Longevity Center, 2004).

11. Cairo, the most populous of all African cities, is overpopulated with its 12 million people (as of 2000) and is among the world's most polluted cities. The ancient Giza Pyramids are engulfed in smog. The U.S. Agency for International Development estimates that pollution is responsible for some

ten thousand to twenty-five thousand deaths a year. Egypt didn't have any environmental laws until 1995.

12. See *The Economist* (Jan. 7, 2006): "Some governments hate the idea of a shrinking population because the absolute size does matter for "great power" status. The bigger the economy, the bigger the military, the greater the geopolitical clout.... People should not mind, though, for *what matters for economic welfare is GDP per person*" (Italics are mine).

13. March 8, 1999.

14. By 2001, Singapore was offering baby bonuses of more than $800 for a second child and $1,600 for a third. The Singapore *Straits Times* called this "subsidizing the stork." The government has endeavored to be a match-maker, producing a dating guide and conducting dating services!

15. Alfred Sauvy, *General Theory of Population* (New York: Basic Books, 1969).

16. Metchnikoff wrote *The Nature of Man: Studies in Optimistic Philosophy* (1904) and *The Prolongation of Life: Optimistic Studies* (1907), in which he formulated the concept of "orthobiosis," which signifies healthy longevity and a natural death. (Reprinted by Springer in 2004 in collaboration with the International Longevity Center, as the second publication in its series, Classics in Longevity and Aging.) Metchnikoff envisioned the scientific transformation of certain "disharmonies" that occurred in the human evolutionary process into "harmonies."

17. President Theodore Roosevelt expressed such concerns.

18. France provides bonuses for third children. To encourage child bearing and family values, a French mother of three receives about $300 monthly from the French government until her children are eighteen. The benefit rises to $450 if there is a fourth child (1996 data). But France has not achieved a level replacement ratio despite these efforts. France is the European nation with the most explicit pronatalist policy. The jury is out as to whether it has produced more children. Although Sweden has no explicit pronatalist policy, in a sense it has the equivalent, because it has such a supportive family policy.

19. The United States provides tax deductions for dependent children as part of national policy.

20. This has been called the economic birthrate and is one price of German unification. In 1994 the German state of Brandenburg decided to pay parents $650 for every new child, and a monthly allowance, called *kindergeld* (child money), has been distributed in West Germany since 1955. In the former communist East Germany, there had been jobs, day care, health coverage, etc. Other former communist nations, such as Hungary and Poland, also provide

payments to families that have new babies, since they too have experienced falls in birthrates. Belgium, Luxembourg, and Portugal also give money to families of newborns.

21. "On Golden Pond," *New York Review of Books,* May 6, 1999.

22. It is said, "The poorer you are, the more children you have. The more children you have, the poorer you are."

23. See United Nations "Replacement Migration. Is it a Solution to Declining and Aging Populations?" 2001. Also see Gary S. Becker, "How Rich Nations Can Defuse the Population Bomb," *Business Week,* May 28, 2001. Of course, immigration brings many problems. Urbanization, one of the concomitants of the Industrial Revolution, reflects an internal migration, leaving, for example, ghost villages in Greece and nearly deserted ones in Africa.

24. In Australia, too, an "us first" message has led to a 20 percent reduction in immigration quotas in 1996–1998. This marks a change from the end of the "White Australia" policy in 1973 and has led to one of the most culturally and ethnically diverse societies in the world.

25. Of course, there are nations with many distinct ethnic groups that are all "native." Ivory Coast in West Africa has nearly sixty such groups. Nonetheless, under President Felix Houphouët-Boigny, there was a strong sense of nationhood, which he instilled. Will this survive his 1993 death?

26. Perhaps external forces can manage such internal ethnic and/or religious tensions. But the UN, NATO, the United States, and European states stood by for several painful years until 1995 and did not intervene in the horrors of Bosnia directly either through peacekeeping troops or effective diplomacy.

27. One of the problems of assimilation or integration in society is seen in the issue of the genital mutilation of girls by African immigrants. The larger issue, of course, is how to deal with the customs and values of immigrants unacceptable in the new country. This particular example of sexual mutilation affects millions of Muslim women in more than twenty-five countries in Africa and the Middle East. The World Health Organization estimates as many as 114 million women worldwide have been mutilated in an erroneous interpretation of the Koran. It is essentially an African practice and occurs among various religious, socioeconomic, and cultural groups. Western critics are seen as cultural imperialists by some Africans and Middle Easterners. Britain, Sweden, and Switzerland have passed specific laws banning the procedure.

28. This also applies to the global economy and global health—i.e., the ease with which infectious disease can spread worldwide.

29. Developed nations promote the brain drain when it suits their interest; when, for example, they seek the technically skilled, nurses, etc.

30. Total fertility rate (TFR) indicates the mean number of children a woman would have, given the current age-specific birthrate. The total fertility rate is influenced by urbanization, rising education, social security for old age, outside employment of women, fewer marriages, more divorce, less affordable housing space, the desire for higher standards of living, and general economic conditions, as well as the availability of more effective means of contraception and legal abortion; in short, by the growing freedom of choices by women. The birth rate is the number of births per 1,000 inhabitants.

31. Hideo Ibe, *Aging in Japan*, (New York: International Longevity Center, 2000).

32. There are thousands of ethnic Japanese from Brazil and Peru.

33. Mexico accounts for more than a quarter of all the foreign-born residents in the United States This is the largest of any country since the 1890 census, when about 30 percent of the country's foreign-born population was from Germany. About one-tenth of Mexico's population lives in America. The Mexican government is trying to formalize its relationship with these exiles. About $9.2 billion in remittances were sent from America to Mexico in 2001.

34. This is the middle series. The lowest series population is 282,706,000 and the highest is 1,182,390,000. (United States Census Bureau).

35. There is some evidence that the decline in birthrate predated the one-child-per-family policy.

36. Until the middle of the nineteenth century, no Western city matched ancient Rome in population.

Chapter 18: Worldwide Democratization of Longevity

1. Calculated up to age sixty-five. Christopher J.L. Murray and Alan D. Lopez (eds.), *The Global Burden of Disease* (Harvard School of Public Health on behalf of the World Health Organization and World Bank, 1996). (See summary.)

2. Ibid.

3. Etienne G. Krug and others, "The World Report on Violence and Health," *The Lancet* (2002) *360*: 1083–1088.

4. Anxiety, too, can be healthy and protective (called signal anxiety) or unhealthy and negative (called traumatic).

5. This can be done. Fortification of sugar in Central America has reduced the prevalence of Vitamin A deficiency.

6. The Montreal Protocol to the Vienna Convention for the Protection of the Ozone Layer.

7. *Racism and the Evolutionary Time Clock.* Geneticists can now map ancient migrations and secure a better understanding of human evolution. L.L. Cavalli-Sforza, P. Menozzi, and A. Piazza were the movers behind the Human Genome Diversity Project (HGDP), an effort to collect DNA samples and anthropological information from twenty-five individuals in each of one hundred targeted populations around the world. They created the first genetic atlas, based on the fact that genes in modern populations carry the encoded history of the remote past of human migrations. This gives us a record of the world's diverse populations. Their book, *The History and Geography of Human Genes* (1994), became an instant classic of genetic archeology. It combined the study of genetic differences between populations with linguistic studies, conventional archeology, history, and prehistory. This book eliminates any serious basis for the concepts of race and therefore any basis for racial superiority and racism (Princeton, NJ: Princeton University Press).

8. Source: Population Division, DESA, United Nations. *World Population Ageing 1950–2050.* (Geneva: United Nations, 2002). http://www.un.org/esa/population/publications/worldageing19502050/pdf/90chapteriv.pdf.

9. *War and Public Health,* Barry S. Levy and Victor W. Sidel (eds.) (New York: Oxford University Press, 1977).

10. Organisation for Economic Cooperation and Development (OECD). April 11, 2005. Table: Net Official Development Assistance in 2004 (Paris: Organisation for Economic Cooperation and Development).

11. It is noteworthy that forty nations committed military forces voluntarily to provide food, water, and shelter to the 2004 tsunami survivors.

12. The annual regular UN budget (2005) is $1.3 billion, with sixty-one thousand employees. http://www.un.org/geninfo/ir/ch5/ch5.htm, http://www.un.org/geninfo/ir/ch6/ch6_txt.htm.

13. The European Union as a whole is outdistancing the United States in support of the United Nations.

14. Jean-Marie Guehenno, *The End of the Nation-State* V. Elliott (trans.) (Minneapolis: University of Minnesota Press, 1996).

15. The original fifteen member nations of the European Union gave up monetary sovereignty, agreeing upon the euro as a common currency.

16. Eric Dash, "World Bank's Former Chief Will Be Hired by Citigroup," *New York Times,* Nov. 4, 2005.

17. See *The Gerontologist* (2002) *42*: 1–2. Also see the Declaration of Human Rights of Older Persons at the end of this chapter.
18. The United Nations mandates retirement at age 60!
19. When a country's percentage of older people, sixty and above, reaches 7 percent the UN calls it an "aging society" and when that reaches 14 percent it is considered an "aged society." Europe is the oldest region of the world, projected to have 25 percent of its population over sixty-five in 2020.
20. U.N. Population Fund. *State of the World Population 2007: Unleashing the Potential of Urban Growth,* 2007.
21. Agence-France Press, "Worries Raised about Exodus of Doctors and Nurses," *New York Times,* Nov. 27, 2005.
22. Willy Brandt, *North-South: A Program for Survival* (Cambridge, MA: MIT Press, 1980).
23. Robert William Fogel, *The Escape from Hunger and Premature Death, 1700–2100: Europe, America, and the Third World* (Cambridge, U.K.: Cambridge University Press, 2004).
24. Hanover, NH: University Press of New England, 1996.

Chapter 19: Threats to Longevity

1. TB, diphtheria, polio, AIDS, cholera, and typhoid were some of the diseases.
2. Carbon dioxide is not covered in the U.S. Clean Air Act. Also see Philip Abelson, "Limiting Atmosphere CO_2," *Science* (2000) *289*: 1293.
3. The United States emits the most carbon dioxide, and the industrialized nations account for 48 percent of the world's CO_2 emissions. But as China, Indonesia, and India become economic giants burning coal, this percentage will change. Already, China is the third largest contributor to global climate changes after the United States and Russia. By 2050, it could be the leader.
4. Established by the World Meteorological Organization and the UN Environmental Program. The panel is made up of over 2,000 scientists.
5. William D. Nordhaus, "Global Warming Economics. Policy Forum: Climate Change," *Science* (2001) *294*: 1283–1284.
6. Biofuels include cellulose, sugarcane, and corn ethanols.
7. Kenneth Bradshaw, "The Price of Keeping Cool in Asia," *New York Times,* Feb. 23, 2007.
8. The Japanese are among the world's longest-living populations. Japanese sixty-five years or older, the survivors of the postwar days when tuberculo-

sis was the country's leading cause of death, account for about 17 percent of the population. Many survivors carry the tubercle bacillus. While the disease remains dormant in some 90 percent of its hosts, it can activate and become contagious as the immune system of the carrier declines with age.

9. Marcos A. Espinal and others, "Global Trends in Resistance to Antituberculosis Drugs," *New England Journal of Medicine* (2001) *344*: 1294–1303.

10. It costs about $10 to cure standard TB, $20,000 to treat antibiotic-resistant TB, and not always successfully.

11. Joint United Nations Programme on HIV/AIDS (UNAIDS), *2004 Report on the Global AIDS Epidemic* (Geneva: Joint United Nations Programme on HIV/AIDS (UNAIDS), June 2004.

12. Tropical rain forests are probably not pristine, and they have been altered both by human agriculture and natural cycles of destruction and reconstruction over the centuries. Cleared forest shows up on satellite pictures, for example, in Amazonia.

13. In 2007 Indonesia planted twenty-nine million trees in one day.

14. Wilson states that the vast majority of life forms remain undiscovered. He supports the Encyclopedia of Life Project. See Edward O. Wilson, "That's Life," *New York Times,* Sept. 6, 2007.

15. The world's first seed bank was established by Nikolai I. Vavilov, a Russian biologist, botanist and geneticist who traveled five continents in the 1920s and 1930s for potato tubers, beans, grains, fodder, fruit and vegetable seeds, and wild and cultivated corn. The Vavilov collection in St. Petersburg, Russia, contains more than 10 percent of the earth's cultivated plant life and represents true wealth. There are now some 120 germ plasma banks in the world. U.S. collections are kept in Fort Collins, Colorado, and Beltsville, Maryland. Seed banks make possible development of blight- and drought-resistant potatoes and increased yields of grains and other crops. Historically, humans have saved seeds for up to ten thousand years, since the beginnings of agriculture. Preserved seeds must be regularly germinated.

16. Owen B. Toon, Alan Robock, Richard P. Turco, Charles Bardeen, Luke Oman, and Georgiy L. Stenchikov, "Consequences of Regional-Scale Nuclear Conflicts," *Science* (2007) *315*: 1224–1225.

17. In *The Antibiotic Paradox* (Cambridge, MA: Da Capo Press, 1992).

18. Gordon Ada, "HIV and Pandemic Influenza Virus: Two Great Infectious Disease Challenges," *Virology* (2000) *268*: 227–230.

19. Such as variants of the fungus that caused the Irish potato blight in the 1840s. Such variants again threatened potato crops around the world in 1993.

20. See Peter Kareiva, Sean Watts, Robert McDonald, and Tim Boucher, "Domesticated Nature: Shaping Landscapes and Ecosystems for Human Welfare," *Science* (2007) *316*: 1866–1869, which noted that by 1995 "only 17 percent of the world's land area had escaped direct influence by humans."

21. Institute of Medicine, "Emerging infections. Microbial Threats to Health in the United States." Joshua Lederberg, Robert E. Shope, and Stanley C. Oakes Jr. (eds.), 1992.

22. This also ties in with the asymmetry between humanity and resources discussed in Chapter 17.

23. J. Lederberg, "Crowded at the Summit: Emerging Infections and the Global Food Chain." Unpublished manuscript, January 1993.

24. Warfarin is an anticoagulant that thins the blood; it also kills rodents by causing hemorrhage.

25. In 2005, the World Bank estimated that a global human flu pandemic could cost $800 billion in economic losses.

26. Such as the feared A(H5NI) strain of Asian bird flu should it commingle with the human flu virus or mutate so human-to-human transmission would occur. H and N refer to genes that can combine to create an antigen shift. Since we do not know what genetic variant could cause and sustain human-to-human transmission, it is impossible to produce a vaccine in advance.

Chapter 20: The Good Life

1. The title of this chapter is ambiguous, for it refers both to quality of life and the moral life, which hopefully are indistinguishable. A moral life can be secular as well as religious. The chapter is written from the perspective of one who enjoys the advantages that the industrialized world can provide. For the millions of individuals who live in poverty in the developed and developing worlds, these thoughts will have little resonance. This is all the more reason the rich nations must assume more responsibility for overcoming the extreme economic, and longevity, divide that currently defines our world.

2. There is an International Society for the Quality of Life and, at least since the 1960s, efforts have been under way to define and measure quality of life. Of course, from the beginning of civilization, there have been concerns about what constitutes "the good life."

3. Joel Tsevat and others, "Health Values of Hospitalized Patients 80 Years or Older," *Journal of the American Medical Association* (1988) *279*: 371–375.

4. Konstantin Arbeev and others, "Disability Trends in Gender and Race Groups of Early Retirement Ages in the USA," *Sozial- und Präventivmedizin/Social and Preventive Medicine* (2004) *49*: 142–151.

5. Sidney Katz, Thomas D. Downs, Helen R. Cash, and Robert C. Grotz, "Progress in Development of the Index of ADL," *The Gerontologist* (1970) *10*: 20–30.

6. Bernice L. Neugarten, Robert J. Havighurst, and Sheldon S. Tobin, "The Measurements of Life Satisfaction," *Journal of Gerontology* (1961) *16*: 134–143. It seems to me that an unspecifiable number of people must claim satisfaction with their lives; to feel or believe otherwise is unbearable. This is another challenge to measurement.

7. Thomas H. Holmes and Richard H. Rahe, "The Social Adjustment Scale," *Journal of Psychosomatic Research* (1969) *11*: 213–218.

8. Bruce S. McEwen, "Allostasis and Allostatic Load: Implications for Neuropsycho-pharmacology," *Neuropsychopharmacology* (2000) *22*: 108–124.

9. National Center for Health Statistics, U.S. Department of Health and Human Services, Centers for Disease Control and Prevention. *National Health and Nutrition Examination Survey.* (Hyattsville, MD: 2005). www.cdc.gov/nchs/nhanes.htm.

10. Such as people who have a sunny disposition.

11. The neurobiological ingredients explaining pleasure, such as the roles of dopamine and endorphins, contribute to understanding the nature of addiction.

12. *New York Times*, July 16, 1996.

13. Daniel Kahneman, "Maps of Bounded Rationality. A Perspective on Intuitive Judgment and Choices," in T. Frangsmyr (ed.), *Les Prix Nobel* (Stockholm: Sweeden, Alm Quist & Wiksell International, 2002).

14. The French workers work less but have a higher rate of productivity than American workers!! See Paul Krugman, "French Family Values," *New York Times*, July 29, 2005.

15. Juliet B. Schor, *The Overworked American: The Unexpected Decline of Leisure* (New York: Basic Books, 1993).

16. Bernard Berenson, *Sunset and Twilight, From the Diaries of 1947–1958* (New York: Harcourt, Brace & World, 1963).

17. Waldamer Nielsen, *The Big Foundations* (New York: Columbia University Press, 1972).

18. In 2004, donations to the poor hit a new low! Yet concern for the poor dates to early Christianity in the West. What is the nation receiving in return for tax write-offs? Mostly to the well-to-do?

19. By 2002, the Russian Federation and Ukraine had both fallen twenty places since the collapse of Communism. The UN notes that only eighty-two of some 190 countries in the world can be described as full-fledged democracies that recognize basic human rights, such as freedom of expression.

20. James E. Birren and others, *Human Aging I: A Biological and Behavioral Study*. Public Health Service Publication No. 986. Washington, DC: U.S. Government Printing Office, 1963. (Reprinted 1971 and 1974.)

21. *The Myth of Sisyphus* (New York: Vantage Books, 1955) (originally 1942).

22. Robert William Fogel, *The Escape from Hunger and Premature Death, 1700–2100: Europe, America, and the Third World* (Cambridge, U.K.: Cambridge University Press, 2004).

Chapter 21: Imagining Longevity

1. Robert N. Butler, Huber Warner, T. Franklin Williams, et al., "The Aging Factor in Health and Disease," *Aging Clinical and Experimental Research* (2002) *16*: 104–112.

2. Gregory Stock, director of the program on medicine, technology, and science at the UCLA School of Medicine. Actively explores the possibilities for extended life.

3. Kira S. Birditt and Karen L. Fingerman, "Do We Get Better at Picking Our Battles? Age Group Differences in Descriptions of Behavioral Reactions to Interpersonal Tensions," *Journals of Gerontology: Psychological Sciences* (2006) *60B*: 121–P128.

4. Laura L. Carstensen, Helene Hoi Fung, and Susan T. Charles, "Socioemotional Selectivity Theory and the Regulation of Emotion in the Second Half of Life," *Motivation and Emotion* (2003) *27*: 103–123.

5. See Michael Ignatieff, "The Gods of War," a review he wrote in *The New York Review of Books*, Oct. 9, 1997, especially his discussion of Barbara Ehrenreich's *Blood Rites: Origins and History of the Passions of War* (New York: Metropolitan Books/Henry Holt, 1997) and Philippe Delmas, *The Rosy Future of War* (New York: Free Press, 1997). Also see Samuel Hynes, *Soldiers' Tale: Bearing Witness to Modern War* (New York: Allen Lane/Penguin Press, 1997).

6. Henry Sigerest, *Early Greek, Hindu, and Persian Medicine,* (Vol. II), Oxford University Press, 1961, p. 111, 182.

7. We have Aristotle in the *Rhetoric*, Cicero's majestic oration *De Senectute*, some of Montaigne's essays, notions concerning the stages of life in Shake-

speare and Rousseau, Francis Bacon's essay *Of Youth and Age*, Goethe's *Faust* (Part II), Schopenhauer's *Ages of Man*, the thinking of Tolstoy in *The Death of Ivan Illych* and his great novel *Resurrection*, the books of Elie Metchnikoff, Marcel Proust's *Remembrance of Things Past*, *Back to Methuselah* by George Bernard Shaw, some of Carl Gustave Jung's writings, the life force against old age of Nikos Kazantzakis' *Zorba the Greek* and the great twentieth-century novel of Garcia-Marquez, *Love in the Time of Cholera*, as well as Western religious and ethical thinking in such works as Ecclesiastes and Job.

8. James F. Fries, *Compression of Morbidity* (Alliance for Health & the Future at the International Longevity Center, 2005).

BIBLIOGRAPHY

Chapter 1

General References

Allard, Michel, Victor Libre, and Jean-Marie Robine. *Jeanne Calment from Van Gogh's Time to Ours.* New York: W.H. Freeman, 1994.

Austed, Steven N. *Why We Age.* New York: Wiley, 1997.

Bertillon, Jacques. *La dépopulation de la France et des remèdes à apporter.* Nancy: Imp. Berger-Levrault, 1896.

Birren James E., Robert N. Butler, Samuel W. Greenhouse, Louis Sokoloff, and Marvin Yarrow, (eds.). *Human Aging I: A Biological and Behavioral Study.* Public Health Service Publication No. 986. Washington, DC: GPO, 1963. Reprinted 1971 and 1974.

Bourdelais, Patrice. *L'âge de la vieillesse histoire du vieillissement de la population.* Paris: Odile Jacob, 1963.

Braudel, Fernand. *The Civilization and Capitalism, 15th–18th Century: The Structures of Everyday Life.* Vol. 1. New York: Harper & Row, 1981 (English translation).

Busse, Ewald and Eric Pfeiffer. *Behavior and Adaptation in Late Life,* Boston: Little, Brown, 1997. 2nd edition.

Butler, Robert N. "The Façade of Chronological Aging." *American Journal of Psychiatry* (1963) *119*: 721–728.

——. *The Longevity Revolution, Policy Forum on Aging Report: Your Aging Future, 1985–2030.* Racine, Wis.: Wingspread, 1985.

——. "Medicine's Responsibility to the Longevity Revolution." *Medical Times* (1986) *114*: 25–26.

Butler, Robert N., Richard Sprott, Huber Warner, et al. "Biomarkers of Aging: From Primitive Organisms to Man." *Journal Gerontological Biological Science* (2004) *59*: B560–B567.

Cantor, Norman F. *In the Wake of the Plague: The Black Death and the World It Made.* New York: Free Press, 2001.

Carey, James R. *Longevity: The Biology and Demography of the Lifespan.* Princeton, NJ: Princeton University Press, 2003.

Cook, Michael. *A Brief History of the Human Race.* Boston: W.W. Norton, 2003.

Fogel, Robert William. *The Escape From Hunger and Premature Death, 1700–2100. Europe, America and the Third World,* Cambridge, U.K.: Cambridge University Press, 2004.

Fogel, Robert W. and Donna L. Costa. "A Theory of Technophysio Evolution, with Some Implications for Forecasting Population, Health Care Costs, and Pension Costs." *Demography* (1997) *34*: 49–66.

Fries, James F. "Aging, Natural Death, and the Compression of Morbidity." *New England Journal of Medicine* (1980) *303*: 130–135.

Gompertz, Benjamin. "On the Nature of the Function Expressive of the Law of Human Mortality and on a New Mode of Determining Life Contingencies." *Philos Trans R. So. Land* (1825) *115*: 513–585.

Gruman, Gerald J. "A History of Ideas about the Prolongation of Life. The Evolution of Prolongevity Hypotheses to 1800." *Transactions of the American Philosophical Society* (2003) *56*: 1–102. Reprint. New York: Springer Publishing and the International Longevity Center–USA.

Institute of Medicine. *Extending Life, Enhancing Life: A National Research Agenda on Aging.* Washington, DC: National Academies Press, 1991.

Jones, Landon Y. *Great Expectations: America and the Baby Boom Generation.* New York: Ballantine Books, 1980.

Kent, Mary M. and Carl Haub. "Global Demographic Divide." *Population Bulletin* (2005) *60*: 4.

Kinsella, Kevin and Cynthia M. Taeuber. *An Aging World II.* Center for International Research, U.S. Bureau of the Census, 1993.

Kinsella, Kevin and David Phillips. *Global Aging: The Challenge of Success.* Population Reference Bureau, 2005.

Kirkwood, Thomas B.L. "Evolution of Ageing." *Nature* (1997) *270*: 301–304.

Kirkwood, Thomas B.L. and Michael R. Rose. "Evolution of Senescence: Late Survival Sacrificed for Reproduction." *Philosophical Transactions of the Royal Society of London,* Series B (1991) *332*: 15–24.

Kitagawa, Evelyn M. and Philip Hauser. *Differential Mortality in the United States: A Study in Socio-Economic Epidemiology.* Cambridge, MA: Harvard University Press, 1973.

"Living Longer and Doing Worse? Present and Future Trends in the Health of the Elderly." Special issue of *Journal of Aging and Health,* Vol. 3, no. 2, 1991.

Makeham, William M. "On the Law of Mortality." *Journal Inst. Actuaries* (1867) *13*: 325–358.

Marmot, Michael. *The Status Syndrome: How Social Standing Affects Our Health and Longevity.* New York: Times Books, 2004.

McKeown, Thomas. *The Modern Rise of Population.* New York: Academic Press, 1976.

McNeill, William H. *Plagues and Peoples.* Garden City, NY: Anchor Press/Doubleday, 1976.

Medawar, Peter B. *An Unsolved Problem of Biology.* London: Lewis, 1952.

Metchnikoff, Elie. *The Nature of Man. Studies in Optimistic Philosophy.* Translation by P. Chalmers Mitchell. London: Heinemann, 1903.

——. *The Prolongation of Life: Optimistic Studies.* Translated by P. Chalmers Mitchell. New York: Springer Publishing and the International Longevity Center–USA, 2003 (1907).

Miller, Lucasta. *The Brontë Myth.* New York: Alfred A. Knopf, 2004.

Mithen, Steven. *After the Ice: A Global Human History, 20000–5000 BC.* London: Weidenfeld & Nicolson, 2003.

Neugarten, Bernice L. *The Meanings of Age.* Selected Papers of Bernice L. Neugarten. Ed., Dale A. Neugarten. Chicago: University of Chicago Press, 1996.

Oeppen, Jim and James W. Vaupel. "Broken Limits to Life Expectancy." *Science* (2002) *296*: 1029–1031.

Olshansky, S. Jay, Bruce A. Carnes, and Aline Désesquelles. "Prospects for Human Longevity." *Science* (2001) *291*: 1491–1492.

Olshansky, S. Jay, Bruce A. Carnes, and Christine Cassell. "In Search of Methuselah: Estimating the Upper Limits to Human Longevity." *Science* (1990) *250*: 634–640.

Olshansky, S. Jay, Dan Perry, Richard A. Miller, and Robert N. Butler. "In Pursuit of the Longevity Dividend." *The Scientist* (2006) *20*: 28–36.

Rowe, John W. and Robert L. Kahn. *Successful Aging.* New York: Pantheon Books, 1998.

Sacher, George A. "Longevity, Aging and Death: An Evolutionary Perspective." *The Gerontologist* (1977) *18*: 112–122.

Sauvy, Alfred. *General Theory of Population.* New York: Basic Books, 1969.

Shryock, Henry S. and Jacob S. Siegel. *The Methods and Materials of Demography*, Vol. 2. Washington, DC: U.S. Department of Commerce, Bureau of the Census, 1975.

Tucker, Jonathan B. *Scourge: The Once and Future Threat of Smallpox.* New York: Atlantic Monthly Press, 2001.

University of California, Berkeley and Max Planck Institute for Demographic

Research (Germany). *Human Mortality Database* available at www
.mortality.org or www.humanmortality.de. (Accessed July 2004.)

U.S. Census Bureau. *65+ in the United States 2005.* Washington, DC: GPO, 2006.

Zinzer, Hans. *Rats, Lice and History.* Boston: Little Brown, 1984.

References: Evolution of Social Protections

Abel, Emil K. *Hearts of Wisdom: American Women Caring for Kin, 1850–1940.*
Cambridge, MA: Harvard University Press, 2000.

Beer, Samuel. "The Roots of New Labour, Liberalism Rediscovered." *The Economist,* 7 Feb. 1998, 23–25.

Berkman, Lisa R. and Ichiro Kawachi. *Social Epidemiology.* New York: Oxford
University Press, 2000.

Beveridge, William H. *Social Insurance and Allied Services.* London, HMSO: 1942.

Bourdieu, Pierre. *Distinction: A Social Critique of the Judgment of Taste,* 1979.

Brinkley, Alan. *The End of Reform, New Deal Liberalism in Recession and War.*
New York: Alfred A. Knopf, 1995.

Butler, Robert N. and Kenzo Kiikuni. *Who Is Responsible for My Old Age?* New
York: Springer, 1993.

Coyle, Diane. *Paradoxes of Prosperity: Why the New Capitalism Benefits Us All.*
Texere, 2001.

Decker, Sandra L. and Carol Rappaport. "Medicare and Inequalities in Health
Outcomes: The Case of Breast Cancer." *Contemporary Economic Policy*
(2002) *20*: 1–11.

Dworkin, R. *Sovereign Virtue, The Theory and Practice of Equality.* Cambridge,
MA: Harvard University Press, 2000.

Ehrenreich, Barbara. *Nickel and Dimed: On (Not) Getting By in America.* New
York: Henry Holt, 2001.

Fogel, Robert W. *The Fourth Great Awakening and the Future of Equalitarianism.*
Chicago: University of Chicago Press, 2000.

Folbre, Nancy. *The Invisible Heart, Economics and Family Values.* New York: The
New Press, 2001.

Friedman, Milton. *Capitalism and Freedom.* Chicago: University of Chicago Press,
1962.

Giddens, Anthony. *The Third Way: The Renewal of Social Democracy.* Cambridge, U.K.: Polity Press, 1995.

Gilleard, Chris and Paul Higgs. *Cultures of Ageing: Self, Citizen and the Body.*
Harlow, U.K.: Pearson, 2000.

Guillemard, Anne-Marie. *Aging and the Welfare-State Crisis.* Newark, DE: University of Delaware Press, 2000.

Kawachi, Ichiro, Bruce P. Kennedy, and Richard G. Wilkinson. *The Society and Population Health Reader: Income Inequality and Health.* New York: New Press, 2000.

Kennedy, David M. *Freedom from Fear: The American People in Depression and War, 1929–1945.* New York: Oxford University Press, 1999.

Keyes, Maynard. *The General Theory of Employment, Interest and Money.* London: MacMillan, 1973.

Kitagawa, Evelyn M. and Philip M. Hauser. *Differential Mortality in the United States: A Study in Socio-Economic Epidemiology.* Cambridge, MA: Harvard University Press, 1973.

Lindert, Peter H. *Growing Public, Social Spending and Economic Growth Since the Eighteenth Century.* Cambridge, U.K.: Cambridge University Press, 2004. Vols. 1 and 2.

Marin, Linda G. and Beth J. Soldo, eds. *Racial and Ethnic Differences in the Health of Older Americans.* Washington, DC: National Academy Press, 1997.

Marmot, Michael. *The Status Syndrome: How Social Standing Affects Our Health and Longevity.* New York: Times Books, 2004.

National Commission on Excellence in Education. *A Nation at Risk: The Imperative for Education Reform: A Report to the Nation and the Secretary of Education.* U.S. Department of Education, Washington, DC: GPO, 1983.

Phillips, Kevin. *Wealth and Democracy: A Political History of the American Rich.* New York: Broadway Books, 2002.

Pierson, Christopher. *Beyond the Welfare State.* Cambridge, U.K.: Polity Press, 1991.

Rawls, John. *A Theory of Justice.* Oxford, U.K.: Oxford University Press, 1973.

Rothschild, Emma. *Economic Sentiments, Adam Smith, Condorcet, and the Enlightenment.* Cambridge, MA: Harvard University Press, 2001.

Smith, Adam. *An Inquiry into the Nature and Causes of the Wealth of Nations.* Ed., Edwin Canaan. New York: Random House, 1977.

Soros, George. *The Crisis of Global Capitalism.* Open Society Institute. New York: Public Affairs, 1998.

Titmuss, Richard M. *The Gift Relationship.* London: Allen & Unwin, 1970.

Tocqueville, Alexis de. *Democracy in America.* Trans., Harvey C. Mansfield and Delba Winthrop. Chicago: University of Chicago Press, 2001.

Young, Michael and Peter Willmott. *Family and Kinship in East London.* Labor Party Platform, 1945.

Chapter 2

Bryson, Kenneth R. and Lynne M. Casper. "Coresident Grandparents and Grandchildren." *U.S. Census Bureau, Current Population Reports, Special Studies* P23–P198. Washington, DC: GPO, May 1999.

Callahan, Daniel. *Setting Limits: Medical Goals in an Aging Society.* Washington, DC: Georgetown University Press, 1995.

Civic Ventures. *The New Face of Retirement: Older Americans, Civic Engagement, and the Longevity Revolution.* San Francisco: Civic Ventures, 1999.

Commonwealth Fund. *The Nation's Great Overlooked Resource: The Contributions of Americans 55+.* New York: The Commonwealth Fund, 1992.

Cutler, David and Mark McClellan. "Is Technological Change in Medicine Worth It?" *Health Affairs* (2001) *20*: 11–29.

Cutler, David, James Poterba, Louise Sheiner, and Lawrence Summers. "An Aging Society: Opportunity or Challenge?" *Brookings Papers on Economic Activity* (1990) *1*: 1–56.

Jacobzone, Stéphane. "Ageing and Care for Frail Elderly Persons: An Overview of International Perspectives," *Labour Market and Social Policy Occasional Paper No. 38.* Paris: Organization for Economic Cooperation and Development, 1999.

Jacobzone, Stéphane, Emmanuelle Cambois, Emmanuel Chaplain, and Jean-Marie Robine. "The Health of Older Persons in OECD Countries: Is It Improving Fast Enough to Compensate for Population Ageing?" *Labour Market and Social Policy Occasional Paper No. 37.* Paris: Organization for Economic Cooperation and Development, 1998. Also see: http://www.oecd.org/LongAbstract/0,2546,en_2649_34587_2732535_119666_1_1_1,000.html.

Nordhaus, William. *The Health of Nations: The Contribution of Improved Health to Living Standards.* National Bureau of Economic Research Working Paper No. 8818, 2002.

Peterson, Peter G. *Gray Dawn: How the Coming Age Wave Will Transform America—and the World.* New York: Times Books, 1999.

Smith, James P. "Healthy Bodies and Thick Wallets: The Dual Relation between Health and Economic Status." *Journal of Economic Perspectives* (1999) *13*: 145–166.

World Bank (Estelle James). *Averting the Old Age Crisis.* Oxford, U.K.: Oxford University Press, 1994.

Chapter 3

Arber, Sara and Claudine Attias-Donfut (eds.). *The Myth of Generational Conflict: The Family and State in Ageing Societies.* London: Routledge, 2000.

Auerbach, Alan J., Jagadeesh Gokhale, and Laurence J. Kotlikoff. "Generational Accounting: A Meaningful Way to Evaluate Fiscal Policy." *Journal of Economic Perspectives* (1994) *8, 1*:73–94.

Auerbach, Alan, Laurence Kotlikoff, and Willi Leibfritz (eds.). *Generational Accounting around the World.* Chicago: University of Chicago Press, 1999.

Baier, Annette. "The Rights of Past and Future Persons," in E. Partridge (ed.), *Responsibilities to Future Generations.* Prometheus Books, 1980, 171–186.

Beckerman, Wilfred and Joanna Pasek. *Justice, Posterity and the Environment.* Oxford: Oxford University Press, 2001.

Bengtson, Vern L. and W. Andrew Achenbaum (eds.). *The Changing Contract across Generations.* New York: Aldine de Gruyter, 1993.

Harden, Garrett. "The Tragedy of the Commons," *Science* (1968) *162*:1243–1248.

Kotlikoff, Laurence J. *Generational Accounting—Knowing Who Pays and When, for What We Spend.* New York: Free Press, 1992.

Laslett, Peter and James S. Fishkin (eds.). *Justice Between Age Groups and Generations.* New Haven: Yale University Press, 1992.

U.S. Census Bureau. *65+ in the United States,* 2005.

Walker, Alan (ed.). *The New Generational Contract, Intergenerational Relations, Old Age and Welfare.* London: University College, London Press, 1996.

Wessel, David. "Moving up: Challenges to the American Dream," *The Wall Street Journal,* May 13, 2005, A1.

Zinn, Howard. *A People's History of the United States 1492–Present.* New York: Harper Perennial, 2003 (1980).

Chapter 4

Achenbaum, W. Andrew. *Old Age in the New Land: The American Experience Since 1790.* Baltimore: Johns Hopkins Press, 1978.

Bonadonna, Giovanni, et al. "Adjuvant Cyclophosphamide-methotrexate and Fluorouracil in Node-positive Breast Cancer." *New England Journal of Medicine* (1995) *332*: 901–906.

Butler, Robert N. "Ageism: Another Form of Bigotry." *The Gerontologist* (1969) *4*: 243–246.

——. *Why Survive? Being Old in America.* New York: Harper & Row, 1975.

Bytheway, Bill. *Ageism.* Buckingham, U.K.: Open University Press, 1995.

Cohen, Eli S. "The Complex Nature of Ageism: What Is It? Who Does It? Who Perceives It?" *The Gerontologist* (2001) *41*: 576–577.

Cole, Thomas R. *The Journey of Life: A Cultural History of Aging in America.* New York: Cambridge University Press, 1992.

Daniels, Norman. *Am I My Parents Keeper?* New York: Oxford University Press, 1987.

Fischer, David Hackett. *Growing Old in America.* The Bland-Lee lectures delivered at Clark University. New York: Oxford University Press, 1978.

Goffman, Erving. *The Asylums.* Essays on the social situation of mental patients and other inmates. Garden City, NY: Doubleday, 1961.

Gruman, Gerald J. "Cultural Origins of Present-Day "Age-ism": The Modernization of the Life Cycle." In *Aging and the Elderly: Humanisitc Perspectives in Gerontology,* Stuart Spicker, Kathleen Woodward, and David Van Tassel (eds.). Atlanta Highlands, NJ: Humanities Press, 1978.

International Longevity Center. *Ageism in America,* 2006.

Lindau, Stacy Tessler, et al. "A Study of Sexuality and Health among Older Adults in the United States," *New England Journal of Medicine* (2007) *357*: 762–764.

McMullen, Julie A. and Victor W. Marshall. "Ageism, Age Relations, and Garment Industry Work in Montreal." *The Gerontologist* (2001) *41*: 111–122.

Metchnikoff, Elie. *The Nature of Man, Studies in Optimistic Philosophy.* New York: Putnam, 1903.

——. *The Prolongation of Life, Optimistic Studies.* Chalmers Mitchell (tr.). New York: Springer with the International Longevity Center, 2003 (London: William Heinemann, 1907).

Miller, Arthur. *The Death of a Salesman.* New York: Viking, 1949.

Moss, Howard B. "Old Age: Not a Barrier to Treatment." *The New England Journal of Medicine* (2001) *345*: 1128–1129.

Palmore, Erdman B. *Ageism: Negative and Positive,* 2nd ed. New York: Springer, 1999.

——. *Encyclopedia of Ageism.* New York: Haworth Press, 2005.

Rosenthal, Evelyn (ed.). "Women and Varieties of Ageism," *Journal of Women and Aging* (1990) *2*:1–164.

Simmons, Leo. *The Role of the Aged in Primitive Society.* North Haven, CT: Shoe String Press, 1970.

Spicker, Stuart, Kathleen Woodward, and David Van Tassel (eds.). *Aging and the Elderly: Humanistic Perspectives in Gerontology.* Atlantic Highlands, NJ: Humanities Press, 1978.

The New York Times Magazine. "The Age Boom." March 8, 1997.

United Nations. *Human Rights and Older Persons*. Geneva, Switzerland: UN Information Service.

Woodward, Kathleen. *Figuring Age, Women, Bodies, Generations*. Bloomington: Indiana University Press, 1999.

Chapter 5

Aaron, Henry J., Timothy Taylor, and Thomas E. Mann (eds.). *Values and Public Policy*. Washington, DC: Brookings Institution, 1993.

Aries, Philippe. *Centuries of Childhood: A Social History of Family Life*. New York: Alfred A. Knopf, 1962.

Bengtson, Vern L. and William Achenbaum (eds.). *The Changing Contract Across Generations*. Hawthorne, NY: Aldine de Gruyter, 1993.

Brody, Elaine. *Women in the Middle: Their Parent-Care Years*. New York: Springer, 1990.

Coontz, Stephanie. *The Way We Never Were: American Families and the Nostalgia Trap*. New York: Basic Books, 1992.

———. *Marriage, a History: From Obedience to Intimacy or How to Have Conquered Marriage*. New York: Viking, 2005.

Elder, Glen H. Jr. *Children of the Great Depression: Social Change in Life Experience* (25th anniversary edition). Chicago: University of Chicago Press, 1999.

Hareven, Tamara. "The History of the Family and the Complexity of Social Change." *American Historical Review* (1991) 96:95–124.

Shanas, Ethel and Gordon F. Streib. *Social Structure and the Family: Generational Relations*. Englewood Cliffs, NJ: Prentice-Hall, 1965.

Uhlenberg, Peter. *Implications of Increasing Divorce for the Elderly*. Paper presented at the United Nations International Conference on Aging Populations in the Context of the Family, Ketakyushu, Japan.

Chapter 6

Barker, David J.P. *Fetal and Infant Origins of Adult Disease*. London: British Medical Journal Publishing Group, 1992.

———, ed. *Mothers, Babies and Diseases in Later Life*. London: British Medical Journal Publishing Group, 1994.

Barlow, Brian K., Rebecca C. Brown, Robert N. Butler, Philip Landrigan, et al.

"Early Environmental Origins of Neurodegenerative Disease in Late Life: Research and Risk Assessment." *Environmental Health Perspectives* (2005) *113*: 1230–1270.

Baue, Arthur E., Eugen Faist, and Donald E. Fry (eds.). *Multiple Organ Failure: Pathophysiology, Prevention, and Therapy.* New York: Springer-Verlag, 2000.

Birren, James E., Robert N. Butler, Samuel W. Greenhouse, Louis Sokoloff, and Marian R. Yarrow (eds.). *Human Aging I: Biological and Behavioral Study.* Public Health Service Publication No. 986. Washington, DC: GPO, 1963. (Reprinted 1971 and 1974.)

Brody, Jacob A. "Postponement or Prevention in Aging," in *Delaying the Onset of Late-Life Dysfunction,* Robert N. Butler and Jacob A. Brody (eds.). New York: Springer, 1995.

Brody, Jacob A. and Edward L. Schneider. "Diseases and Disorders of Aging: An Hypothesis," *Journal of Chronic Diseases* (1986) *39*: 871–876.

Butler, Robert N., Huber Warner, T. Franklin Williams, et al. "The Aging Factor in Health and Disease," *Aging Clinical and Experimental Research* (2004) *16*:104–112.

Dubos, Rene. *Mirage of Health: Utopias, Progress and Biological Change.* New York: Harper, 1959.

Felitti, Vincent J., Robert F. Anda, Dale Nordenberg, David F. Williamson, et al. "Relationship of Childhood Abuse and Household Dysfunction to Many of the Leading Causes of Death in Adults." The Adverse Childhood Experiences (ACE) Study. *American Journal of Preventive Medicine* (1998) *14*: 245–258.

Fogel, Robert W. Can We Afford Longevity? In *Longevity and Quality of Life,* Robert N. Butler and Claude Jasmin (eds.). New York: Kluwer Academic/Plenum Publishers, 2000.

——. "Economic Growth, Population Theory, and Physiology: The Bearing of Long-term Processes on the Making of Economic Policy." *American Economic Review* (1994) *84*: 369–395.

Fried, Linda P., Catherine M. Tangen, Jeremy Walston, et al. "Frailty in Older Adults: Evidence for a Phenotype." *Journals of Gerontology* (2001) *56*: M146–M156.

Fries, James F. "Aging, Natural Death, and the Compression of Morbidity." *New England Journal of Medicine* (1980) *303*: 130–135.

Gluckman, Peter D. and Marka Hanson. "Living with the Past: Evolutionary Development and Patterns of Disease." *Science* (2004) *305*:1733–1736.

Gompertz, Benjamin. "On the Nature of the Function Expressive of the Law of Human Mortality and on a New Mode of Determining Life Contingencies." *Philosophical Transactions of the Royal Society of London* (1825) *115*: 513–585.

Gruenberg, Ernest M. "The Failures of Success." *Milbank Memorial Fund Quarterly/Health and Society* (1977) *55*:3–24.

Guralnik, Jack M., Luigi Ferrucci, Eleanor M. Simonsick, et al. "Lower-extremity Function in Persons over the Age of 70 Years as a Predictor of Subsequent Disability." *New England Journal of Medicine* (1995) *332*: 556–561.

Hayflick, Leonard. *Has Anyone Ever Died of Old Age?* New York: International Longevity Center–USA, 2003.

———. *How and Why We Age.* New York: Ballantine Books, 1994.

Holtzman, Neil A. and Theresa M. Marteau. "Will Genetics Revolutionize Medicine?" *New England Journal of Medicine* (2002) *343*:141–144.

Hoover, Robert N. "Cancer Nature, Nurture or Both." *New England Journal of Medicine* (2000) *343*: 135–136.

Inouye, Sharon K., Stephanie Studenski, Mary E. Tinetti, and George A. Kucher. "Geriatric Syndromes: Clinical, Research and Policy Implications of a Core Geriatric Concept." *Journal of the American Geriatrics Society* (2007) *55*: 780–791.

Katz, Sidney, Amasa B. Ford, Roland W. Moskowitz, et al. "Studies of Illness in the Aged. The Index of ADL: A Standardized Measure of Biological and Psychosocial Function." *Journal of the American Medical Association* (1963) *185*: 914–919.

Kiberstis, Paula and Leslie Roberts. "It's Not Just the Genes." *Science* (2002) *296*: 685.

Kitagawa, Evelyn M. and Philip M. Hauser. *Differential Mortality in the United States: A Study in Socioeconomic Epidemiology.* Cambridge, MA: Harvard University Press, 1973.

Kohn, Robert R. "Cause of Death in Very Old People." *Journal of the American Medical Association* (1982) *247*: 2793–2797.

Landrigan, Philip J., Babasaheb Sonawane, Robert N. Butler, Leonardo Trasande, Richard Callan, and Michael J. Droller. "Early Environmental Origins of Neurodegenerative Disease in Late Life: Research and Risk Assessment," *Environmental Health Perspectives* (2005) *113*: 1230–1233.

Lawton, Powell M. and Elaine Brody. "Assessment of Older People: Self-maintaining and Instrumental Activities of Daily Living." *The Gerontologist* (1969) *9*: 179–186.

Lichtenstein, Paul, et al. "Environmental and Heritable Factors in the Causation of Cancer." *New England Journal of Medicine* (2000) *343*: 78–85.

Manton, Kenneth G., Larry S. Corder, and Eric Stallard. "Estimates of Change in Chronic Disability and Institutional Incidence and Prevalence Rates in the

U.S. Older Population from 1982, 1984, and 1989." National Long-term Case Survey. *Journal of Gerontology: Social Sciences* (1993) *48*: S153–S166.

Manton, Kenneth G., Eric Stallard, and Larry Corder. "Education-specific Estimates of Life Expectancy and Age-specific Disability in the U.S. Elderly Population: 1982 to 1991." *Journal of Aging and Health* (1997) *9*: 419–450.

McEwen, Bruce S. "Allostasis and Allostatic Load: Implications for Neuropsychopharmacology." *Neuropsychopharmacology* (2000) *22*: 108–124.

Medawar, Peter Brian. *An Unsolved Problem of Biology.* London: Lewis, 1952.

Murray, Christopher J.L. and Alan D. Lopez (eds.). *The Global Burden of Disease.* Boston: Harvard University Press, 1996.

Oeppen, James and James Vaupel. "Enhanced: Broken Limits to Life Expectancy." *Science* (2002) *296*:1029–1031.

Olshansky, S. Jay and Bruce A. Carnes. "Prospect for Extended Survival: A Critical Review of the Biological Evidence." In *Health and Mortality Among Elderly Populations,* Graziella Caselli and Alan D. Lopez (eds.). Oxford: Clarendon Press, 1996.

Olshansky, S. Jay, Bruce Carnes, and Robert N. Butler. "If Humans Were Built to Last." *Scientific American* (2002) *284*: 50–55.

Olshansky, S. Jay, Bruce Carnes, and Catherine Cassel. "In Search of Methuselah: Estimating the Upper Limits to Human Longevity." *Science* (1990) *250*: 634–640.

Olshansky, S. Jay, Bruce A. Carnes, and Aline Desesquelles. "Prospects for Human Longevity." *Science* (2001) *291*: 1491–1492.

Omran, Abdel R. "The Epidemiologic Transition: A Theory of the Epidemiology of Population Change." *Milbank Memorial Fund Quarterly* (1971) *49*: 509–538.

——. "The Epidemiologic Transition Theory Revisited Thirty Years Later." *World Health Statistics Quarterly/Rapport trimestriel de statisiques sanitaires mondiales* (1998) *51*(nos. 2–4): 99–119.

Palmore, Erdman. 1972. "Gerontophobia versus Ageism." *The Gerontologist* (1972) *12*: 213.

Payer, Lynn. *Medicine and Culture: Varieties of Treatment in the United States, England, West Germany and France.* New York: Holt, 1988.

Pope, Harrison G. Jr., Katharine A. Phillips, and Roberto Olivardia. *Adonis Complex: The Secret Crisis of Male Body Obsession.* New York: Free Press, 2001.

Reaven, Gerald. *Syndrome X, the Silent Killer: The New Heart Disease Risk.* New York: Simon & Schuster, 2001.

Riley, James C. *Rising Life Expectancy: A Global History.* Cambridge, U.K.: Cambridge University Press, 2001.

Schwartz, William B. *Life Without Disease: The Pursuit of Medical Utopia.* Berkeley: University of California Press, 1998.

Scitovsky, Anne A. "The High Cost of Dying: What Do the Data Show?" *Milbank Memorial Fund Quarterly* (1984) *62*: 591–608.

Svanborg, Alvar, Gunilla Bergstrom, and Dan Mellstrom. "Epidemiological Studies on Social and Medical Conditions of the Elderly. Report on a Survey." (Euro Reports and Studies 12.) Copenhagen, Denmark: World Health Organization, 1982.

Temple, Larissa K.F., Robin S. McLeod, Steven Gallinger, and James G. Wright. "Defining Disease in the Genomics Era." *Science* (2001) *293*: 807–808.

Thomas, Lewis. "Medical Lessons from History." In *The Medusa and the Snail, More Notes of a Biology Watcher.* New York: Viking Press, 1979.

Tinetti, Mary E. and Terri Fried. "The End of the Disease Era." *The American Journal of Medicine* (2004) *116*: 179–185.

Wallace, Douglas C. "Mitochondrial Genetics: A Paradigm for Aging and Degenerative Diseases?" *Science* (1992) *256*: 628–632.

Weinberg, Robert A. *One Renegade Cell: How Cancer Begins.* Science Masters Series. New York: Basic Books, 1998.

Whorton, James C. *Inner Hygiene: Constipation and the Pursuit of Health in Modern Society.* New York: Oxford University Press, 2000.

Williams, George C. and Randolph M. Nesse. "The Dawn of Darwinian Medicine." *The Quarterly Review of Biology* (1991) *66*: 1–22.

Wilmoth, John R. "The Future of Human Longevity: A Demographer's Perspective." *Science* (1998) *280*: 395–397.

Wilmoth, John R., Leo J. Deegan, Hans Lundström, and Shiro Horiuchi. "Increase of Maximum Lifespan in Sweden, 1861–1999." *Science* (2000) *289*: 2366–2368.

Zimmer, Carl. "Do Chronic Diseases Have an Infectious Root." *Science* (2001) *293*: 1974–1977.

Chapter 7

Association of American Medical Colleges (AAMC). *Protecting Subjects, Preserving Trust, Promoting Progress–Policy and Guidelines for the Oversight of Individual Interests in Human Subjects Research,* 2002.

Bloom, David E. and David Canning. "The Health and Wealth of Nations," *Science* (2000) *287*: 1207–1209.

Bloom, David E. and Pia Malaney. "Macroeconomic Consequences of the Russian Mortality Crisis." *World Development* (1998) *26*: 2073–2085.

Bloom, David E., David Canning, and Jaypee Sevilla. *Health, Human Capital and Economic Growth.* World Health Organization Commission on Macroeconomics and Health Working Paper No. WG1:8, 2001.

Bok, Derek. *Universities in the Marketplace: The Commercialization of Higher Education.* Princeton, NJ: Princeton University Press, 2003.

Brundtland, Gro Harlem. "Better Health Stokes Productivity." *International Herald-Tribune,* March 2, 1999.

Butler, Robert N. "Do Longevity and Health Generate Wealth?" *Cambridge Handbook of Aging.* Cambridge, U.K.: Cambridge University Press, 2005.

Callahan, Daniel. *What Price Better Health?* Berkeley: University of California Press, 2003.

Commission on Macroeconomics and Health. *Macroeconomics and Health: Investing in Health for Economic Development.* Geneva: World Health Organization, 2001.

Comroe, Julius H. Jr. "The Road from Research to New Diagnosis and Therapy." *Science* (1978) *200*: 931–937.

Comroe, Julius H. Jr. and Robert D. Dripps. "Scientific Basis for the Support of Biomedical Science." *Science* (1976) *192*: 105–111.

Cutler, David M. *Your Money or Your Life: Strong Medicine for America's Health Care System.* New York: Oxford University Press, 2004.

Cutler, David, Mark McClellan, and Joseph Newhouse. "The Costs and Benefits of Intensive Treatment for Cardiovascular Disease." In *Measuring the Prices of Medical Treatments,* Jack E. Triplett (ed.). Washington, DC: Brookings Institution Press, 1999.

Federman, Daniel D., Kathi E. Hanna, and Laura Lyman Rodriquez, eds. *Responsible Research: A Systems Approach to Protecting Research Subjects.* Washington, DC: National Academy Press, 2003.

Finch, Caleb E. and Thomas B.L. Kirkwood. *Chance, Development and Aging.* New York: Oxford University Press, 2000.

Fogel, Robert W. *The Fourth Great Awakening and the Future of Egalitarianism.* Chicago: Chicago University Press, 2000.

——. "New Findings on Secular Trends in Nutrition and Mortality: Some Implications for Population Theory." In *Handbook of Population and Family Economics,* Mark Rosenzweig and Oded Stark (eds.). Amsterdam: Elsevier, 1997.

Jones, Steve. *Genetics in Medicine: Real Promises, Unreal Expectations.* New York: Milbank Memorial Fund, 2000.

Kalemi-Ozcan, Sebnem. "Does the Mortality Decline Promote Economic Growth?" *Journal of Economic Growth* (2002) *7*: 411–439.

Kass, Leon. *Beyond Therapy, Biotechnology and the Pursuit of Happiness.* New York: Harper Collins, 2003.

Keller, Evelyn F. *The Century of the Gene.* Cambridge, MA: Harvard University Press, 2000.

Krimsky, Sheldon. *Science in the Private Interest: Has the Lure of Profits Corrupted/ Biomedical Research?* Latham, MD: Rowman & Little, 2003.

Lee, Chulhee. *Health and Wealth Accumulation: Evidence from Nineteenth-Century America.* National Bureau of Economic Research Working Paper No. 10035, 2003.

Lewontin, Richard C. *Triple Helix: Gene, Organism, and Environment.* Cambridge, MA: Harvard University Press, 2000.

Murphy, Kevin and Robert Topel (eds.). *Exceptional Returns: The Economic Value of America's Investment in Medical Research.* Funding First, 2000.

Olshansky, S. Jay and Bruce Carnes. *The Quest for Immortality Science and the Frontiers of Aging,* New York: W.W. Norton, 2001; see also http://www.medhist.ac.uk. (Wellcome Library for the History and Understanding of Medicine, United Kingdom.)

Webb, Anthony. "Do Health and Longevity Create Wealth?" International Longevity Center, 2006.

Chapter 8

Alzheimer, Alois. Ueber eine eigneartige Erkrankung der Hirnrinde. *Centralblatt fur Nervenheilkunde und Psychiatrie* (1907) *30*: 177–179.

Andreasen, Nancy C. *Brave New Brain, Conquering Mental Illness in the Era of the Genome.* New York: Oxford University Press, 2001.

Bachman, David L., P.A. Wolf, R.T. Linn, J.E. Knoefel, J.L. Cobb, A.J. Belanger, L.R. White, and R.B. D'Agostino. "Incidence of Dementia and Probable Alzheimer's Disease in a General Population: The Framingham Study." *Neurology* (1993) *43*: 515–519.

Butler, Robert N. How Alzheimer's became a public issue. *Generation Quarterly, Journal of the Western Gerontological Society* (1984) *9*: 33–35.

Butler, Robert N., Myrna I. Lewis, and Trey Sunderland. *Aging and Mental Health: Psychosocial and Biomedical Approaches.* Fifth edition. Boston: Allyn & Bacon, 1998.

Fillit, Howard M., Robert N. Butler, Alan W. O'Connell, et al. "Achieving and Maintaining Cognitive Vitality with Aging." *Mayo Clinic Proceedings* (2002) *77*: 681–696.

Fratiglioni L., M. Grut, Y. Forsell, M. Viitanen, M. Grafstrom, K. Holmen, K. Ericsson, L. Backman, A. Ahlbom, B. and Winblad. "Prevalence of Alzheimer's Disease and Other Dementias in an Elderly Urban Population: Relationship with Age, Sex and Education." *Neurology* (1991) *41*: 1886–1892.

Freedman, Vicki, K. Aykan, and Linda G. Martin. "Aggregate Changes in Severe Cognitive Impairment among Older Americans, 1993 and 1998." *Journal of Gerontology* (2001) *56B*: S100–S111.

Gazzaniga, Michael S. (ed.). *Cognitive Neuroscience.* Cambridge, MA: MIT Press, 1999.

Glenner, George G. and C.W. Wong. "Alzheimer's Disease and Down's Syndrome: Sharing of a Unique Cerebrovascular Amyloid Fibril Protein." *Biochemical and Biophysical Research Communications* (1984) *122*: 1131–1135.

Goedert, Michel and Maria Grazia Spillantini. "A Century of Alzheimer's Disease." *Science* (2006) *314*: 777–781.

Hsiao, Karen, et al. "Correlative Memory Deficits, AB elevation, and Amyloid Plaques in Transgenic Mice." *Science* (1996) *274*: 99–103.

International Longevity Center, *Achieving and Maintaining Cognitive Vitality with Aging.* New York: ILC-USA, 2000.

International Longevity Center. *Alzheimer's—The Disease of the Century,* 2006.

Irizarry, Michael C. and Bradley T. Hyman. Chapter 5 in *Principles of Neuroepidemiology,* T. Batchelor and M.E. Cudkowicz (eds.). Boston: Butterworth-Heineman, 2001.

Katzman, Robert and Kathleen Bick. *Alzheimer's Disease: The Changing View.* San Diego, CA: Academic Press, 2000.

Kempermann, G. and Fred H. Gage. "New Nerve Cells for the Adult Brain." *Scientific American,* May 1999.

Kandel, Erik, James H. Schwartz, and Thomas M. Jessel. *Principles of Neural Science.* New York: McGraw Hill, 1991.

Mace, Nancy L. and Peter V. Rabins. *The 36-Hour Day, Guide to Caring for Persons with Alzheimer's Disease, Related Dementing Illness, and Memory Loss in Later Life.* New York: Warner Books, 1994.

Matsuo, E.S., R.W. Shin, M.L. Billingsley, A.V. deVoorde, M. O'Connor, J.Q. Trojanowski, and V.M.Y. Lee. "Biopsy-derived Adult Human Brain tau Is Phosphorylated at Many of the Same Sites as Alzheimer's Disease Paired Helical Filament tau." *Neuron* (1994) *13*: 989–1002.

Roberson, Erik D. and Lennart Mucke. "100 Years and Counting: Prospects for Defeating Alzheimer's disease." *Science* (2006) *314*: 781–784.

Scheibel, Arnold B. "Microvascular Changes in Alzheimer's Disease." In *The Biological Substrates of Alzheimer's Disease*, Arnold B. Scheibel, et al. (eds.). New York: Academic Press, 1996, 177–192.

Selkoe, Dennis J. "Alzheimer's Disease: Genotypes, Phenotype, and Treatments." *Science* (1997) *275*: 630–631.

——. "Cell Biology of the Amyloid B-protein Precursor and the Mechanism of Alzheimer's disease." *Annual Review of Cell Biology* (1994) *10*: 373–403.

Seubert, P., C. Vigo-Pelfrey, F. Esch, M. Lee, J. Whaley, C. Seindlehurst, R. McCormack, R. Wolfert, D. Selkoe, L. Leiderburg, and D. Schenk. "Isolation and Quantification of Soluble Alzheimer's B-peptide from Biological Fluids." *Nature* (1992) *359*: 325–327.

Schenk, D. *The Forgetting. Alzheimer's: Portrait of an Epidemic*. New York: Doubleday, 2001.

Strittmatter, W.J., A.M. Saunders, D. Schechel, M. Pericak-Vance, J. Enghild, G.S. Salvesen, and A.D. Roses. "Apolipoprotein E: High-avidity Binding to 6-amyloid and Increased Frequency of Type 4 Allele in Late-onset Familial Alzheimer's Disease." *Proceedings of the National Academy of Sciences of the United States of America* (1993) *90*: 1977–1981.

Chapter 9

Andersen-Ranberg, Karen, Marianne Schroll, and Bernard Jeune. "Healthy Centenarians Do Not Exist, but Autonomous Centenarians Do: A Population-Based Study of Morbidity Among Danish Centenarians." *Journal of the American Gerontological Society* (2001) *49*: 900–908.

Butler, Robert N., Richard L. Sprott, Steven N. Austad, Nir Barzilai, A. Braun, and S. Helfand. *Longevity Genes: From Primitive Organisms to Humans*. International Longevity Center, New York, 2002.

Carnes, Bruce A. and S. Jay Olshansky. "Heterogeneity and its Biodemographic Implications for Longevity and Mortality." *Journal of Experimental Gerontology* (2001) *36*: 419–430.

Carnes, Bruce A., S. Jay Olshansky, and Douglas Grahn. "Biological Evidence for Limits to the Duration of Life." *Biogerontology* (2003) *4*: 31–45.

Fabrizo, Paola, Plether Fabiola Pozza, D. Scott, Christi M. Gendrom, D. Valter, and D. Longo. "Regulation of Longevity and Stress Resistance by Sch9 in Yeast." *Science* (2001) *292*: 288–290.

Finch, Caleb E. and Thomas B.L. Kirkwood. *Chance, Development, and Aging*. New York: Oxford University Press, 2000.

Frisoni, Giovanni B., Jukka Louhija, Cristina Geroldi, and Marco Trabucchi. "Longevity and the Allele of APOE: The Finnish Centenarians Study," *Journal of Gerontological Medicine Science, 56A*: M75–M78.

Gompertz, Benjamin. "On the Nature of the Function Expressive of the Law of Human Mortality and on a New Mode of Determining Life Contingencies." *Philosophical Transactions of the Royal Society of London* (1825) *115*: 513–585.

Grant, Peter R. and B. Rosemary Grant. "Unpredictable Evolution in a 30-year Study of Darwin's Finches." *Science* (2002) *296*: 707–711.

Hayflick, Leonard. *How and Why We Age.* New York: Ballantine, 1991.

Herskind, Anne Maria, Matthew McGue, Niels V. Holm, Thorkild I.A. Sorensen, Bent Harvald, and James W. Vaupel. "The Heritability of Human Longevity: A Population-based Study of 2872 Danish Twin Pairs Born 1870–1900," *Human Genetics* (1996) *97*: 319–323.

Institute of Medicine. *Exploring the Biological Contributions of Human Health, Does Sex Matter?* Teresa M. Wizemann and Mary-Lou Pardue. Washington, DC: National Academy Press, 2001.

International Longevity Center. *Longevity Genes, from Primitive Organisms to Humans*, 2002.

Johanson, Donald and Blake Edgar. *From Lucy to Language.* New York: Peter A. Neuraumont, 1996.

Johnson, Thomas E. "Increased Life-span of *age–1* Mutants in *Caenorhabditis elegans* and Lower Gompertz Rate of Aging." *Science* (1990) *249*: 908–912.

Kenyon, Cynthia, Jean Chang, Gensch Erin, Adam Rudner, and Ramon Tabtiang. "C. *Elegans* Mutant Lives Twice as Long as Wild Type." *Nature* (1993) *366*: 461–464.

Kerber, Richard A., Elizabeth O'Brien, Ken R. Smith, and Richard M. Cawthon. "Familial Excess Longevity in Utah Genealogies." *Journal of Gerontology: Biological Sciences* (2001) *56A*: B130–B139.

Kirkwood, Thomas B.L. and Robin Holliday. "The Evolution of Ageing and Longevity." *Proceedings of the Royal Society of London, Series B, Biological Sciences* (1979) *205*: 531–546.

Makeham, William M. "On the Law of Mortality." *Journal of the Institute of Actuaries* (1867) *13*: 325–358.

Martin, George M. "Genetics and the Pathobiology of Aging." *Philosophical Transactions Royal Report* (1997) *352*: 1773–1780.

Mayr, Ernst. *What Evolution Is.* New York: Basic Books, 2001.

Medawar, Peter B. *An Unsolved Problem of Biology.* London: Lewis, 1952.

Oppen, Jim and James W. Vaupel. "Broken Limits to Life Expectancy." *Science* (2002) *296*: 1029–1031.

Palumbi, Stephen R. "Humans as the World's Greatest Evolutionary Force." *Science* (2001) *293*: 1786–1790.

Pearl, Raymond. "Studies on Human Longevity IV The Inheritance of Longevity, Preliminary Report." *Human Biology* (1931) *3*: 245–269.

Perls, Thomas T. "The Oldest Old." *Scientific American* (1995) *272*: 70–75.

Robine, Jean-Marie and Michel Allard. "The Oldest Human." *Science* (1998) *279*: 1834.

Rose, Michael R. *Evolutionary Biology of Aging*. New York: Oxford University Press.

Rowe, John W. and Robert L. Kahn. *Successful Aging*. New York: Random House, 1998.

Schächter, François, Laurence Faure-Delanef, Frédérique Guénot, Hervé Rouger, Philippe Froguel, Laurence Lesueur-Ginot, and Daniel Cohen. "Genetic Associations with Human Longevity at the *APOE* and *ACE* loci." *Nature Genetics* (1994) *6*(1): 29–32.

Stock, Gregory. *Redesigning Humans: Our Inevitable Genetic Future*. Boston: Houghton Mifflin, 2002.

Strechler, Bernard L. and A.S. Mildvan. "General Theory of Mortality and Aging." *Science* (1960) *132*: 14–21.

Strechler, Bernard L. *Time, Cells and Aging*. New York: Academic Press, 1962.

Tomita-Mitchell, Aoy, Brindha P. Muniappan, Pablo Herrero-Jimenez, Helmut Zarbl, and William G. Thilly. "Single Nucleotide Polymorphism Spectra in Newborns and Centenarians: Identification of Genes Coding for Rise of Mortal Disease." *Gene* (1998) *223*: 381–391.

Wallace, Douglas C. "Mitochondrial Genetics: A Paradigm for Aging and Degenerative Diseases?" *Science* (1992) *256*: 628–631.

Watson, James D. and Andrew Berry. *DNA: The Secret of Life*. New York: Alfred A. Knopf, 2003.

Weismann, August. *Essays Upon Heredity and Kindred Biological Problems*. Oxford: Clarendon Press, 1981.

Williams, George C. "Pleiotropy, Natural Selection and the Evolution of Senescence." *Evolution* (1957) *11*: 398–411.

Wilmoth, John R., Leo J. Deegan, Hans Lundström, and Shiro Horiuchi. "Increase of Maximum Lifespan in Sweden, 1861–1999." *Science* (2000) *289*: 2366–2368.

Chapter 10

Austad, Steven N. *Why We Age.* John Wiley and Sons, 1997.

Beckman, Kenneth B. and Bruce N. Ames. "The Free Radical Theory of Aging Matures." *Physiological Reviews* (1998) *78*: 547–581.

Birren, James E., Robert N. Butler, Sam W. Greenhouse, Louis Sokoloff, and Marian R. Yarrow (eds.). *Human Aging I: A Biological and Behavioral Study.* Public Health Service Publication No. 986 Washington, DC: U.S. Government Printing Office, 1963. (Reprinted 1971 and 1974.)

Brody, Jacob A. and Edward L. Schneider. "Diseases and Disorders of Aging: An Hypothesis." *Journal of Chronic Diseases* (1986) *39*: 871–876.

Busse, Edward W. and George L. Maddox. *The Duke Longitudinal Studies of Normal Aging 1955–1980: An Overview of History, Design and Findings.* New York: Springer, 1985.

Butler, Robert N., Steven N. Austad, and Nir Barzilai. "Longevity Genes: From Primitive Organisms to Humans." *Journal of Gerontology,* A *Biological Science and Medical Science* (2000) *58*: B581–B584.

Butler, Robert N., Huber R. Warner, T. Franklin Williams, et al. "The aging factor in health and disease: The promise of basic research on aging." *Aging Clinical and Experimental Research* (2004) *16*: 2.

Cohn, L., A.G. Feller, M.W. Draper, I.W. Rudman, and D. Rudman. "Carpal Tunnel Syndrome and Gynaecomastia during Growth Hormone Treatment of Elderly Men with Low Circulatory IGF–1 Concentrations." *Clinical Endocrinology* (OXF) (1993) *39*: 417–425.

Cornoni-Huntley, Joan C., Dwight B. Brock, Adrian M. Ostfeld, James O. Taylor, and Robert B. Wallace (eds.). *Established Population for Epidemiologic Studies of the Elderly: Resource Data Book.* (NIH Publication No. 8602443.) Bethesda, MD: Department of Health and Human Services, 1986.

Finch, Caleb E. *The Biology of Human Longevity.* Amsterdam: Elsevier, Academic Press, 2007.

——. *Longevity, Senescence, and the Genome.* Chicago: University of Chicago Press, 1990.

Finch, Caleb E. and Thomas B.L. Kirkwood. *Chance, Development, and Aging.* New York: Oxford University Press, 2000.

Fries, James F. "Aging; Natural Death and the Compression of Morbidity." *New England Journal of Medicine* (1980) *303*: 130–135.

Greider, Carol W. and Elizabeth H. Blackburn. "Telomeres, Telomerase and Cancer." *Scientific American* (1996) *274*: 92–97.

Hajnóczky, György and Jan B. Hoek. "Cell Signaling: Mitochondrial Longevity Pathways." *Science* (2007) *315*: 607–609.

Harman, Denham. *Increasing Healthy Life Span: Conventional Measures and Slowing the Innate Aging Process.* Annals of the New York Academy of Sciences, *959*, 2002.

Hayflick, Leonard. *How and Why We Age.* New York: Ballantine Books, 1996.

Herskind, Anne Maria, Matthew McGue, Niels V. Holm, Thorkild I.A. Sorensen, Bent Harvald, and James W. Vaupel. "The Heritability of Human Longevity: A Population-based Study of 2872 Danish Twin Pairs Born 1870–1900." *Human Genetics* (1996) *97*: 319–323.

Holloszy, John O. and Wendy M. Kohrt. *Handbook of Physiology: Aging* (Section 11). New York: Oxford University Press, 1995.

Holmes, Donna and Steven N. Austad. "Birds as Animal Models for the Comparative Biology of Aging: A Prospectus." *Journal of Gerontology* (1995) *50An2*: B59–B66.

Ingram, Donald K., Richard G. Cutler, Richard Weidruch, David M. Renquist, Joseph J. Knapka, Milton April, Claude Belcher, et al. "Dietary Restriction and Aging: The Initiation of a Primate Study." *Journal of Gerontology* (1990) *45*(5): B148–B163.

International Longevity Center. *Biomarkers of Aging: From Primitive Organisms to Man*, 2000.

Jeanne, Bernard and James W. Vaupel (eds.). *Exceptional Longevity: From Prehistory to the Present.* Berlin: Springer, 1995.

Kirkwood, Tom. *Time of Our Lives: The Science of Human Aging.* Oxford, U.K.: Oxford University Press, 1999.

Kuiken, Thijs, Frederick A. Leighton, Ron A.M. Fouchier, James W. LeDuc., J.S. Malik Peiris, Alejandro Schudel, Klaus Stohr, and Albert Osterhaus. "Pathogen Surveillance in Animals." *Science* (2005) *309*: 1680–1681.

Manton, Kenneth G., XiLian Gu, and Vicki L. Lamb. "Change in Chronic Disability from 1982 to 2004/2005 as Measured by Long-Term Changes in Function and Health in the U.S. Elderly Population." Proceedings of the National Academy of Science (2006) *103*: 18374–18379.

Masore, Edward. "Caloric Restriction: A Key to Understanding and Modulating Aging." *Research Problems in Aging.* Amsterdam: Elsevier, 2002.

McCay, Clive, M. Crowel, and L. Maynard. "The Effect of Retarded Growth upon the Length of Lifespan and upon the Ultimate Body Size." *Journal of Nutrition* (1935) *10*: 63–79.

McClearn, Gerald E. "Heterogeneous Reference Populations in Animal Model

Research in Aging." *Institute for Laboratory Animal Research (ILAR) Journal* (1997) *38*: 119–123.

Medawar, Peter B. *An Unsolved Problem of Biology.* London: Lewis, 1952.

Metchnikoff, Elie. *The Prolongation of Life: Optimistic Studies.* International Longevity Center Classics in Longevity and Aging Series. New York: Springer, 2003.

Olshansky, S. Jay, Leonard Hayflick, Cristine Cassel, and Thomas T. Perls. "Introduction: Anti-Aging Medicine: The Hype and the Reality—Part I." *Journal of Gerontology and Biological Science* (2004) *59*: B513–B514.

Olshansky, S. Jay, Leonard Hayflick, and Thomas T. Perls. "Introduction: Anti-Aging Medicine: The Hype and the Reality—Part II." *Journal of Gerontology and Biological Science* (2004) *59*: B649–B651.

Olshansky, S. Jay, Daniel Perry, Richard A. Miller, and Robert N. Butler. "In Pursuit of the Longevity Dividend." *The Scientist* (2006) *20*: 28–36.

Palmore, Erdman B., Ewald Busse, Ilene C. Siegler, John B. Nowlin, and George L. Maddox (eds.). *Normal Aging III: Reports from the Duke Longitudinal Studies 1975–1984.* Durham, NC: Duke University Press, 1985.

Perls, Thomas T. and Margery Hutter Silver, with John F. Lauerman. *Living to 100: Lessons in Living to Your Maximum Potential at Any Age.* New York: Basic Books, 1999.

Rose, Michael R. "Laboratory Evolution of Postponed Senescence in *Drosophila melanogaster.*" *Evolution* (1984) *38*: 1004–1010.

Rudman, Daniel, Axel G. Feller, Hoskote S. Nagraj, Gregory A. Gergans, Pardee Y. Lalitha, Allen F. Goldberg, Robert A. Schlenker, Lester Cohn, Inge W. Rudman, and Dale E. Mattson. "Effects of Human Growth Hormone in Men Over 60 Years Old." *The New England Journal of Medicine* (1990) *323*: 1–6.

Sacher, George A. "Longevity, Aging, and Death: An Evolutionary Perspective." *The Gerontologist* (1978) *18*: 112–119.

Sapolsky, Robert M., Lewis C. Krey, and Bruce S. McEwen. "The Neuroendocrinology of Stress and Aging: The Glucocorticoid Cascade Hypothesis." *Endocrinology Review* (1986) *7*: 284.

Shock, Nathan W., Richard C. Greulich, Reubin Andres, David Arenberg, Paul T. Costa, Edward G. Lakatta, and Jordan D. Tobin. *Normal Human Aging: The Baltimore Longitudinal Study of Aging,* U.S. Department of Health and Human Services, NIH Publication No. 84–2450, 1984.

Suzman, Richard M., David P. Willis, and Kenneth G. Manton. *The Oldest Old.* New York: Oxford University Press, 1992.

Swerdlow, Anthony J., Craig D. Higgins, Peter Adlard, and Mike A. Preece. "Risk of Cancer in Patients Treated with Human Pituitary Growth Hormone in the U.K., 1959–85: A Cohort Study." *The Lancet* (2002) *360*, Issue 9329: 273–277.

Walford, Roy L. *The Immunologic Theory of Aging.* Copenhagen: Williams & Wilkins, 1969.

Wallace, Douglas C. "Mitochondrial Genetics: A Paradigm for Aging and Degenerative Diseases." *Science* (1992) *256*: 628–632.

Weismann, August. "The Duration of Life." In *Collected Essays Upon Heredity and Kindred Biological Problems.* E.B. Poulton (ed.). Oxford: Clarendon Press, 1889.

Wizemann, Theresa M. and Mary-Lou Pardue (eds.). *Exploring the Biological Contributions to Human Health: Does Sex Matter?* Institute of Medicine, Washington DC: National Academy Press, 2001.

Chapter 11

Ames, Bruce N., Mark K. Shigenaga, and Tory M. Hagen. "Oxidants, Antioxidants, and the Degenerative Diseases of Aging." *Proceedings of the National Academy of Sciences* (1933) *90*: 7915–7922.

Avorn, Jerry. *Powerful Medicine: The Benefits, Risks and Costs of Prescription Drugs.* New York: Knopf, 2004.

Boston Women's Health Book Collective. *The New Our Bodies, Ourselves: A Book By and For Women.* New York: Simon & Shuster, 1996.

Cornaro, Luigi. *The Art of Living.* International Longevity Center in collaboration with Springer Publishing Company, 2005.

De Smet, Peter A.G.M. "Herbal Remedies: Drug Therapy," *New England Journal of Medicine* (2002) *347*: 2046–2056.

Diabetes Prevention Program Research Group. "Reduction in the Incidence of Type 2 Diabetes with Lifestyle Intervention or Metformin." *The New England Journal of Medicine* (2002) *346*: 393–403.

Fillit, Howard. *Treating Alzheimer's: Accelerating Drug Discovery.* International Longevity Center issue brief, March–April 2003.

Fillit, Howard M. and Robert N. Butler (eds.). *Cognitive Decline: Strategies for Prevention.* London: Greenwich Medical Media, 1997.

Foster, Steven and Varro E. Tyler. *Tyler's Honest Herbal: A Sensible Guide to the Use of Herbs and Related Remedies.* 4th ed. New York: Haworth, 1999.

Fries, James F. "Aging, Natural Death and the Compression of Morbidity." *New England Journal of Medicine* (1980) *303*: 130–135.

Fries, James F. and Lawrence M. Crapo. *Vitality and Aging: Implications of the Rectangular Curve.* San Francisco: W.H. Freeman, 1981.

Fries, James F. *Compression of Morbidity in Retrospect and in Prospect.* Alliance for Health and the Future at the International Longevity Center, 2005.

Gruman, Gerald J. *A History of Ideas about the Prolongation of Life.* International Longevity Center in collaboration with Springer, 2005.

Guide to Clinical Preventive Services. 2nd ed. Report of the U.S. Preventive Services Task Force, 1995.

Guralnik, Jack M., Eleanor M. Simonsick, Luigi Ferrucci, Robert J. Glynn, Lisa F. Berkman, Dan G. Blazer, Paul A. Scherr, and Robert A. Wallace. "A Short Physical Performance Battery Assessing Lower Extremity Function: Association with Self-reported Disability and Prediction of Mortality and Nursing Home Admission." *Journal of Gerontology: Medical Sciences* (1994) *49*: M85–M94.

Healthy People 2000: National Health Promotion and Disease Prevention Objectives. Washington DC: Government Printing Office (DHHS Publication No. [PHS] 91–50213), 1991.

Heywood, Vernon H., R.T. Watson, and I. Baste, et al. (eds.). *Global Biodiversity Assessment Summary for Policy Makers.* Cambridge, U.K.: Cambridge University Press, 1995.

Hodgson, Thomas A. "Cigarette Smoking and Lifetime Medical Expenditures." *The Milbank Quarterly* (1992) *70*: 81–125.

Kennedy, D. "Drug Discovery." *Science* (2004) *303*: 1729, 1795–1822.

Levy, Daniel and Susan A. Brink. *A Change of Heart: How the Framingham Heart Study Helped Unravel the Mysteries of Cardiovascular Disease.* New York: A.A. Knopf, 2005.

Marcus, Donald M. and Arthur P. Grollman. "Botanical Medicines—The Need for New Regulations." *New England Journal of Medicine* (2002) *347*: 2073–2076.

McGinnis, J. *Michael Chair, Food Marketing and the Diets of Children and Youth.* Institute of Medicine, 2005.

Metchnikoff, Elie. *The Prolongation of Life: Optimistic Studies.* International Longevity Center in collaboration with Springer, 2004.

Micozzi, Mark S. (ed.). *Fundamentals of Complementary and Alternative Medicine.* 2nd ed. New York: Churchill Livingstone.

Olshansky, S. Jay, Leonard Hayflick, and Thomas T. Perls (eds.). *Anti-Aging Medicine: The Hype and the Reality.* Special publication of the *Journal of Gerontology: Biological Sciences,* 2004.

Olshansky, S. Jay, Douglas J. Passaro, Ronald C. Hershow, Jennifer Layden, Bruce A. Carnes, Jacob Brody, Leonard Hayflick, Robert N. Butler, David B. Allison, and David S. Ludwig. "Potential Decline in Life Expectancy in the United States in the 21st Century." *New England Journal of Medicine* (2005) *352*: 1138–1145.

Paffenbarger, Ralph S. Jr., Robert T. Hyde, I-Min Lee Wing, and C. Heich. "Physical Activity, All-cause Mortality, and Longevity of College Alumni." *New England Journal of Medicine* (1986) *314*: 605–613.

Robbers, James E. and Varro E. Tyler. *Tyler's Herbs of Choice: The Therapeutic Use of Phytomedicinals.* New York: Haworth, 1999.

Russell, Louise B. "The Role of Prevention in Health Reform." *New England Journal of Medicine* (1993) *329*: 352–354.

Seeman, Teresa E., Peter A. Chapentier, Lisa F. Berkman, Mary E. Tinetti, Jack M. Guralnik, Marilyn Albert, Dan Blazer, and John W. Rowe. "Predicting Changes in Physical Performance in a High-functioning Elderly Cohort." MacArthur Studies of Successful Aging. *Journal of Gerontology: Medical Sciences* (1994) *49*: M97–M108.

Stewart, Paul M. "Aging and Fountain of Age Hormones." *New England Journal of Medicine* (2006) *355*: 1724–1726.

U.S. Department of Health, Education and Welfare, Public Health Service, and the Office of the Assistant Secretary for Health and Surgeon General. *Healthy People—The Surgeon General's Report on Health Promotion and Disease Prevention.* Washington, DC: U.S. Government Printing Office, 1979.

U.S. Surgeon General. *Smoking and Health: Report of the Advisory Committee to the Surgeon General of the Public Health Service,* 1964.

Woolf, Steven H., Robert S. Lawrence, and Steven Jonas. *Health Promotion and Disease Prevention in Clinical Practice.* Complement to *Guide to Clinical Preventive Services.* 2nd ed., 1995.

Chapter 12

Anderson, Gerard F. "Medicare and Chronic Conditions." *New England Journal of Medicine* (2005) *353*: 305–309.

Boockvar, Kenneth S., Alan Litke, Joan D. Penrod, et al. "Patient Relocation in the 6 Months after Hip Fracture: Risk Factors for Fragmented Care." *Journal of the American Geriatric Society* (2004) *53*: 1826–1831.

Brocklehurst, John Charles, Raymond Tallis, and Howard Fillit (eds.). *Brocklehurst's Textbook of Geriatric Medicine and Gerontology.* 5th ed. New York: Churchill-Livingston, 1998.

Burgio, Kathryn L., Julie L. Locher, Patricia S. Goode, Michael Hardin, Joan McDowell, Marianne Dombrowski, and Dorothy Candib. "Behavioral vs. Drug Treatment for Urge Urinary Incontinence in Older Women: A Randomized Controlled Trial." *Journal of the American Geriatrics Society* (2000) *48*: 370–374.

Butler, R.N. "The Teaching Nursing Home." *Journal of the American Medical Association* (1981) *245*: 1435–1437.

Cohen, Havey J., John R. Feussner, Morris Weinberger, Molly Carnes, Ronald C. Hamdy, Frank Hsieh, Ciaran Phibbs, Donald Courtney, Kenneth W. Lyles, Conrad May, Cynthia McMurtry, Leslye Pennypacker, David M. Smith, Nina Ainslie, Thomas Hornick, Kayla Brodkin, and Philip Lavori. "A Controlled Trial of Inpatient and Outpatient Geriatric Evaluation and Management." *New England Journal of Medicine* (2002) *346*: 905–912.

Creditor, Morton C. "Hazards of Hospitalization of the Elderly." *Annals of Internal Medicine* (1993) *118*: 219–223.

Engel, George L. "The Need for a New Medical Model: A Challenge for Biomedicine." *Science* (1977) *196*:129–136.

Finucane, Thomas E., Colleen Christmas, and Kathy Travis. "Tube Feeding in Patients with Advanced Dementia." *Journal of the American Medical Association* (1999) *282*: 1365–1370.

Freeman, Joseph T. "The History of Geriatrics." *Annals of Medical History* (1938) *10*: 324–355.

Goffman, Erving. *Asylums.* New York: Anchor, 1961.

Inouye, Sharon K., Sidney T. Bogardus Jr., Dorothy I. Baker, Linda Leo-Summers, and Leo M. Cooney Jr. "The Hospital Elder Life Program: A Model of Care to Prevent Cognitive and Functional Decline in Older Hospitalized Patients." *Journal of the American Geriatrics Society* (2000) *48*: 1697–1706.

Inouye, Sharon, Sidney T. Bogardus, Peter A. Charpentier, Linda Leo-Summers, Denise Acampora, Theodore R. Holford, and Theodore R. Cooney. "A Multicomponent Intervention to Prevent Delirium in Hospitalized Older Patients." *New England Journal of Medicine* (1999) *340*: 669–676.

Inouye, Sharon K., Stephanie Studenski, Mary E. Tinetti, and George A. Kucher. "Geriatric Syndromes: Clinical, Research and Policy Implications of a Core Geriatric Concept." *Journal of the American Geriatrics Society* (2007) *55*: 780–791.

Institute of Medicine. *To Err Is Human: Building a Safer Health System.* Washington, DC: National Academy Press, 2000.

International Longevity Center. *Myths of the High Medical Costs of Old Age and Dying,* 2007.

———. *Redesigning Health Care for an Older America, Seven Guiding Principles,* 2004.

Kozak, Lola J., Maria F. Owings, and Margaret J. Hall. "National Discharge Survey: Annual Summary with Detailed Diagnosis and Procedure Data." National Center for Health Statistics. *Vital Health Stat.* (2002) *13*: 1–199.

Morin, Charles M., Cheryl Colecchi, Jackie Stone, Rakesh Sood, and Douglas Brink. "Behavioral and Pharmacological Therapies for Late-life Insomnia; A Randomized Controlled Trial." *Journal of the American Medical Association* (1999) *281*: 991–999.

Morrison, R. Sean and Diane E. Meier. "Clinical Practice. Palliative Care." *New England Journal of Medicine* (2004) *350*: 2582–2590.

Nascher, Ignaz L. *Geriatrics: The Diseases of Old Age and Their Treatment.* New York: Arnow Press, 1979.

Tew, James D. Jr., Benoit H. Mulsant, Roger F. Haskett, Joan Prudic, Michael E. Thase, Raymond R. Crowe, Diane Dolata, Amy E. Begley, Charles F. Reynolds III, and Harold A. Sackeim. "Acute Efficacy of ECT in the Treatment of Major Depression in the Old-Old." *American Journal of Psychiatry* (1999) *156*: 1865–1870.

Tinetti, Mary E., Dorothy I. Baker, Gail McAvay, Elizabeth B. Claus, Patricia Garrett, Margaret Gottschalk, Marie L. Koch, Kathryn Trainor, and Ralph I. Horwitz. "A Multifactorial Intervention to Reduce the Risk of Falling among Elderly People Living in the Community." *New England Journal of Medicine* (1994) *331*: 821–827.

Tinetti, Mary E. and Terri Fried. "The End of the Disease Era." *The American Journal of Medicine* (2004) *116*: 179–185.

Zarit, Steven H., Leonard I. Pearlin, and Jon Hendricks. "Health Inequalities across the Life Course." Special issue *Journal of Gerontology: Social Sciences,* Series B (2005) *60B*: 1–139.

Chapter 13

AARP. *The Business Case for Workers Age 50+.* A Report for AARP prepared by Towers Pervin, December 2005.

Bass, Scott A. (ed.). *Aging and Active: How Americans Over 55 Are Contributing to Society.* Americans Over 55 at Work Program, 1995.

Bass, Scott A., Francis G. Caro, and Yung-Ping Chen (eds.). *Achieving a Productive Aging Society.* Aubern House Paperback, June 1993.

Bond, Lynne A., Stephen J. Cutler, and Armin Grams (eds.). *Promoting Successful and Productive Aging.* Thousand Oaks, CA: Sage: 1995.

Butler, Robert N. *Why Survive? Being Old in America.* New York: Harper & Row,

1975. (See especially "The Right to Work" and "Loosening Up Life" chapters. Reissued by the Johns Hopkins Press, 2002.)

Butler, Robert N. and Herbert P. Gleason. *Productive Aging, Enhancing Vitality in Later Life.* New York: Springer, 1985.

Butler, Robert N., Malvin Schechter, and Mia Oberlink (eds.). *The Promise of Productive Aging: From Biology to Social Policy.* New York: Springer, 1990.

Costa, Dora (ed.). *The Evolution of Retirement: An American Economic History, 1880– 1990: Health and Labor Force Participation Over the Life Cycle: Evidence from the Past.* National Bureau of Economic Research Conference, 2003.

DeLong, David W. *Lost Knowledge, Confronting the Threat of an Aging Workforce.* New York: Oxford University Press, 2004.

Freedman, Marc. *Prime Time: How Baby Boomers Will Revolutionize Retirement and Transform America.* New York: Public Affairs, 1999.

Fogel, Robert William. *The Fourth Great Awakening and the Future of Egalitarianism.* Chicago: The University of Chicago Press, 2002.

Gallo, William T., Elizabeth H. Bradley, Michele Siegel, and Stanislav V. Kasl. "Health Effects of Involuntary Job Loss among Older Workers: Findings from the Health and Retirement Survey." *Journal of Gerontological Social Sciences* (2000) 55B: 5131–5141.

Gardner, John W. *Self-Renewal: The Individual and the Innovative Society.* New York: W.W. Norton, 1963.

Kumashiro, Masaharu. *Paths to Productive Aging.* Proceedings of the XIVth University of Occupational and Environmental Health (UOEH) and Institute of Industrial Ecological Sciences International Symposia and International Ergonomics Association Technical Group. Taylor & Francis, 1995.

McCann, Lauri A. "Age Discrimination in Employment Legislation: the United States Experience." *Age Dissemination in Employment Research Report*, April 2002 (www.aarp.org/research/work/agediscrim, accessed September 2005).

Metchnikoff, Elie. *The Prolongation of Life: Optimistic Studies.* Republished by the International Longevity Center and Springer, 2004.

Morrow-Howell, Nancy, James Hinterlong, and Michael Sherraden (eds.). *Productive Aging Concepts and Challenges.* Baltimore: The Johns Hopkins University Press, 2001.

Muller, Charlotte and Marjorie Honig. *Charting the Productivity and Independence of Older Persons.* A study of six industrialized nations. International Longevity Center, 2000.

Munnell, Alicia H. and Jerilyn Libby. *Will People Be Healthy Enough to Work Longer? An Issue in Brief.* Center for Retirement Research at Boston College, No. 2007–3, March 2007.

Rix, Sara. *Update on the Older Worker Research Report.* AARP, 2005.

Uchitelle, Louis. *The Disposable American: Layoffs and Their Consequences.* New York: A.A. Knopf, 2006.

Chapter 14

Ackerman, Bruce and Anne Alstoot. *The Stakeholder Society.* New Haven, CT: Yale University Press, 1999.

Baker, Dean. *Social Security: The Phony Crisis.* Chicago: University of Chicago Press, 1999.

Ball, Robert. *Essentials: Bob Ball on Social Security.* New York: Century Foundation Press, 2004.

Burkhauser, Richard V. and Timothy M. Smeeding. *Social Security Reform: A Budget-Neutral Approach to Reducing Older Women's Disproportionate Risk of Poverty.* Syracuse, NY: Syracuse University, Center for Policy Research, 1994.

Butler, Robert N. "An Overview of Research on Aging and the Status of Gerontology Today." *Milbank Memorial Fund Quarterly Health and Society* (1983) *61*: 351–361.

——. "The Relation of Extended Life to Extended Employment Since the Passage of Social Security in 1935," *Milbank Memorial Fund Quarterly Health and Society* (1983) *61*: 420–429.

Butler, Robert N. and Herbert P. Gleason. *Productive Aging, Enhancing Vitality in Later Life.* New York: Springer, 1985.

Center for Women Policy Studies. *Earnings Sharing in Social Security: A Model for Reform.* Washington, DC: Center for Women Policy Studies, 1988.

Commonwealth Fund Commission on Elderly People Living Alone. *Old, Alone and Poor.* Baltimore, MD: 1987.

Diamond, Peter A. *Privatization of Social Security: Lessons from Chile.* NBER Working Paper No. 4510. Cambridge, MA: National Bureau of Economic Research, 1993.

——. "A Framework for Social Security Analysis." *Journal of Public Economics* (1977) *8*: 275–298.

Diamond, Peter A. and Peter R. Orszag. *Saving Social Security, a Balanced Approach.* Washington, DC: Brookings Institution Press, 2004.

Employee Benefits Research Institute. *Baby Boomers in Retirement: What Are Their Prospects?* Issue Brief No. 151. Washington, DC: EBRI, 1999.

Gruber, Jonathan and David A. Wise, eds. *Social Security and Retirement around the World: Micro-Estimation.* Chicago: University of Chicago Press, 1999.

Hurd, Michael D. and Michael J. Boskin. "The Effects of Social Security on Retirement in the Early 1970s." *Quarterly Journal of Economics* (1984) 99: 767–790.

International Longevity Center. *The Consequences of Population Aging for Society.* New York: International Longevity Center, 2000.

——. *Old and Poor in New York City.* Issue Brief, September–October 2002. New York: International Longevity Center, 2002.

Kotlikoff, Laurence J. and Scott Burns. *The Coming Generational Storm: What You Need to Know about America's Economic Future.* Cambridge, MA: MIT Press, 2004.

Leone, Richard C. "Stick with Public Pensions." *Foreign Affairs* (1997) 76: 39–53.

Levitt, Arthur with Paula Dwyer. *Take on the Street: What Wall Street and Corporate America Don't Want You to Know, What You Can Do to Fight Back.* New York: Pantheon, 2002.

Peterson, Peter A. *Running on Empty: How the Democratic and Republican Parties Are Bankrupting Our Future and What Americans Can Do about It.* New York: Farrar, Straus & Giroux, 2004.

Schiller, Robert J. *Irrational Exuberance.* Princeton, NJ: Princeton University Press, 2000.

Smeeding, Timothy M. *Social Security Reform: Improving Benefit Adequacy and Economic Security for Women.* Policy Brief No. 16. Syracuse, NY: Center for Policy Research, Syracuse University, 1999.

U.S. Bureau of the Census. *Poverty in the United States: 1996.* Current Population Reports, Series P–60, No. 198. Washington, DC: USGPO, 1997.

Wasow, Bernard. *Setting the Record Straight: Social Security Work for Latinos.* New York: Century Foundation, 2002.

Wolman, William and Anne Colamosca. *The Great 401(k) Hoax: Why Your Family's Financial Security Is at Risk and What You Can Do about It.* New York: Perseus, 2002.

Chapter 15

Butler, Robert N. and Herbert P. Gleason (eds.). *Productive Aging—Enhancing Vitality in Later Life.* New York: Springer, 1985.

Costa, Dora L. *The Evolution of Retirement: An American Economic History 1880–1990.* Chicago: University of Chicago Press, 1998.

Gill, Indermit S., Truman Parkard, and Juan Yermo. *Keeping the Promise of Old*

Age Income Security in Latin America. Washington, DC: World Bank, 2004.

Graham, Benjamin. *The Intelligent Investor* (rev. ed). New York: Harper Business Essentials, 2003.

Gruber, Jonathan and David A. Wise (eds.). *Social Security and Retirement around the World.* Chicago: University of Chicago Press, 1999.

Fischer, David Hackett. *Growing Old in America.* New York: Oxford University Press.

Kotlikoff, Laurence J. and Scott Burns. *The Coming Generational Storm: What You Need to Know about America's Economic Future.* Cambridge, MA: MIT Press, 2004.

Munnell, Alicia Haydock and Annika Sunden. *Coming Up Short: The Challenge of 401(k) Plans.* Washington, DC: Brookings Institution Press, 2004.

Schiller, Robert J. *Irrational Exuberance.* Princeton, NJ: Princeton University Press, 2000.

Siegel, Jeremy J. *Stocks for the Long Run: The Definitive Guide to Financial Market Returns and Long-Term Investment Strategies* (3rd ed.). New York: McGraw-Hill, 2002.

Wolff, Edward N. *Retirement Insecurity: The Income Shortfalls Awaiting the Soon-to-Retire.* Washington, DC: The Economic Policy Institute, May 2002.

Chapter 16

Achenbaum, W. Andrew. *Social Security: Visions and Revisions: A Twentieth Century Fund Study.* New York: Cambridge University Press, 1986.

——. *Old Age in the New Land: The American Experience Since 1790.* Baltimore, MD: Johns Hopkins University Press, 1978.

Agee, James and Walker Evans. *Let Us Now Praise Famous Men.* Boston: Houghton Mifflin, 2000 (1939).

Brinkley, Alan. *End of Reform: New Deal Liberalism in Recession and War.* New York: Knopf, 1996.

——. *Voices of Protest: Huey Long, Father Coughlin, and the Great Depression.* New York: Knopf, 1983.

Butler, Robert N. "The Pacification and the Politics of Aging." Chapter 11 in Robert N. Butler, *Why Survive? Being Old in America.* New York: Harper & Row, 1975. (Reissued by Johns Hopkins Press, 2002.)

Corak, Miles (ed.). *Generational Income Mobility in North America and Europe.* Cambridge, U.K.: Cambridge University Press, 2004.

Dahl, Robert. *How Democratic Is the American Constitution?* New Haven, CT: Yale University Press, 2002.

Howe, Neil and William Strauss. *Millennials Rising: The Next Great Generation.* New York: Vintage Books, 2000.

Kennedy, David M. and E. Vann Woodward. *Freedom from Fear: The American People in Depression and War, 1929–1945.* New York: Oxford University Press, 1999.

Lewis, Sinclair. *Babbitt.* Dover, 2003 (1922).

Nozick, Robert. *Anarchy, State and Utopia.* New York: Basic Books, 1974.

Phillips, Kevin. *American Theocracy.* New York: Penguin Books, 2007.

——. *Wealth and Democracy: A Political History of the American Rich.* New York: Broadway Books, 2002.

Schlesinger, Arthur M. *Crisis of the Old Order, 1919–1933.* 3 Volumes. Vol. 1, *The Age of Roosevelt.* Boston: Houghton Mifflin, 1950.

Sunstein, Cass R. *Designing Democracy: What Constitutions Do.* New York: Oxford University Press, 2001.

Tocqueville, Alexis de. *Democracy in America.* Translated by H.C. Mansfield and D. Winthrop. Chicago: University of Chicago Press, 2001.

Torres-Gil, Fernando M. and TsuAnn Kuo. "Social Policy and the Politics of Hispanic Aging." *Journal of Gerontological Social Work* (1998) *30*: 143–158.

Turkel, Studs. *Hard Times: An Oral History of the Great Depression.* New York: Pantheon, 1970.

Chapter 17

Braudel, Fernand. *Civilization and Capitalism,* 15th–18th Century, Vols. 1–3. New York: Harper & Row, 1979.

Cohen, Joel E. "How Many People Can the Earth Support?" *The Sciences,* November/December 1995.

——. "Human Population: The Next Half Century." *Science* (2003) *302*: 1175–1177.

Hardin, Garret. "The Tragedy of the Commons." *Science* (1968) *162*: 1243–1248.

Hareven, Tamara. "The History of the Family and the Complexity of Social Change," *American Historical Review* (1991) *96*: 95–124.

Harrison, Paul. *AAAS Atlas of Population and Environment.* Berkeley: University of California Press, 2001.

Hesketh, Therese, Lu Li, and Wei Kin G. Zku. "The Effect of China's One-child Family Policy after 26 years." *New England Journal of Medicine* (2005) *353*: 1171–1176.

James, Estelle. *Averting the Old Age Crisis: Policies to Protect the Old and Promote Growth.* A World Bank Policy Research Report, November 1993.

Marks, Laura V. *Sexual Chemistry: A History of the Contraceptive Pill.* New Haven, CT: Yale University Press, 2001.

McNeill, John R. *Something New Under the Sun: An Environmental History of the Twentieth Century World.* New York: W.W. Norton, 2000.

Meadows, Donella H., Jørgen Randers, and Dennis L. Meadows. *The Limits of Growth.* A Report for the Club of Rome's Project on the Predicament of Mankind, 1972. Updated to *Beyond the Limits: Confronting Global Collapse, Envisioning a Sustainable Future,* 1992.

Metchnikoff, Elie Illyich. *The Prolongation of Life: Optimistic Studies.* Classics of Longevity Series. New York: International Longevity Center and Springer, 2004.

Peterson, Peter. *Gray Dawn.* New York: Times Books, 1999.

Tone, Andrea. *Devices and Desires. A History of Contraceptives in America.* New York: Hill & Wang, 2001.

United Nations Population, State of World Population: *Unleashing the Potential of Urban Growth,* 2007.

Chapter 18

Athanasiou, Tom. *Divided Planet: The Ecology of Rich and Poor.* Boston: Little, Brown, 1996.

Barraclough, Geoffrey. *An Introduction to Contemporary History.* New York: Basic Books, 1964.

Bloom, David E. and David Canning. "Health as Human Capital and Its Impact on Economic Performance." *The Geneva Papers on Risk and Insurance* (2003) *28*: 304–315.

Bloom, David E., David Canning, and Jaypee Sevilla. *Health, Human Capital and Economic Growth.* World Health Organization Commission on Macroeconomics and Health Working Paper No. WG1:8, 2001.

Bloom, David E. and Pia Malaney. "Macroeconomic Consequences of the Russian Mortality Crisis." *World Development* (1998) *26*: 2073–2085.

The Center for Strategic and International Studies (CSID). *The Global Retirement Crisis: The Threat to World Stability and What to Do about It,* 2002.

Central Intelligence Agency, National Foreign Intelligence Board. *Global Trends 2015: A Dialogue about the Future with Non-Governmental Experts.* Washington, DC, 2000. (Unclassified Report.)

Coyle, Diane. *Paradoxes of Prosperity: Why the New Capitalism Benefits All.* Texere, 2001, 2002.

Ehrlich, Paul R. and Anne H. Ehrlich. *One with Nineveh: Politics, Consumption, and the Human Future.* Washington, DC: Island, 2004.

Farmer, Paul and Nicole Gastineau Campos. "Rethinking Medical Ethics: A View from Below." *Developing World Bioethics* (2005) 4: 17–41.

The Fourth World Conference on Women, Beijing China, Sept. 4–5, 1995. (Document A/CONF.177/20, UN, New York, 1995.)

Garcia-Moreno, Claudia, Lori Heise, Henrica A.F.M. Jansen, Mary Ellsberg, and Charlotte Watts. "Violence Against Women," *Science* (2005) *310:* 1282–1283.

Heise, Lori, Mary M. Ellsberg, Megan Gottemoeller, *Ending Violence Against Women.* Baltimore, MD: Johns Hopkins University Press, 1999.

Maritain, Jacques. *Man and the State.* Chicago: University of Chicago Press, 1951.

Prabhat, Jha, Anne Mills, Kara Hanson, Lilani Kumaranayake, Lesong Conteh, Christoph Kurowski, Son Name Nguyen, Valerie Cruz, Kent Ranson, Lara M.E. Vaz, Shengchao Yu, Oliver Morton, and Jeffrey D. Sachs. "Improving the Health of the Global Poor," *Science* (2002) *295:* 2036.

Sachs, Jeffrey D. *The End of Poverty: Economic Possibilities for Our Time.* New York: Penguin Press, 2005.

Sen, Amartya K., *Development as Freedom.* New York: Knopf, 1999.

Shabecoff, Philip. *A New Name for Peace: International Environmentalism, Sustainable Development, and Democracy.* Hanover, NH: University Press of New England, 1996.

Simmons, Leo W. *The Role of the Aged in Primitive Societies.* New Haven, CT: Yale University Press, 1945.

Stiglitz, Joseph E. *Globalization and Its Discontents.* New York: W.W. Norton, 2002.

Tout, Ken. *Aging in Developing Countries.* Oxford University Press, 1989.

United Nations. *Agenda for Development,* 1997.

UN World Assembly on Aging. *International Plan of Action on Aging.* Vienna: UN Aging Unit, 1982. (In various languages.)

Chapter 19

Ada, Gordon. "Vaccines and Vaccination: Advances in Immunology." *The New England Journal of Medicine* (2001) *345:* 1042–1053.

Diamond, Jared. *Guns, Germs and Steel.* New York: W.W. Norton, 1997.

Dubos, Rene. *The White Plague: Tuberculosis, Man, and Society.* Boston: Little, Brown, 1952.

Fenn, Elizabeth Anne, *Pox America: The Great Smallpox Epidemic of 1775–1782.* New York: Hill & Wang, 2001.

Flannery, Tim. *The Weather Makers: How Man Is Changing the Climate and What It Means for Life on Earth.* New York: Atlantic Monthly Press, 2006.

Garwin, Richard L. and Georges Charpak. *Megawats and Megatons. A Turning Point in the Nuclear Age.* New York: Knopf, 2001. (Originally published in France, 1997.)

Goldenberg, Jose (ed.). *World Emerging Assessment, Energy and the Challenge of Sustainability.* United Nations Development Programme. New York: United Nations Department of Economic and Social Affairs and World Energy Council, 2001.

Kuiken, T., F.A. Leighton, A.M. Fouchler, L.W. LeDuc, J.S.M. Peirls, A. Schuden, K. Stöhr, A.D.M.E. Osterhaus. "Pathogen Surveillance in Animals." *Science* (2005) *309*: 1680–1681.

Lederberg, Joshua. "Crowded at the Summit: Emerging Infections and the Global Food Chain." Unpublished manuscript, January 1993.

Lederberg, Joshua, Robert E. Shope, Stanley C. Oakes Jr. *Emerging infections: Microbial Threats to Health in the United States.* Report of the Institute of Medicine, 1992.

McMichael, Tony. *Human Frontiers, Environmental and Disease: Past Patterns, Uncertain Futures.* Cambridge, UK: Cambridge University Press, 2001.

McNeill, John R. *Something New Under the Sun: An Environmental History of the Twentieth Century World.* New York: W.W. Norton, 2000.

McNeill, William. *Plagues and Peoples.* Garden City, NY: Anchor, 1976.

Miller, Judith, Stephen Engelberg, and William Brood. *Germs,* New York: Simon & Schuster, 2001.

Monto, Arnold S. "The Threat of an Avian Influenza Pandemic." *New England Journal of Medicine* (2005) *352*:4; 323–325.

Murray, Christopher J. and Alan D. Lopez (eds.). *The Global Burden of Disease: A Comprehensive Assessment of Mortality and Disability From Diseases, Injuries and Risk Factors in 1990 and Projected 2020.* Cambridge, MA: Harvard University Press, 1996.

Palumbi, Stephen R. "Humans as the World's Greatest Evolutionary Force." *Science* (2001) *293*: 1786–1790.

Reichman, Lee B. with Janice Hopkins Tanne. *Time Bomb: The Global Epidemic of Multi-Drug-Resistant Tuberculosis.* New York: McGraw-Hill, 2002.

Speilman, Andrew and Michael D'Antonio. *Mosquito: A Natural History of Our Most Persistent and Deadly Foe.* New York: Hyperion, 2002.

Tucker, Jonathan B. *Scourge: The Once and Future Threat of Smallpox*. New York: Atlantic Monthly Press, 2001.

Wilson, Eduard O. *The Future of Life*. New York: Knopf, 2002.

Zinsser, Hans. *Rats, Lice and History*. Boston: Little, Brown, 1984.

Chapter 20

Adler, Mortimer J. *We Hold These Truths*. New York: Macmillan, 1987.

Bakija, Jon M. and William G. Gale. "Tax Break." *Tax Notes*, June 23, 2003.

Berenson, Bernard. *Sunset and Twilight, From the Diaries of 1947–1958*. New York: Harcourt, Brace & World, 1963.

Blazer, Dan G. *The Age of Melancholy: "Major Depression" and its Social Origins*. New York: Routledge, 2005.

Butler, Robert N. "The Life Review: An Interpretation of Reminiscence in the Aged." *Psychiatry* (1963) *26*: 65–76.

Butler, Robert N. and Claude Jasmin. *Longevity and Quality of Life: Opportunities and Challenges*. New York: Kluwer Academic/Plenum, 1999.

Campbell, Angus, Phillip E. Converse, and Willard L. Rodger. *The Quality of American Life*. New York: Russell Sage, 1976.

Cohen, Wilbur J. *Toward a Social Report*. Ann Arbor, MI: Russell Sage, 1969.

Federal Interagency Forum on Aging Related Statistics. *Older Americans: Key Indicators of Well-Being*. Washington, DC: U.S. Government Printing Office, 2004.

Frankl, Viktor. *Man's Search for Meaning*. Boston: Beacon Press, 1959.

George, Linda K. and Lucille B. Bearson. *Quality of Life in Older Persons: Meaning and Measurement*. New York: Human Sciences Press, 1980.

Gurland, Barry J. and Roni V. Gurland. "Choices, Choosing and the Nature of Quality of Life." Unpublished manuscript, 2006.

Holmes, Thomas H. and Richard H. Rahe. "The Social Adjustment Scale." *Journal of Psychosomatic Research* (1969) *11*: 213–218.

International Longevity Center. *Longevity and Quality of Life*, 1998.

Jung, Carl G. *Modern Man in Search of a Soul*. New York: Harcourt, Brace & World, 1933.

Lewis, Sinclair. *Babbitt*. New York: Harcourt, Brace & World, 1922.

Neugarten, Bernice L. *The Meanings of Age*. Selected Papers of Bernice L. Neugarten. Dale A. Neugarten (ed.). Chicago: University of Chicago Press, 1996.

Nielsen, Waldamar. *The Big Foundations*. New York: Columbia University Press, 1972.

Olshansky, S.J., B.A. Carnes, and R.N. Butler. "If Humans Were Built to Last," *Scientific American* (2001) *284*: 42–47.

Phillips, Kevin. *Wealth and Democracy: A Political History of the American Rich.* New York: Broadway Books, 2002.

Ryff, Carol. "Psychosocial Well-being in Adult Life." *Current Directions in Psychological Science* (1995) *4*: 99–104.

Seligman, Martin E.P. *Authentic Happiness: Using the New Positive Psychology to Realize Your Potential for Lasting Fulfillment.* New York: Free Press, 2002.

Chapter 21

Baltes, Paul B. and Ursula M. Staudinger. "Wisdom: A Metaheuristic (pragmatic) to Orchestrate Mind and Virtue Toward Excellence." *American Psychologist* (2000) *55*: 122–136.

Buber, Martin. *The William Alanson White Memorial Lectures.* 4th Series. *Psychiatry* (1957) *20*: 95–113.

Butler, R.N. "The Life Review: An Interpretation of Reminiscence in the Aged." *Psychiatry* (1963) *26*: 65–76.

Darwin, Charles. *On the Origin of Species by Means of Natural Selection.* London: John Murray, 1859.

Fortey, Richard. *Life: A Natural History of the First Four Billion Years of Life on Earth.* New York: Alfred A. Knopf, 1998.

Frankl, Viktor E. *Man's Search for Meaning.* New York: Washington Source Press, 1963.

Gruman, Gerald J. *A History of Ideas about the Prolongation of Life.* The International Longevity Center in collaboration with Springer, 2004.

Harris, John. "Intimations of Immortality: The Ethics and Justice of Life Extending Therapies." In Michael Freedman (ed.), *Current Legal Problems.* New York: Oxford University Press, 2002, and the International Longevity Center, 2007.

Lehman, Harvey. *Age and Achievement.* Princeton, NJ: Princeton University Press, 1953.

Mason, Herbert. *Gilgamesh: A Verse Narrative.* New York: Penguin Books, 1972.

McKee, Patrick L. (ed.). *Philosophic Foundation of Gerontology.* New York: Human Sciences Press, 1982.

Metchnikoff, Elie. *Prolongation of Life: Optimistic Studies.* Springer Publishing in cooperation with the International Longevity Center, 2004.

Nef, Evelyn Stefansson. *Finding My Way: The Autobiography of an Optimist.* The Francis Press, 2002.

Olshansky, S. Jay, Daniel Perry, Richard A. Miller, and Robert N. Butler. "In Pursuit of the Longevity Dividend." *The Scientist* (2006) *20*: 28–36.

Randall, John H. *The Making of the Modern Mind.* New York, 1926, 1976.

Schopenhauer, Arthur. "The Ages of Life" in *Counsels and Maxims.* Trans. T. Bailey Saunders. London: Swan Sonnenschein & Co., 1890.

Sigerest, Henry. *Primitive and Archaic Medicine.* Volume I of complete 8-volume *History of Medicine.* Oxford University Press, 1951.

Smith, Houston. *The World's Great Religions.* Harper San Francisco, 1991.

Staudinger, Ursula M. "Older and Wiser? Integrating Results on the Relationship Between Age and Wisdom-Related Performance." *International Journal of Behavioral Development* (1999) *23*: 641–664.

———. *The Study of Life Review: An Approach to the Investigation of Intellectual Development across the Lifespan.* Berlin: Sigma, 1989.

Sternberg, Robert J. *Wisdom: Its Nature, Origins & Development.* Cambridge, U.K.: Cambridge University Press, 1990.

Stock, Gregory. *Redesigning Humans: Our Inevitable Genetic Future.* Boston: Houghton Mifflin, 2002.

Warner, Huber, Julie Anderson, Steven Austad, Ettore Bergamni, Dale Bredesen, Robert Butler, et al. "Science Fact and the SENS Agenda." *EMBO Reports* (2005) *6*: 1006.

ACKNOWLEDGMENTS

Myrna Lewis, my extraordinary wife and intellectual partner, and I worked together in the field of aging for some thirty-five years. Her influence is everywhere in my work, including in this book. Brilliant and insightful about the human condition in general, she understood the special issues that affect middle-aged and older women. Without her support I could never have written this book.

I am especially indebted to Lindsay Jones, my editor at PublicAffairs, and to Susan Weinberg for their insights devoted to the final shaping of this book. I am grateful also for the interest of Peter Osnos.

I acknowledge the excellent early editing of both Judith Estrine of the International Longevity Center and also the final efforts of Mindy Werner.

I am deeply grateful to both the Frederick and Amelia Schimper Foundation of New York, which in 1975 also supported *Why Survive? Being Old in America,* and to the Merck Foundation for their generous financial support.

I owe a very special debt of thanks to the always-helpful Milagros Marrero, who was wonderfully patient with the frequent retyping of the manuscript over many years. Kyoung Kim, my skilled research associate, labored very effectively to ensure that I have my facts straight. Morriseen Barmore, my executive assistant, is unflappable. She guarded the gate with diplomacy, efficiency, and good judgment, protecting me so that I could carry on my various activities, including the writing of this book. I am indebted to Megan McIntyre, who has enriched the communications office of the International Longevity Center.

My thanks to Georges Borchardt, who has always supported my work. I appreciate the helpful criticism of certain chapters by the following persons:

William Abrams, Julia Alvarez, Jacob Brody, Christine Butler, Anthony Cerami, Howard Fillit, Philip Gerbino, James Gleason, Gerald Gruman, Robert Katzman, Mark Lebwohl, Charlotte Muller, Robert Myers, David Rothman, Mark Saunders, Mickey Spiegel, Richard Sprott, Kathleen Woodward, and Steven Zweig.

I thank Edward Stolman and Nancy Conrad of Glen Ellen in beautiful Sonoma Valley, California, and Paul and Mira Lehr at the Pritikin Longevity Center, who provided me wonderful retreats and accommodations to complete my manuscript.

INDEX

About the Author

Robert N. Butler, M.D., is President and Chief Executive Officer and Co-Chair of the Alliance for Health and the Future of the International Longevity Center–USA, and professor of geriatrics at the Brookdale Department of Geriatrics and Adult Development at the Mount Sinai Medical Center in New York City.

From 1975 to 1982 he was the founding director of the National Institute on Aging of the National Institutes of Health. In 1982 he founded the first department of geriatrics in a U.S. medical school. In 1990 he founded the International Longevity Center.

In 1976 Butler won the Pulitzer Prize for his book *Why Survive? Being Old in America.* He is co-author (with Dr. Myrna I. Lewis) of the books *Aging and Mental Health* and *Love and Sex After 60.* He is presently working on a book entitled *Life Review.*

Dr. Butler was a principal investigator of one of the first interdisciplinary, comprehensive, longitudinal studies of healthy community-residing older persons, conducted at the National Institute of Mental Health (1955–1966), which resulted in the landmark books *Human Aging I* and *II.* It was found that much attributed to old age is in fact a function of disease, social-economic adversity, and even personality. This research helped establish the fact that senility is not inevitable with aging, but is, instead, a consequence of disease. The NIMH research contributed to a different vision of old age. It set the stage for the later concepts of "healthy aging," "productive aging," and "successful aging."

In 2003 he received the Heinz Award for the Human Condition. The award was established to honor the memory of the late Senator John Heinz.

PublicAffairs is a publishing house founded in 1997. It is a tribute to the standards, values, and flair of three persons who have served as mentors to countless reporters, writers, editors, and book people of all kinds, including me.

I. F. STONE, proprietor of *I. F. Stone's Weekly*, combined a commitment to the First Amendment with entrepreneurial zeal and reporting skill and became one of the great independent journalists in American history. At the age of eighty, Izzy published *The Trial of Socrates*, which was a national bestseller. He wrote the book after he taught himself ancient Greek.

BENJAMIN C. BRADLEE was for nearly thirty years the charismatic editorial leader of *The Washington Post*. It was Ben who gave the *Post* the range and courage to pursue such historic issues as Watergate. He supported his reporters with a tenacity that made them fearless and it is no accident that so many became authors of influential, best-selling books.

ROBERT L. BERNSTEIN, the chief executive of Random House for more than a quarter century, guided one of the nation's premier publishing houses. Bob was personally responsible for many books of political dissent and argument that challenged tyranny around the globe. He is also the founder and longtime chair of Human Rights Watch, one of the most respected human rights organizations in the world.

·   ·   ·

For fifty years, the banner of Public Affairs Press was carried by its owner Morris B. Schnapper, who published Gandhi, Nasser, Toynbee, Truman, and about 1,500 other authors. In 1983, Schnapper was described by *The Washington Post* as "a redoubtable gadfly." His legacy will endure in the books to come.

Peter Osnos, *Founder and Editor-at-Large*